Probability and Its Applications

Published in association with the Applied Probability Trust

Editors: J. Gani, C.C. Heyde, P. Jagers, T.G. Kurtz

Probability and Its Applications

Mu-Fa Chen

Eigenvalues, Inequalities, and Ergodic Theory

 Springer

Mu-Fa Chen
Department of Mathematics, Beijing Normal University, Beijing 100875,
The People's Republic of China

Series Editors

J. Gani
Stochastic Analysis Group CMA
Australian National University
Canberra ACT 0200
Australia

C.C. Heyde
Stochastic Analysis Group, CMA
Australian National University
Canberra ACT 0200
Australia

T.G. Kurtz
Department of Mathematics
University of Wisconsin
480 Lincoln Drive
Madison, WI 53706
USA

P. Jagers
Mathematical Statistics
Chalmers University of Technology
S-41296 Göteborg
Sweden

Mathematics Subject Classification (2000): 60J25, 60K35, 37A25, 37A30, 47A45, 58C40, 34B24, 34L15, 35P15, 91B02

British Library Cataloguing in Publication Data
Chen, Mufa
  Eigenvalues, inequalities and ergodic theory.
  (Probability and its applications)
  1. Eigenvalues 2. Inequalities (Mathematics) 3. Ergodic theory
  I. Title
  512.9′436
ISBN 1852338687

Library of Congress Cataloging-in-Publication Data
Chen, Mu-fa.
  Eigenvalues, inequalities, and ergodic theory / Mu-Fa Chen.
    p. cm. — (Probability and its applications)
  Includes bibliographical references and indexes.
  ISBN 1-85233-868-7 (alk. paper)
    1. Eigenvalues. 2. Inequalities (Mathematics) 3. Ergodic theory. I. Title. II. Probability
  and its applications (Springer-Verlag).
  QA193.C44 2004
  512.9′436—dc22   2004049193

ISBN 1-85233-868-7   Springer-Verlag London Berlin Heidelberg
Springer Science+Business Media
springeronline.com

Typesetting: Camera-ready by author.
12/3830-543210 Printed on acid-free paper   SPIN 10969397

Preface

First, let us explain the precise meaning of the compressed title. The word "eigenvalues" means the first nontrivial Neumann or Dirichlet eigenvalues, or the principal eigenvalues. The word "inequalities" means the Poincaré inequalities, the logarithmic Sobolev inequalities, the Nash inequalities, and so on. Actually, the first eigenvalues can be described by some Poincaré inequalities, and so the second topic has a wider range than the first one. Next, for a Markov process, corresponding to its operator, each inequality describes a type of ergodicity. Thus, study of the inequalities and their relations provides a way to develop the ergodic theory for Markov processes. Due to these facts, from a probabilistic point of view, the book can also be regarded as a study of "ergodic convergence rates of Markov processes," which could serve as an alternative title of the book. However, this book is aimed at a larger class of readers, not only probabilists.

The importance of these topics should be obvious. On the one hand, the first eigenvalue is the leading term in the spectrum, which plays an important role in almost every branch of mathematics. On the other hand, the ergodic convergence rates constitute a recent research area in the theory of Markov processes. This study has a very wide range of applications. In particular, it provides a tool to describe the phase transitions and the effectiveness of random algorithms, which are now a very fashionable research area.

This book surveys, in a popular way, the main progress made in the field by our group. It consists of ten chapters plus two appendixes. The first chapter is an overview of the second to the eighth ones. Mainly, we study several different inequalities or different types of convergence by using three mathematical tools: a probabilistic tool, the coupling methods (Chapters 2 and 3); a generalized Cheeger's method originating in Riemannian geometry (Chapter 4); and an approach coming from potential theory and harmonic analysis (Chapters 6 and 7). The explicit criteria for different types of convergence and the explicit estimates of the convergence rates (or the optimal constants in the inequalities) in dimension one are given in Chapters 5 and 6; some generalizations are given in Chapter 7. The proofs of a diagram of nine types of ergodicity (Theorem 1.9) are presented in Chapter 8. Very often, we deal with one-dimensional elliptic operators or tridiagonal matrices (which can be infinite) in detail, but we also handle general differential and integral operators. To avoid heavy technical details, some proofs are split among several locations in the text. This also provides different views of the same problem at different levels. The topics of the last two chapters (9 and 10) are different but closely related. Chapter 9 surveys the study of a class of interacting particle systems (from which a large part of the problems studied in this book are motivated), and illustrates some applications. In the last chapter, one can see an interesting application of the first eigenvalue, its eigenfunctions, and an ergodic theorem to stochastic models of economics. Some related open problems are included in each chapter. Moreover, an effort is made to make each chapter, except the first one, more or less self-contained. Thus, once one has read about the program in Chapter 1, one may freely go on to the other chapters. The main exception is Chapter 3, which depends heavily on Chapter 2. As usual, a quick way to get an impression about what is done in the book is to look at the summaries given at the beginning of each chapter.

One should not be disappointed if one cannot find an answer in the book for one's own model. The complete solutions to our problems have only recently been obtained in dimension one. Nevertheless, it is hoped that the three methods studied in the book will be helpful. Each method has its own advantages and disadvantages. In principle, the coupling method can produce sharper estimates than the other two methods, but additional work is required to figure out a suitable coupling and, more seriously, a good distance. The Cheeger and capacitary methods work in a very general setup and are powerful qualitatively, but they leave the estimation of isoperimetric constants to the reader. The last task is usually quite hard in higher-dimensional situations.

This book serves as an introduction to a developing field. We emphasize the ideas through simple examples rather than technical proofs, and most of them are only sketched. It is hoped that the book will be readable by nonspecialists. In the past ten years or more, the author has tried rather hard to make acceptable lectures; the present book is based on these lecture notes: Chen (1994b; 1997a; 1998a; 1999c; 2001a; 2002b; 2002c; 2003b; 2004a; 2004b) [see Chen (2001c)]. Having presented eleven lectures in Japan in 2002, the author understood that it would be worthwhile to publish a short book, and then the job was started.

Since each topic discussed in the book has a long history and contains a great number of publications, it is impossible to collect a complete list of references. We emphasize the recent progress and related references. It is hoped that the bibliography is still rich enough that the reader can discover a large number of contributors in the field and more related references.

Beijing, The People's Republic of China　　　　　Mu-Fa Chen, October 2004

Acknowledgments

As mentioned before, this book is based on lecture notes presented over the past ten years or so. Thus, the book should be dedicated, with the author's deep acknowledgment, to the mathematicians and their universities/institutes whose kind invitations, financial support, and warm hospitality made those lectures possible. Without their encouragement and effort, the book would never exist. With the kind permission of his readers, the author is happy to list some of the names below (since 1993), with an apology to those that are missing:

- Z.M. Ma and J.A. Yan, Institute of Applied Mathematics, Chinese Academy of Sciences. D.Y. Chen, G.Q. Zhang, J.D. Chen, and M.P. Qian, Beijing (Peking) University. T.S. Chiang, C.R. Hwang, Y.S. Chow, and S.J. Sheu, Institute of Mathematics, Academy Sinica, Taipei. C.H. Chen, Y.S. Chow, A.C. Hsiung, W.T. Huang, W.Q. Liang, and C.Z. Wei, Institute of Statistical Science, Academy Sinica, Taipei. H. Chen, National Taiwan University. T.F. Lin, Soochow University. Y.J. Lee and W.J. Huang, National University of Kaohsiung. C.L. Wang, National Dong Hwa University.

- D.A. Dawson and S. Feng [McMaster University], Carleton University. G. O'Brien, N. Madras, and J.M. Sun, York University. D. McDonald, University of Ottawa. M. Barlow, E.A. Perkins, and S.J. Luo, University of British Columbia.

- E. Scacciatelli, G. Nappo, and A. Pellegrinotti [University of Roma III], University of Roma I. L. Accardi, University of Roma II. C. Boldrighini, University of Camerino [University of Roma I]. V. Capasso and Y.G. Lu, University of Bari.

- B. Grigelionis, Akademijios, Lithuania.

- L. Stettner and J. Zabczyk, Polish Academy of Sciences.

- W.Th.F. den Hollander, Utrecht University [Universiteit Leiden].

- Louis H.Y. Chen, K.P. Choi, and J.H. Lou, Singapore University.

- R. Durrett, L. Gross, and Z.Q. Chen [University of Washington Seattle], Cornell University. D.L. Burkholder, University of Illinois. C. Heyde, K. Sigman, and Y.Z. Shao, Columbia University.

- D. Elworthy, Warwick University. S. Kurylev, C. Linton, S. Veselov, and H.Z. Zhao, Loughborough University. T.S. Zhang, University of Manchester. G. Grimmett, Cambridge University. Z. Brzezniak and P. Busch, University of Hull. T. Lyons, University of Oxford. A. Truman, N. Jacod, and J.L. Wu, University of Wales Swansea.
- F. Götze and M. Röckner, University of Bielefeld. S. Albeverio and K.-T. Sturm, University of Bonn. J.-D. Deuschel and A. Bovier, Technical University of Berlin.
- K.J. Hochberg, Bar-Ilan University. B. Granovsky, Technion-Israel Institute of Technology.
- B. Yart, Grenoble University [University, Paris V]. S. Fang and B. Schmit, University of Bourgogne. J. Bertoin and Z. Shi, University of Paris VI. L.M. Wu, Blaise Pascal University and Wuhan University.
- R.A. Minlos, E. Pechersky, and E. Zizhina, the Information Transmission Problems, Russian Academy of Sciences.
- A.H. Xia, University of New South Wales [Melbourne University]. C. Heyde, J. Gani, and W. Dai, Australian National University. E. Seneta, University of Sydney. F.C. Klebaner, University of Melbourne. Y.X. Lin, Wollongong University.
- I. Shigekawa, Y. Takahashi, T. Kumagai, N. Yosida, S. Watanabe, and Q.P. Liu, Kyoto University. M. Fukushima, S. Kotani, S. Aida, and N. Ikeda, Osaka University. H. Osada, S. Liang, and K. Sato, Nagoya University. T. Funaki and S. Kusuoka, Tokyo University.
- E. Bolthausen, University of Zurich, P. Embrechts and A.-S. Sznitman, ETH.
- London Mathematical Society for the visit to the United Kingdom during November 4–22, 2003.

Next, the author acknowledges the organizers of the following conferences (since 1993) for their invitations and financial support.

- The Sixth International Vilnuis Conference on Probability and Mathematical Statistics (June 1993, Vilnuis).
- The International Conference on Dirichlet Forms and Stochastic Processes (October 1993, Beijing).
- The 23rd and 25th Conferences on Stochastic Processes and Their Applications (June 1995, Singapore and July 1998, Oregon).
- The Symposium on Probability Towards the Year 2000 (October 1995, New York).
- Stochastic Differential Geometry and Infinite-Dimensional Analysis (April 1996, Hangzhou).
- Workshop on Interacting Particle Systems and Their Applications (June 1996, Haifa).
- IMS Workshop on Applied Probability (June 1999, Hong Kong).

- The Second Sino-French Colloquium in Probability and Applications (April 2001, Wuhan).
- The Conference on Stochastic Analysis on Large Scale Interacting Systems (July 2002, Japan).
- Stochastic Analysis and Statistical Mechanics, Yukawa Institute (July 2002, Kyoto).
- International Congress of Mathematicians (August 2002, Beijing).
- The First Sino-German Conference on Stochastic Analysis—A Satellite Conference of ICM 2002 (September 2002, Beijing).
- Stochastic Analysis in Infinite Dimensional Spaces (November 2002, Kyoto)
- Japanese National Conference on Stochastic Processes and Related Fields (December 2002, Tokyo).

Thanks are given to the editors, managing editors, and production editors, of the Springer Series in Statistics, Probability and Its Applications, especially J. Gani and S. Harding for their effort in publishing the book, and to the copyeditor D. Kramer for the effort in improving the English language.

Thanks are also given to World Scientific Publishing Company for permission to use some material from the author's previous book (1992a, 2004: 2nd edition).

The continued support of the National Natural Science Foundation of China, the Research Fund for Doctoral Program of Higher Education, as well as the Qiu Shi Science and Technology Foundation, and the 973 Project are also acknowledged.

Finally, the author is grateful to the colleagues in our group: F.Y. Wang, Y.H. Zhang, Y.H. Mao, and Y.Z. Wang for their fruitful cooperation. The suggestions and corrections to the earlier drafts of the book by a number of friends, especially J.W. Chen and H.J. Zhang, and a term of students are also appreciated. Moreover, the author would like to acknowledge S.J. Yan, Z.T. Hou, Z.K. Wang, and D.W. Stroock for their teaching and advice.

Contents

Chapter 1

An Overview of the Book

This chapter is an overview of the book, especially of the first eight chapters. It consists of four sections. In the first section, we explain what eigenvalues we are interested in and show the difficulties in studying the first (nontrivial) eigenvalue through elementary examples. The second section presents some new (dual) variational formulas and explicit bounds for the first eigenvalue of the Laplacian on Riemannian manifolds or Jacobi matrices (Markov chains), and explains the main idea of the proof, which is a probabilistic approach: the coupling methods. In the third section, we introduce some recent lower bounds of several basic inequalities, based on a generalization of Cheeger's approach which comes from Riemannian geometry. In the last section, a diagram of nine different types of ergodicity and a table of explicit criteria for them are presented. The criteria are motivated by the weighted Hardy inequality, which comes from harmonic analysis.

1.1 Introduction

Let me now explain what eigenvalue we are talking about.

Definition. The first (nontrivial) eigenvalue

Consider a tridiagonal matrix (or in probabilistic language, a birth–death process with state space $E = \{0, 1, 2, \ldots\}$ and Q-matrix)

$$Q = (q_{ij}) = \begin{pmatrix} -b_0 & b_0 & 0 & 0 & \cdots \\ a_1 & -(a_1 + b_1) & b_1 & 0 & \cdots \\ 0 & a_2 & -(a_2 + b_2) & b_2 & \cdots \\ \vdots & \ddots & & \ddots & \ddots \end{pmatrix},$$

where a_k, $b_k > 0$. Since the sum of each row equals 0, we have $Q\mathbf{1} = \mathbf{0} = 0 \cdot \mathbf{1}$, where $\mathbf{1}$ is the vector having elements 1 everywhere and $\mathbf{0}$ is the zero vector.

This means that the Q-matrix has an eigenvalue 0 with eigenvector **1**. Next, consider the finite case $E_n = \{0, 1, \ldots, n\}$. Then, the eigenvalues of $-Q$ are discrete: $0 = \lambda_0 < \lambda_1 \leqslant \cdots \leqslant \lambda_n$. We are interested in the first (nontrivial) eigenvalue $\lambda_1 = \lambda_1 - \lambda_0 =: \text{gap}\,(Q)$ (also called the *spectral gap* of Q). In the infinite case, $\lambda_1 := \inf\{\{\text{Spectrum of } (-Q)\} \setminus \{0\}\}$ can be 0. Certainly, one can consider a self-adjoint elliptic operator in $\mathbb{R}^d$ or the Laplacian Δ on manifolds or an infinite-dimensional operator as in the study of interacting particle systems.

Since the spectral theory is of central importance in many branches of mathematics and the first nontrivial eigenvalue is the leading term of the spectrum, it should not be surprising that the study of λ_1 has a very wide range of applications.

Difficulties

To get a concrete feeling about the difficulties of the topic, let us look at the following examples with finite state spaces.

When $E = \{0, 1\}$, it is trivial that $\lambda_1 = a_1 + b_0$. Everyone is happy to see this result, since if either a_1 or b_0 increases, so does λ_1. If we go one more step, $E = \{0, 1, 2\}$, then we have four parameters, b_0, b_1 and a_1, a_2. In this case, $\lambda_1 = 2^{-1}\big[a_1 + a_2 + b_0 + b_1 - \sqrt{(a_1 - a_2 + b_0 - b_1)^2 + 4a_1 b_1}\,\big]$. It is disappointing to see this result, since parameters effect on λ_1 is not clear at all. When $E = \{0, 1, 2, 3\}$, we have six parameters: $b_0, b_1, b_2, a_1, a_2, a_3$. The solution is expressed by the three quantities B, C, and D:

$$\lambda_1 = \frac{D}{3} - \frac{C}{3 \cdot 2^{1/3}} + \frac{2^{1/3}\,(3\,B - D^2)}{3\,C},$$

where the quantities D, B, and C are not too complicated:

$$D = a_1 + a_2 + a_3 + b_0 + b_1 + b_2,$$
$$B = a_3\,b_0 + a_2\,(a_3 + b_0) + a_3\,b_1 + b_0\,b_1 + b_0\,b_2 + b_1\,b_2 + a_1\,(a_2 + a_3 + b_2),$$
$$C = \left(A + \sqrt{4(3\,B - D^2)^3 + A^2}\,\right)^{1/3}.$$

However, in the last expression, another quantity, A, is involved. What, then, is A?

$$
\begin{aligned}
A = {}&-2\,a_1^3 - 2\,a_2^3 - 2\,a_3^3 + 3\,a_3^2 b_0 + 3a_3 b_0^2 - 2b_0^3 + 3a_3^2 b_1 - 12\,a_3\,b_0 b_1 + 3b_0^2 b_1 \\
&+ 3\,a_3\,b_1^2 + 3\,b_0\,b_1^2 - 2\,b_1^3 - 6\,a_3^2 b_2 + 6\,a_3 b_0 b_2 + 3\,b_0^2 b_2 + 6\,a_3 b_1 b_2 - 12\,b_0 b_1 b_2 \\
&+ 3b_1^2\,b_2 - 6a_3 b_2^2 + 3b_0\,b_2^2 + 3b_1 b_2^2 - 2b_2^3 + 3a_1^2\,(a_2 + a_3 - 2\,b_0 - 2\,b_1 + b_2) \\
&+ 3\,a_2^2\,[a_3 + b_0 - 2\,(b_1 + b_2)] \\
&+ 3a_2\,[a_3^2 + b_0^2 - 2\,b_1^2 - b_1\,b_2 - 2b_2^2 - a_3(4b_0 - 2b_1 + b_2) + 2b_0(b_1 + b_2)] \\
&+ 3\,a_1\,[a_2^2 + a_3^2 - 2\,b_0^2 - b_0\,b_1 - 2\,b_1^2 - a_2(4\,a_3 - 2\,b_0 + b_1 - 2\,b_2) \\
&+ 2\,b_0\,b_2 + 2\,b_1\,b_2 + b_2^2 + 2\,a_3(b_0 + b_1 + b_2)],
\end{aligned}
$$

computed using Mathematica. One should be shocked, at least I was, to see this result, since the roles of the parameters are completely hidden! Of course, everyone understands that it is impossible to compute λ_1 explicitly when the size of the matrix is greater than five!

Now, how about the estimation of λ_1? To see this, let us consider the perturbation of the eigenvalues and eigenfunctions. We consider the infinite state space $E = \{0, 1, 2, \ldots\}$. Denote by g and Degree(g), respectively, the eigenfunction of λ_1 and the degree of g when g is polynomial. Three examples of the perturbation of λ_1 and Degree(g) are listed in Table 1.1.

Table 1.1 Three examples of the perturbation of λ_1 and Degree(g)

$b_i\ (i \geqslant 0)$	$a_i\ (i \geqslant 1)$	λ_1	**Degree** (g)
$i + c\ (c > 0)$	$2i$	1	1
$i + 1$	$2i + 3$	2	2
$i + 1$	$2i + \left(4 + \sqrt{2}\right)$	3	3

The first line is the well-known linear model, for which $\lambda_1 = 1$, independent of the constant $c > 0$, and g is linear. Next, keeping the same birth rate, $b_i = i + 1$, the perturbation of the death rate a_i from $2i$ to $2i + 3$ (respectively, $2i + 4 + \sqrt{2}$) leads to the change of λ_1 from one to two (respectively, three). More surprisingly, the eigenfunction g is changed from linear to quadratic (respectively, cubic). For the intermediate values of a_i between $2i$, $2i + 3$, and $2i + 4 + \sqrt{2}$, λ_1 is unknown, since g is nonpolynomial. As seen from these examples, the first eigenvalue is very sensitive. Hence, in general, it is very hard to estimate λ_1.

Hopefully, we have presented enough examples to show the extreme difficulties of the topic. Very fortunately, at last, we are able to present a complete solution to this problem in the present context. Please be patient; the result will be given only later.

For a long period, we did not know how to proceed. So we visited several branches of mathematics. Finally, we found that the topic was well studied in Riemannian geometry.

1.2 New variational formula for the first eigenvalue

A story of estimating λ_1 in geometry

Here is a short story about the study of λ_1 in geometry.

Consider the Laplacian Δ on a connected compact Riemannian manifold (M, g), where g is the Riemannian metric. The spectrum of Δ is discrete: $\cdots \leqslant -\lambda_2 \leqslant -\lambda_1 < -\lambda_0 = 0$ (may be repeated). Estimating these eigenvalues λ_k (especially λ_1) is an important chapter in modern geometry. As far as we know, five books, excluding books on general spectral theory, have been devoted to this topic: I. Chavel (1984), P.H. Bérard (1986), R. Schoen and S.T. Yau (1988), P. Li (1993), and C.Y. Ma (1993). About 2000 references are collected in the second quoted book. Thus, it is impossible for us to introduce an overview of what has been done in geometry. Instead, we would like to show the reader ten of the most beautiful lower bounds. For a manifold M, denote its dimension, diameter, and the lower bound of Ricci curvature by d, D, and K (Ricci$_M \geqslant Kg$), respectively. The simplest example is the unit sphere $\mathbb{S}^d$ in $\mathbb{R}^{d+1}$, for which $D = \pi$ and $K = d - 1$. We are interested in estimating λ_1 in terms of these three geometric quantities. It is relatively easy to obtain an upper bound by applying a test function $f \in C^1(M)$ to the classical variational formula

$$\lambda_1 = \inf\left\{ \int_M \|\nabla f\|^2 \mathrm{d}x : f \in C^1(M), \int_M f \mathrm{d}x = 0, \int_M f^2 \mathrm{d}x = 1 \right\}, \quad (1.0)$$

where "$\mathrm{d}x$" is the Riemannian volume element. To obtain the lower bound, however, is much harder. In Table 1.2, we list ten of the strongest lower bounds that have been derived in the past, using various sophisticated methods.

Table 1.2   Ten lower bounds of λ_1

Author(s)	Lower bound	
A. Lichnerowicz (1958)	$\dfrac{d}{d-1}K, \quad K \geqslant 0$	(1.1)
P.H. Bérard, G. Besson, & S. Gallot (1985)	$d\left\{\dfrac{\int_0^{\pi/2}\cos^{d-1}t\,\mathrm{d}t}{\int_0^{D/2}\cos^{d-1}t\,\mathrm{d}t}\right\}^{2/d}, \quad K = d-1 > 0$	(1.2)
P. Li & S.T. Yau (1980)	$\dfrac{\pi^2}{2 \cdot D^2}, \quad K \geqslant 0$	(1.3)
J.Q. Zhong & H.C. Yang (1984)	$\dfrac{\pi^2}{D^2}, \quad K \geqslant 0$	(1.4)
D.G. Yang (1999)	$\dfrac{\pi^2}{D^2} + \dfrac{K}{4}, \quad K \geqslant 0$	(1.5)
P. Li & S.T. Yau (1980)	$\dfrac{1}{D^2(d-1)\exp\left[1 + \sqrt{1 + 16\alpha^2}\right]}, \quad K \leqslant 0$	(1.6)
K.R. Cai (1991)	$\dfrac{\pi^2}{D^2} + K, \quad K \leqslant 0$	(1.7)
D. Zhao (1999)	$\dfrac{\pi^2}{D^2} + 0.52K, \quad K \leqslant 0.$	(1.8)
H.C. Yang (1990) & F. Jia (1991)	$\dfrac{\pi^2}{D^2}e^{-\alpha}, \quad \text{if } d \geqslant 5, \quad K \leqslant 0$	(1.9)
H.C. Yang (1990) & F. Jia (1991)	$\dfrac{\pi^2}{2\,D^2}e^{-\alpha'}, \quad \text{if } 2 \leqslant d \leqslant 4, \quad K \leqslant 0$	(1.10)

In Table 1.2, the two parameters α and α' are defined as

$$\alpha = 2^{-1}D\sqrt{|K|(d-1)} \quad \text{and} \quad \alpha' = 2^{-1}D\sqrt{|K|((d-1)\vee 2)}.$$

The first estimate is due to A. Lichnerowicz 46 years ago. It is very good, since it is indeed sharp for the unit sphere in two or more dimensions. After 27 years, this result was improved by three French mathematicians, given in (1.2). The problem here is that these two estimates become trivial for zero curvature, the unit circle for instance. It is well known that the zero curvature case is harder than that of positive curvature. The first progress was made by Li and Yau (1.3) and improved by Zhong and Yang (1.4), by removing the factor two from (1.3). For the nonexpert, one may think that this is not essential. However, it is regarded as a deep result in geometry, since it is sharp for the unit circle. The fifth estimate is a mixture of the first and the fourth sharp estimates.

We now go to the case of negative curvature. The first result (1.6) is again due to Li and Yau in the same paper quoted above. Combining the two results (1.3) and (1.6), it should be clear that the negative case is much harder than the positive one. Li and Yau's results are improved step by step by many geometers as listed in Table 1.2.

Among these estimates, seven [(1.1), (1.2), (1.4), (1.5), (1.7)–(1.9)], shown in boldface, are sharp. The first two are sharp for the unit sphere in two or higher dimensions but fail for the unit circle; the fourth, the fifth, and the seventh to ninth are all sharp for the unit circle. The above authors include several famous geometers, and several of the results received awards. As seen from the table, the picture is now very complete, due to the efforts of geometers in the past 46 years or more. For such a well-developed field, what can we do now? Our original starting point was to learn from the geometers and to study their methods, especially recent developments. It is surprising that we actually went to the opposite direction, that is, studying the first eigenvalue by using a probabilistic method. At last, we discovered a general formula for λ_1.

New variational formula

Before stating our new variational formula, we introduce two notations:

$$C(r) = \cosh^{d-1}\left[\frac{r}{2}\sqrt{\frac{-K}{d-1}}\right], \ r \in (0, D); \quad \mathscr{F} = \{f \in C[0, D] : f > 0 \text{ on } (0, D)\},$$

where $\cosh^r x = (\cosh x)^r$. Here, we have used all three quantities: the dimension d, the diameter D, and the lower bound K of Ricci curvature. Note that $C(r)$ is always real for any $K \in \mathbb{R}$.

Theorem 1.1 (General formula [Chen and F.Y. Wang, 1997a]).

$$\lambda_1 \geqslant \sup_{f \in \mathscr{F}} \inf_{r \in (0, D)} \frac{4f(r)}{\int_0^r C(s)^{-1}\mathrm{d}s \int_s^D C(u)f(u)\mathrm{d}u} =: \xi_1. \tag{1.11}$$

The variational formula (1.11) has its essential value in estimating the lower bound. It is a dual of the classical variational formula (1.0) in the sense that "inf" in (1.0) is replaced by "sup" in (1.11). The classical formula goes back to Lord S.J.W. Rayleigh (1877) or E. Fischer (1905). Noticing that there are no common points in the two formulas (1.0) and (1.11), this explains the reason why such a formula never appeared before. Certainly, the new formula can produce many new lower bounds. For instance, the one corresponding to the trivial function $f = 1$ is still nontrivial in geometry. It also has a nice probabilistic meaning: the convergence rate of strong ergodicity (cf. Section 5.6). Clearly, in order to obtain a better estimate, one needs to be more careful in choosing the test functions. Applying the general formula (1.11) to the elementary test functions $\sin(\alpha r)$ and $\cosh^{1-d}(\alpha r)\sin(\beta r)$ with $\alpha = 2^{-1}D\sqrt{|K|/(d-1)}$ and $\beta = \pi/(2D)$, we obtain the following corollary.

Corollary 1.2 (Chen and F.Y. Wang, 1997a).

$$\lambda_1 \geqslant \frac{dK}{d-1}\left\{1 - \cos^d\left[\frac{D}{2}\sqrt{\frac{K}{d-1}}\right]\right\}^{-1}, \qquad d > 1, \quad K \geqslant 0. \tag{1.12}$$

$$\lambda_1 \geqslant \frac{\pi^2}{D^2}\sqrt{1 - \frac{2D^2K}{\pi^4}}\cosh^{1-d}\left[\frac{D}{2}\sqrt{\frac{-K}{d-1}}\right], \qquad d > 1, \, K \leqslant 0. \tag{1.13}$$

Applying the formula (1.11) to some very complicated test functions, we can prove, assisted by a computer, the following result.

Corollary 1.3 (Chen, E. Scacciatelli, and L. Yao, 2002).

$$\lambda_1 \geqslant \pi^2/D^2 + K/2, \qquad K \in \mathbb{R}. \tag{1.14}$$

Surprisingly, these two corollaries improve all the estimates (1.1)–(1.10). Estimate (1.12) improves (1.1) and (1.2), estimate (1.13) improves (1.9) and (1.10), and estimate (1.14) improves (1.4), (1.5), (1.7), and (1.8). Moreover, the linear approximation in (1.14) is optimal in the sense that the coefficient $1/2$ of K is exact.

A test function is indeed a mimic eigenfunction of λ_1, so it should be chosen appropriately in order to obtain good estimates. A question arises naturally: does there exist a single representative test function such that we can avoid the task of choosing a different test function each time? The answer is seemingly negative, since we have already seen that the eigenvalue and the eigenfunction are both very sensitive. Surprisingly, the answer is affirmative. The representative test function, though very tricky to find, has a rather simple form: $f(r) = \left(\int_0^r C(s)^{-1}\mathrm{d}s\right)^\gamma (\gamma \geqslant 0)$. This is motivated by a study of the weighted Hardy inequality, a powerful tool in harmonic analysis [cf. B. Muckenhoupt (1972), B. Opic and A. Kufner (1990)]. The lower and the upper bounds of ξ_1, given in (1.15) below, correspond to $\gamma = 1/2$ and $\gamma = 1$, respectively.

Corollary 1.4 (Chen, 2000c). For the lower bound ξ_1 of λ_1 given in Theorem 1.1, we have

$$4\delta^{-1} \geqslant \xi_1 \geqslant \delta^{-1}, \tag{1.15}$$

where

$$\delta = \sup_{r \in (0, D)} \left(\int_0^r C(s)^{-1} \mathrm{d}s \right) \left(\int_r^D C(s) \mathrm{d}s \right), \quad C(s) = \cosh^{d-1} \left[\frac{s}{2} \sqrt{\frac{-K}{d-1}} \right].$$

Theorem 1.1 and its corollaries are also valid for manifolds with a convex boundary endowed with the Neumann boundary condition. In this case, the estimates (1.1)–(1.10) are conjectured by the geometers to be correct. However, as far as we know, only Lichnerowicz's estimate (1.1) was proven by J.F. Escobar in 1990. The others in (1.2)–(1.10) and furthermore in (1.12)–(1.15) are all new in geometry.

Sketch of the main proof (Chen and F.Y. Wang, 1993b)

Here we adopt the language of analysis and restrict ourselves to the Euclidean case. The geometric case will be explained in detail in the next chapter. Our main tool is the coupling methods. Given a self-adjoint second-order elliptic operator L in $\mathbb{R}^d$,

$$L = \sum_{i,\,j=1}^d a_{ij}(x) \frac{\partial^2}{\partial x_i \partial x_j} + \sum_{i=1}^d b_i(x) \frac{\partial}{\partial x_i},$$

an elliptic (usually degenerate) operator $\widetilde{L}$ on the product space $\mathbb{R}^d \times \mathbb{R}^d$ is called a *coupling of L* if it satisfies the following *marginality* condition (Chen and S.F. Li, 1989):

$$\widetilde{L}f(x, y) = Lf(x) \ \big(\text{respectively, } \widetilde{L}f(x, y) = Lf(y)\big), \qquad f \in C_b^2(\mathbb{R}^d), \ x \neq y,$$

where on the left-hand side, f is regarded as a bivariate function.

Denote by $\{P_t\}_{t \geqslant 0}$ the semigroup determined by L: $P_t = e^{tL}$. Corresponding to a coupling operator $\widetilde{L}$, we have $\{\widetilde{P}_t\}_{t \geqslant 0}$. The coupling simply means that

$$\widetilde{P}_t f(x, y) = P_t f(x) \ \big(\text{respectively, } \widetilde{P}_t f(x, y) = P_t f(y)\big) \tag{1.20}$$

for all $f \in C_b^2(\mathbb{R}^d)$ and all (x, y) $(x \neq y)$, where on the left-hand side, f is again regarded as a bivariate function. With this preparation in mind, we can now start our proof.

Step 1. Let g be an eigenfunction of $-L$ corresponding to λ_1. That is, $-Lg = \lambda_1 g$. By the standard differential equation (the forward Kolmogorov equation) of the semigroup, we have

$$\frac{d}{dt} P_t g(x) = P_t Lg(x) = -\lambda_1 P_t g(x).$$

Solving this ordinary differential equation in $P_t g(x)$ for fixed g and x, we obtain

$$P_t g(x) = g(x) e^{-\lambda_1 t}. \qquad (1.21)$$

This expression is very nice, since the eigenvalue, its eigenfunction, and the semigroup are all combined in a simple formula. However, it is useless at the moment, since none of these three things are explicitly known.

Step 2. Consider the case of a compact space. Then g is Lipschitz with respect to the distance ρ. Denote by c_g the Lipschitz constant. Now the main condition we need is the following:

$$\widetilde{P}_t \rho(x, y) \leqslant \rho(x, y) e^{-\alpha t}. \qquad (1.22)$$

This condition is more or less equivalent to

$$\widetilde{L} \rho(x, y) \leqslant -\alpha \rho(x, y), \qquad x \neq y \qquad (1.23)$$

(cf. Lemma A.6 in Appendix A). Setting $g_1(x, y) = g(x)$ and $g_2(x, y) = g(y)$, we obtain

$$
\begin{aligned}
e^{-\lambda_1 t} |g(x) - g(y)| &= \left| P_t g(x) - P_t g(y) \right| \quad \text{(by (1.21))} \\
&= \left| \widetilde{P}_t g_1(x, y) - \widetilde{P}_t g_2(x, y) \right| \quad \text{(by (1.20))} \\
&= \left| \widetilde{P}_t (g_1 - g_2)(x, y) \right| \leqslant \widetilde{P}_t |g_1 - g_2|(x, y) \\
&\leqslant c_g \widetilde{P}_t \rho(x, y) \quad \text{(Lipschitz property)} \\
&\leqslant c_g \rho(x, y) e^{-\alpha t} \quad \text{(by (1.22))}.
\end{aligned}
$$

Since g is not a constant, there exist $x \neq y$ such that $g(x) \neq g(y)$. Letting $t \to \infty$, we must have $\lambda_1 \geqslant \alpha$.   □

The proof is unbelievably straightforward. A good point in the proof is the use of the eigenfunction so that we can achieve sharp estimates. On the other hand, it is crucial that we do not need too much knowledge about the eigenfunction, for otherwise, there is no hope that things will work out in such a general setting, since the eigenvalue and its eigenfunction are either known or unknown simultaneously. Aside from the Lipschitz property of g with respect to the distance, which can be avoided by using a localizing procedure for the noncompact case, the key to the proof is clearly condition (1.23). For this, one needs not only a good coupling but also a good choice of the distance. It is a long journey to solving these two problems. The details will be explained in the next two chapters.

Our proof is universal in the sense that it works for general Markov processes. We also obtain variational formulas for noncompact manifolds, elliptic operators in $\mathbb{R}^d$ (Chen and F.Y. Wang, 1997b), and Markov chains (Chen, 1996). It is more difficult to derive the variational formulas for the elliptic operators and Markov chains due to the presence of infinite parameters in these cases. In contrast, there are only three parameters (d, D, and K) in the geometric case. In fact, with the coupling methods at hand, the formula (1.11) is a particular consequence of our general formula (which is complete in dimension one) for elliptic operators. The general formulas have recently been extended to the Dirichlet eigenvalues by Chen, Y.H. Zhang, and X.L. Zhao (2003).

To conclude this section, we return to the matrix case introduced at the beginning of the chapter.

Tridiagonal matrices (birth–death processes)

To answer the question just posed, we need some notation. Define

$$\mu_0 = 1, \quad \mu_i = \frac{b_0 \cdots b_{i-1}}{a_1 \cdots a_i}, \quad i \geqslant 1.$$

Assume that the process is nonexplosive:

$$\sum_{k=0}^{\infty} \frac{1}{b_k \mu_k} \sum_{i=0}^{k} \mu_i = \infty, \quad \text{and moreover} \quad Z = \sum_{i} \mu_i < \infty. \tag{1.24}$$

Then the process is ergodic (positive recurrent). The corresponding Dirichlet form is

$$D(f) = \sum_{i} \pi_i b_i (f_{i+1} - f_i)^2, \quad \mathscr{D}(D) = \{f \in L^2(\pi) : D(f) < \infty\}.$$

Here and in what follows, only the diagonal elements $D(f)$ are written, but the nondiagonal elements can be computed from the diagonal ones using the quadrilateral role. We then have the *classical variational formula*

$$\lambda_1 = \inf \{D(f) : \pi(f) = 0, \pi(f^2) = 1\},$$

where $\pi(f) = \int f \mathrm{d}\pi$. Define

$$\mathscr{W} = \{w : w_0 = 0, \ w \text{ is strictly increasing}\},$$
$$\widetilde{\mathscr{W}} = \{w : w_0 = 0, \text{there exists } k : 1 \leqslant k \leqslant \infty \text{ such that } w_i = w_{i \wedge k}$$
$$\text{and } w \text{ is strictly increasing in } [0, k]\},$$
$$I_i(w) = \frac{1}{\mu_i b_i (w_{i+1} - w_i)} \sum_{j \geqslant i+1} \mu_j w_j.$$

Note that $\widetilde{\mathscr{W}}$ is simply a modification of $\mathscr{W}$. Hence, only the two notations $\mathscr{W}$ and $I(w)$ are essential here.

Theorem 1.5 (Chen (1996; 2000c; 2001b)). Let $\bar{w} = w - \pi(w)$. For ergodic birth–death processes (i.e., (1.24) holds), we have

(1) Dual variational formulas:

$$\inf_{w \in \overline{\mathcal{W}}} \sup_{i \geqslant 1} I_i(\overline{w})^{-1} = \lambda_1 = \sup_{w \in \mathcal{W}} \inf_{i \geqslant 0} I_i(\overline{w})^{-1}.$$

(2) Explicit bounds and an approximation procedure: Two explicit sequences $\{\eta_n\}$ and $\{\tilde{\eta}_n\}$ are constructed such that

$$Z\delta^{-1} \geqslant \tilde{\eta}_n^{-1} \geqslant \lambda_1 \geqslant \eta_n^{-1} \geqslant (4\delta)^{-1},$$

where $\delta = \sup\limits_{i \geqslant 1} \sum\limits_{j \leqslant i-1} (\mu_j b_j)^{-1} \sum\limits_{j \geqslant i} \mu_j.$

(3) Explicit criterion: $\lambda_1 > 0$ iff $\delta < \infty$.

Here the word "dual" means that the upper and lower bounds in part (1) of the theorem are interchangeable if one exchanges "sup" and "inf." Certainly, with slight modifications, this result is also valid for finite matrices; refer to Chen (1999a). Starting from the examples given in Section 1.1, could you have expected such a short and complete answer?

Theorem 1.1 and the second formula in Theorem 1.5 (1) will be proved in Chapter 3, for which the coupling tool is prepared in Chapter 2. An analytic proof of the second formula in Theorem 1.5 (1) is also presented in Chapter 3. Further results are presented in Chapters 5 and 6.

1.3   Basic inequalities and new forms of Cheeger's constants

Basic inequalities

We now go to a more general setup. Let $(E, \mathcal{E}, \pi)$ be a probability space satisfying $\{(x, x) : x \in E\} \in \mathcal{E} \times \mathcal{E}$. Denote by $L^p(\pi)$ the usual real L^p-space with norm $\| \cdot \|_p$. Write $\| \cdot \| = \| \cdot \|_2$.

For a given Dirichlet form $(D, \mathcal{D}(D))$, the classical variational formula for the first eigenvalue λ_1 can be rewritten in the form (1.25) below with optimal constant $C = \lambda_1^{-1}$. From this point of view, it is natural to study other inequalities. Here are two additional basic inequalities, (1.26) and (1.27):

$$\textit{Poincaré inequality}: \quad \mathrm{Var}(f) \leqslant CD(f), \qquad f \in L^2(\pi), \tag{1.25}$$

$$\textit{Logarithmic Sobolev inequality}: \int f^2 \log \frac{f^2}{\|f\|^2} \mathrm{d}\pi \leqslant CD(f), \tag{1.26}$$
$$f \in L^2(\pi),$$

$$\textit{Nash inequality}: \quad \mathrm{Var}(f) \leqslant CD(f)^{1/p}\|f\|_1^{2/q}, \quad f \in L^2(\pi), \tag{1.27}$$

where $\mathrm{Var}(f) = \pi(f^2) - \pi(f)^2$, $\pi(f) = \int f\mathrm{d}\pi$, $p \in (1, \infty)$, and $1/p + 1/q = 1$. The last two inequalities are due to L. Gross (1976) and J. Nash (1958), respectively.

Our main object is a symmetric (not necessarily Dirichlet) form $(D, \mathscr{D}(D))$ on $L^2(\pi)$, corresponding to an integral operator (or symmetric kernel) on $(E, \mathscr{E})$:

$$D(f) = \frac{1}{2} \int_{E \times E} J(\mathrm{d}x, \mathrm{d}y)[f(y) - f(x)]^2,$$
$$\mathscr{D}(D) = \{f \in L^2(\pi) : D(f) < \infty\}, \tag{1.28}$$

where J is a nonnegative, symmetric measure having no charge on the diagonal set $\{(x, x) : x \in E\}$. A typical example in our mind is the reversible jump process with q-pair $(q(x), q(x, \mathrm{d}y))$ and reversible measure π. Then $J(\mathrm{d}x, \mathrm{d}y) = \pi(\mathrm{d}x)q(x, \mathrm{d}y)$.

For the remainder of this section, we restrict our discussion to the symmetric form of (1.28).

Status of the research

An important topic in this research area is to study under what conditions on the symmetric measure J the above inequalities (1.25)–(1.27) hold. In contrast with the probabilistic method used in Section 1.2, here we adopt a generalization of Cheeger's method (1970), which comes from Riemannian geometry. Naturally, we define $\lambda_1 := \inf\{D(f) : \pi(f) = 0, \|f\| = 1\}$. For bounded jump processes, the fundamental known result is the following. Write $x \wedge y = \min\{x, y\}$ and similarly, $x \vee y = \max\{x, y\}$.

Theorem 1.6 (G.F. Lawler and A.D. Sokal, 1988). $\lambda_1 \geqslant \dfrac{k^2}{2M}$, where

$$k = \inf_{\pi(A) \in (0,1)} \frac{\int_A \pi(\mathrm{d}x) q(x, A^c)}{\pi(A) \wedge \pi(A^c)} \qquad \text{and} \qquad M = \sup_{x \in E} q(x) < \infty.$$

In the past years, the theorem has appeared in six books: Chen (1992a), A.J. Sinclair (1993), F.R.K. Chung (1997), L. Saloff-Coste (1997), Y. Colin de Verdière (1998), D.G. Aldous, and J.A. Fill (2004). From the titles of the books, one can see a wide range of the applications. However, this result fails for an unbounded operator (i.e., $\sup_x q(x) = \infty$). It was a challenging open problem for ten years (until 1998) to handle the unbounded case.

As for the logarithmic Sobolev inequality, there have been a large number of publications in the past twenty years for differential operators. For a survey, see D. Bakry (1992), L. Gross (1993), or A. Guionnet and B. Zegarlinski (2003). Still, there are very limited results for integral operators.

New results

Since the symmetric measure can be very unbounded, we choose a symmetric, nonnegative function $r(x, y)$ such that

$$J^{(\alpha)}(\mathrm{d}x, \mathrm{d}y) := I_{\{r(x,y)^\alpha > 0\}} \frac{J(\mathrm{d}x, \mathrm{d}y)}{r(x, y)^\alpha}, \qquad \alpha > 0,$$

satisfies $J^{(1)}(\mathrm{d}x, E)/\pi(\mathrm{d}x) \leqslant 1$, π-a.s. For convenience, we use the convention $J^{(0)} = J$. Corresponding to the three inequalities above, we introduce some new forms of Cheeger's constants, listed in Table 1.3. Now our main result can be easily stated as follows.

Theorem 1.7. $k^{(1/2)} > 0 \Longrightarrow$ the corresponding inequality holds.

In short, we use $J^{(1/2)}$ and $J^{(1)}$ to handle an unbounded J. The use of the first two kernels comes from the Schwarz inequality. The result is proven in four papers quoted in Table 1.3. In these papers, some estimates, which can be sharp or qualitatively sharp, for the upper or lower bounds are also presented.

Table 1.3 New forms of Cheeger's constants

Inequality	Constant $k^{(\alpha)}$	
Poincaré	$\displaystyle\inf_{\pi(A)\in(0,1)} \frac{J^{(\alpha)}(A \times A^c)}{\pi(A) \wedge \pi(A^c)}$	(Chen and F.Y. Wang, 1998)
Log. Sobolev	$\displaystyle\lim_{r\to 0}\inf_{\pi(A)\in(0,r]} \frac{J^{(\alpha)}(A \times A^c)}{\pi(A)\sqrt{\log[e + \pi(A)^{-1}]}}$	(F.Y. Wang, 2001)
Log. Sobolev	$\displaystyle\lim_{\delta\to\infty}\inf_{\pi(A)>0} \frac{J^{(\alpha)}(A \times A^c) + \delta\pi(A)}{\pi(A)\sqrt{1 - \log\pi(A)}}$	(Chen, 2000b)
Nash	$\displaystyle\inf_{\pi(A)\in(0,1)} \frac{J^{(\alpha)}(A \times A^c)}{[\pi(A) \wedge \pi(A^c)]^{(2q-3)/(2q-2)}}$	(Chen, 1999b)

A presentation of Cheeger's technique is the aim of Chapter 4 where the closely related first Dirichlet eigenvalue is also studied.

1.4 A new picture of ergodic theory and explicit criteria

Importance of the inequalities

Let $(P_t)_{t\geqslant 0}$ be the semigroup determined by a Dirichlet form $(D, \mathscr{D}(D))$. Then, various applications of the inequalities are based on the following results.

Theorem 1.8 (T.M. Liggett (1989), L. Gross (1976), and Chen (1999b)).

(1) Poincaré inequality $\Longleftrightarrow$ L^2-exponential convergence:
$\|P_t f - \pi(f)\|^2 = \mathrm{Var}(P_t f) \leqslant \mathrm{Var}(f)\exp[-2\lambda_1 t]$.

(2) Logarithmic Sobolev inequality $\Longrightarrow$ exponential convergence in entropy:
$\mathrm{Ent}(P_t f) \leqslant \mathrm{Ent}(f)\exp[-2\sigma t]$, where $\mathrm{Ent}(f) = \pi(f\log f) - \pi(f)\log\|f\|_1$ and $2/\sigma$ is the optimal constant C in (1.26).

(3) Nash inequality $\Longleftrightarrow \mathrm{Var}(P_t f) \leqslant C\|f\|_1^2/t^{q-1}$.

In the context of diffusions, one can replace "$\Longrightarrow$" by "$\Longleftrightarrow$" in part (2). Therefore, the above inequalities describe some type of L^2-ergodicity for the semigroup $(P_t)_{t \geqslant 0}$. These inequalities have become powerful tools in the study of infinite-dimensional mathematics (phase transitions, for instance) and the effectiveness of random algorithms.

Three traditional types of ergodicity

The following three types of ergodicity are well known for Markov processes:

Ordinary ergodicity : $\displaystyle\lim_{t\to\infty} \|p_t(x,\cdot) - \pi\|_{\mathrm{Var}} = 0,$

Exponential ergodicity : $\|p_t(x,\cdot) - \pi\|_{\mathrm{Var}} \leqslant C(x)e^{-\alpha t}$ for some $\alpha > 0,$

Strong ergodicity : $\displaystyle\lim_{t\to\infty} \sup_x \|p_t(x,\cdot) - \pi\|_{\mathrm{Var}} = 0$

$$\Longleftarrow \lim_{t\to\infty} e^{\beta t} \sup_x \|p_t(x,\cdot) - \pi\|_{\mathrm{Var}} = 0 \text{ for some } \beta > 0,$$

where $p_t(x, dy)$ is the transition function of the Markov process and $\|\cdot\|_{\mathrm{Var}}$ is the total variation norm. They obey the following implications:

Strong ergodicity $\Longrightarrow$ Exponential ergodicity $\Longrightarrow$ Ordinary ergodicity.

It is natural to ask the following question: does there exist any relation between the above inequalities and the three traditional types of ergodicity?

A new picture of ergodic theory

Theorem 1.9 (Chen (1999c), et al).   Let $(E, \mathscr{E})$ be a measurable space with countably generated $\mathscr{E}$. Then, for a Markov process with state space $(E, \mathscr{E})$, reversible and having transition probability densities with respect to a probability measure, we have the diagram shown in Figure 1.1.

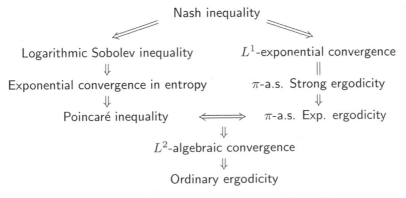

Figure 1.1   Diagram of nine types of ergodicity

In Figure 1.1, L^2-*algebraic convergence* means that $\mathrm{Var}(P_t f) \leqslant CV(f)t^{1-q}$ ($t > 0$) holds for some V having the properties that V is homogeneous of degree two in the sense that

$$V(cf + d) = c^2 V(f)$$

for any constants c and d, and $V(f) < \infty$ for all functions f with finite support. We will come back to this topic in Section 7.6. As usual, L^p ($p \geqslant 1$)-exponential convergence means that

$$\|P_t f - \pi(f)\|_p \leqslant \|f - \pi(f)\|_p\, e^{-\varepsilon t}, \qquad t \geqslant 0,\ f \in L^p(\pi),$$

for some $\varepsilon > 0$.

The diagram is complete in the following sense. Each single implication cannot be replaced by a double one. Moreover, strongly ergodic convergence and the logarithmic Sobolev inequality (respectively, exponential convergence in entropy) are not comparable. With the exception of the equivalences, all the implications in the diagram are suitable for more general Markov processes. Clearly, the diagram extends the ergodic theory of Markov processes.

The application of the diagram is obvious. For instance, from the well-known criteria for exponential ergodicity, one obtains immediately some criteria (which are indeed new) for the Poincaré inequality. On the other hand, by using the estimates obtained from the study of the Poincaré inequality, one may estimate the exponentially ergodic convergence rate (for which knowledge is still very limited).

The diagram was presented in Chen (1999c), stated mainly for Markov chains. Recently, the equivalence of L^1-exponential convergence and strong ergodicity was proved by Y.H. Mao (2002c). A counterexample of diffusion which shows that strongly ergodic convergence does not imply exponential convergence in entropy is constructed by F.Y. Wang (2002). For L^2-algebraic convergence, refer to T.M. Liggett (1991), J.D. Deuschel (1994), Chen and Y.Z. Wang (2003), and references therein.

Detailed proofs of the diagram with some additional results are presented in Chapter 8.

Explicit criteria for several types of ergodicity

As an application of the diagram in Figure 1.1, we obtain a criterion for the exponential ergodicity of birth–death processes, as listed in Table 1.4. To achieve this, we use the equivalence of exponential ergodicity and Poincaré inequality, as well as the explicit criterion for the Poincaré inequality given in part (3) of Theorem 1.5. This solves a long—standing open problem in the study of Markov chains [cf. W.J. Anderson (1991, Section 6.6), Chen (1992a, Section 4.4)].

Next, it is natural to look for some criteria for other types of ergodicity. To do so, we consider only the one-dimensional case. Here we focus on birth–death processes, since one-dimensional diffusion processes are in parallel. A criterion for strong ergodicity was obtained recently by H.J. Zhang, X. Lin and Z.T. Hou (2000), and extended by Y.H. Zhang (2001), using a different approach, to a larger class of Markov chains. The criteria for the logarithmic Sobolev and Nash inequalities and the *discrete spectrum* (the continuous spectrum is empty and all eigenvalues have finite multiplicity) were obtained by S.G. Bobkov and F. Götze (1999a; 1999b), and Y.H. Mao (2004, 2002a,b), respectively, based on the weighted Hardy inequality [see also L. Miclo (1999a,b), F.Y. Wang (2000a,b), F.Z. Gong and F.Y. Wang (2002)]. It is understood now that the results can also be deduced from generalizations of the variational formulas discussed in this chapter [cf. Chen (2002a, 2003a,b) and Chapter 6]. Finally, we summarize these results in Theorem 1.10 and Table 1.4. The first three criteria in the table are classical, but all the others are very recent. The table is arranged in such an order that the property in each line is stronger than the property in the previous line. The only exception is that even though strong ergodicity is often stronger than the logarithmic Sobolev inequality, they are not comparable in general, as mentioned in Section 1.3.

Recall the sequence (μ_n) defined above (1.24) and set $\mu[i,k] = \sum_{i \leqslant j \leqslant k} \mu_j$.

Table 1.4 Ten criteria for birth–death processes

Property	Criterion
Uniqueness	$\sum_{n \geqslant 0} \dfrac{1}{\mu_n b_n} \mu[0,n] = \infty \quad (*)$
Recurrence	$\sum_{n \geqslant 0} \dfrac{1}{\mu_n b_n} = \infty$
Ergodicity	$(*) \ \& \ \mu[0,\infty) < \infty$
Exponential ergodicity L^2-exp. convergence	$(*) \ \& \ \sup_{n \geqslant 1} \mu[n,\infty) \sum_{j \leqslant n-1} \dfrac{1}{\mu_j b_j} < \infty$
Discrete spectrum	$(*) \ \& \ \lim_{n \to \infty} \sup_{k \geqslant n+1} \mu[k,\infty) \sum_{n \leqslant j \leqslant k-1} \dfrac{1}{\mu_j b_j} = 0$
Log. Sobolev inequality	$(*) \ \& \ \sup_{n \geqslant 1} \mu[n,\infty)\log[\mu[n,\infty)^{-1}] \sum_{j \leqslant n-1} \dfrac{1}{\mu_j b_j} < \infty$
Strong ergodicity L^1-exp. convergence	$(*) \ \& \ \sum_{n \geqslant 0} \dfrac{1}{\mu_n b_n} \mu[n+1,\infty) = \sum_{n \geqslant 1} \mu_n \sum_{j \leqslant n-1} \dfrac{1}{\mu_j b_j} < \infty$
Nash inequality	$(*) \ \& \ \sup_{n \geqslant 1} \mu[n,\infty)^{(q-2)/(q-1)} \sum_{j \leqslant n-1} \dfrac{1}{\mu_j b_j} < \infty \ (\varepsilon)$

Theorem 1.10 (Chen, 2001a).   For birth–death processes with birth rates $b_i(i \geqslant 0)$ and death rates $a_i(i \geqslant 1)$, ten criteria are listed in Table 1.4, in which the notation "$(*)$ & $\cdots$" means that one requires the uniqueness condition in the first line plus the condition "$\cdots$". The notation "(ε)" in the last line means that for small q, $1 < q \leqslant 2$, a criterion for the Nash inequality is still unknown.

The proofs of the criteria will be started in Chapter 5 and continued in Chapter 6. In Chapter 5, both the coupling method and an analytic method are used to prove the criteria for exponential or strong ergodicity. In Chapter 6, most of the remaining criteria are proved in terms of a generalization of the second variational formula stated in Theorem 1.5 (1) to some Orlicz spaces. Further generalization to the higher-dimensional case in terms of capacity is left to Chapter 7.

A large part of the author's original research papers are collected in Chen (2001c).

Summary

In conclusion, we have discussed in the chapter three levels of problems, three methods, and mainly four results. According to the range of problems, the principal eigenvalues, the basic inequalities, and the ergodic theory, each has a wider range than the previous one. We have used the coupling method from probability theory, Cheeger's approach from Riemannian geometry, and the weighted Hardy inequality from harmonic analysis. Finally, we have presented some variational formulas for the exponentially ergodic rates, new forms of Cheeger's constants, a comparison diagram, and a table of explicit criteria for several types of ergodicity.

Chapter 2

Optimal Markovian Couplings

This chapter introduces our first mathematical tool, the coupling methods, in the study of the topics in the book, and they will be used many times in the subsequent chapters. We introduce couplings, Markovian couplings (Section 2.1), and optimal Markovian couplings (Sections 2.2 and 2.3), mainly for time-continuous Markov processes. The study emphasizes analysis of the coupling operators rather than the processes. Some constructions of optimal Markovian couplings for Markov chains and diffusions are presented, which are often unexpected. Two general results of applications to the estimation of the first eigenvalue are proved in Section 2.4. Furthermore, some typical applications of the methods are illustrated through simple examples.

2.1 Couplings and Markovian couplings

Let us recall the simple definition of couplings.

Definition 2.1. Let μ_k be a probability on a measurable space $(E_k, \mathscr{E}_k)$, $k = 1, 2$. A probability measure $\tilde{\mu}$ on the product measurable space $(E_1 \times E_2, \mathscr{E}_1 \times \mathscr{E}_2)$ is called a *coupling of μ_1 and μ_2* if the following *marginality* condition holds:

$$\begin{aligned}\tilde{\mu}(A_1 \times E_2) &= \mu_1(A_1), & A_1 \in \mathscr{E}_1,\\ \tilde{\mu}(E_1 \times A_2) &= \mu_2(A_2), & A_2 \in \mathscr{E}_2.\end{aligned} \tag{M}$$

Example 2.2 (Independent coupling $\tilde{\mu}_0$). $\tilde{\mu}_0 = \mu_1 \times \mu_2$. That is, $\tilde{\mu}_0$ is the independent product of μ_1 and μ_2.

This trivial coupling has already a nontrivial application. Let $\mu_k = \mu$ on $\mathbb{R}$, $k = 1, 2$. We say that μ satisfies the *FKG inequality* if

$$\int_{\mathbb{R}} fg\,\mathrm{d}\mu \geqslant \int_{\mathbb{R}} f\,\mathrm{d}\mu \int_{\mathbb{R}} g\,\mathrm{d}\mu, \qquad f, g \in \mathscr{M}, \tag{2.1}$$

where $\mathscr{M}$ is the set of bounded monotone functions on $\mathbb{R}$. Here is a one-line proof based on the independent coupling:

$$\iint \tilde{\mu}_0(\mathrm{d}x, \mathrm{d}y)[f(x) - f(y)][g(x) - g(y)] \geqslant 0, \qquad f, g \in \mathscr{M}.$$

We mention that a criterion of FKG inequality for higher-dimensional measures on $\mathbb{R}^d$ (more precisely, for diffusions) was obtained by Chen and F.Y. Wang (1993a). However, a criterion is still unknown for Markov chains.

Open Problem 2.3. What is the criterion of FKG inequality for Markov jump processes?

We will explain the meaning of the problem carefully at the end of this section and explain the term "Markov jump processes" soon. The next example is nontrivial.

Example 2.4 (Basic coupling $\tilde{\mu}_b$). Let $E_k = E$, $k = 1, 2$. Denote by Δ the diagonals in E: $\Delta = \{(x, x) : x \in E\}$. Take

$$\tilde{\mu}_b(\mathrm{d}x_1, \mathrm{d}x_2) = (\mu_1 \wedge \mu_2)(\mathrm{d}x_1)I_\Delta + \frac{(\mu_1 - \mu_2)^+(\mathrm{d}x_1)(\mu_1 - \mu_2)^-(\mathrm{d}x_2)}{(\mu_1 - \mu_2)^+(E)}I_{\Delta^c},$$

where $\nu^\pm$ is the Jordan–Hahn decomposition of a signed measure ν and $\nu_1 \wedge \nu_2 = \nu_1 - (\nu_1 - \nu_2)^+$.

Note that one may ignore I_{Δ^c} in the above formula, since $(\mu_1 - \mu_2)^+$ and $(\mu_1 - \mu_2)^-$ have different supports.

Actually, the basic coupling is optimal in the following sense. Let ρ be the discrete distance: $\rho(x, y) = 1$ if $x \neq y$, and $= 0$ if $x = y$. Then a simple computation shows that

$$\tilde{\mu}_b(\rho) = \frac{1}{2}\|\mu_1 - \mu_2\|_{\mathrm{Var}}.$$

Thus, by Dobrushin's theorem (see Theorem 2.23 below), we have

$$\tilde{\mu}_b(\rho) = \inf_{\tilde{\mu}} \tilde{\mu}(\rho),$$

where $\tilde{\mu}$ varies over all couplings of μ_1 and μ_2. In other words, $\tilde{\mu}_b(\rho)$ is a ρ-*optimal coupling*. This indicates an optimality for couplings that we are going to study in this chapter.

Similarly, we can define a coupling process of two stochastic processes in terms of their distributions at each time t for fixed initial points. Of course, for given marginal Markov processes, the resulting coupled process may not be Markovian. Non-Markovian couplings are useful, especially in the time-discrete situation. However, in the time-continuous case, they are often not practical. Hence, we now restrict ourselves to the Markovian couplings.

Definition 2.5. Given two Markov processes with semigroups $P_k(t)$ or transition probabilities $P_k(t, x_k, \cdot)$ on $(E_k, \mathscr{E}_k)$, $k = 1, 2$, a *Markovian coupling* is a Markov process with semigroup $\widetilde{P}(t)$ or transition probability $\widetilde{P}(t; x_1, x_2; \cdot)$ on the product space $(E_1 \times E_2, \mathscr{E}_1 \times \mathscr{E}_2)$ having the *marginality*

$$
\begin{aligned}
\widetilde{P}(t; x_1, x_2; A_1 \times E_2) &= P_1(t, x_1, A_1), \\
\widetilde{P}(t; x_1, x_2; E_1 \times A_2) &= P_2(t, x_2, A_2), \quad t \geqslant 0, x_k \in E_k, A_k \in \mathscr{E}_k, k = 1, 2.
\end{aligned} \tag{MP}
$$

Equivalently,

$$
\begin{aligned}
\widetilde{P}(t) f(x_1, x_2) &= P_1(t) f(x_1), \\
\widetilde{P}(t) f(x_1, x_2) &= P_2(t) f(x_2), \quad t \geqslant 0, \ x_k \in E_k, \ f \in {}_b\mathscr{E}_k, \ k = 1, 2,
\end{aligned} \tag{MP}
$$

where ${}_b\mathscr{E}$ is the set of all bounded $\mathscr{E}$-measurable functions. Here, on the left-hand side, f is regarded as a bivariate function.

We now consider Markov jump processes. For this, we need some notation. Let $(E, \mathscr{E})$ be a measurable space such that $\{(x, x) : x \in E\} \in \mathscr{E} \times \mathscr{E}$ and $\{x\} \in \mathscr{E}$ for all $x \in E$. It is well known that for a given sub-Markovian transition function $P(t, x, A) (t \geqslant 0, x \in E, A \in \mathscr{E})$, if it satisfies the *jump condition*

$$
\lim_{t \to 0} P(t, x, \{x\}) = 1, \qquad x \in E, \tag{2.2}
$$

then the limits

$$
q(x) := \lim_{t \to 0} \frac{1 - P(t, x, \{x\})}{t} \quad \text{and} \quad q(x, A) := \lim_{t \to 0} \frac{P(t, x, A \setminus \{x\})}{t} \tag{2.3}
$$

exist for all $x \in E$ and $A \in \mathscr{R}$, where

$$
\mathscr{R} = \left\{ A \in \mathscr{E} : \lim_{t \to 0} \sup_{x \in A} \left[1 - P(t, x, \{x\}) \right] = 0 \right\}.
$$

Moreover, for each $A \in \mathscr{R}$, $q(\cdot)$, $q(\cdot, A) \in \mathscr{E}$, for each $x \in E$, $q(x, \cdot)$ is a finite measure on $(E, \mathscr{R})$, and $0 \leqslant q(x, A) \leqslant q(x) \leqslant \infty$ for all $x \in E$ and $A \in \mathscr{R}$. The pair $(q(x), q(x, A)) (x \in E, A \in \mathscr{R})$ is called a *q-pair* (also called the transition intensity or transition rate). The q-pair is said to be *totally stable* if $q(x) < \infty$ for all $x \in E$. Then $q(x, \cdot)$ can be uniquely extended to the whole space $\mathscr{E}$ as a finite measure. Next, the q-pair $(q(x), q(x, A))$ is called *conservative* if $q(x, E) = q(x) < \infty$ for all $x \in E$ (Note that the conservativity here is different from the one often used in the context of diffusions). Because of the above facts, we often call the sub-Markovian transition $P(t, x, A)$ satisfying (2.3) a *jump process* or a *q-process*. Finally, a q-pair is called *regular* if it is not only totally stable and conservative but also determines uniquely a jump process (nonexplosive).

When E is countable, conventionally we use the matrix $Q = (q_{ij} : i, j \in E)$ (called a *Q-matrix*) and $P(t) = (p_{ij}(t) : i, j \in E)$,

$$
p'_{ij}(t)\big|_{t=0} = q_{ij},
$$

instead of the q-pair and the jump process, respectively. Here $q_{ii} = -q_i$, $i \in E$. We also call $P(t) = (p_{ij}(t))$ a *Markov chain* (which is used throughout this book only for a discrete state space) or a Q-*process*.

In practice, what we know in advance is the q-pair $(q(x), q(x, \mathrm{d}y))$ but not $P(t, x, \mathrm{d}y)$. Hence, our real interest goes in the opposite direction. How does a q-pair determine the properties of $P(t, x, \mathrm{d}y)$? A large part of the book (Chen, 1992a) is devoted to the theory of jump processes. Here, we would like to mention that the theory now has a very nice application to quantum physics that was missed in the quoted book. Refer to the survey article by A.A. Konstantinov, U.P. Maslov, and A.M. Chebotarev (1990) and references within.

Clearly, there is a one-to-one correspondence between a q-pair and the operator Ω:

$$\Omega f(x) = \int_E q(x, \mathrm{d}y)[f(y) - f(x)] - [q(x) - q(x, E)]f(x), \qquad f \in {}_b\mathscr{E}.$$

Because of this correspondence, we will use both according to our convenience. Corresponding to a coupled Markov jump process, we have a q-pair $(\tilde{q}(x_1, x_2), \tilde{q}(x_1, x_2; \mathrm{d}y_1, \mathrm{d}y_2))$ as follows:

$$\tilde{q}(x_1, x_2) = \lim_{t \to 0} \frac{1 - \widetilde{P}(t; x_1, x_2; \{x_1\} \times \{x_2\})}{t}, \qquad (x_1, x_2) \in E_1 \times E_2,$$

$$\tilde{q}(x_1, x_2; \tilde{A}) = \lim_{t \to 0} \frac{\widetilde{P}(t; x_1, x_2; \tilde{A})}{t}, \qquad (x_1, x_2) \notin \tilde{A} \in \widetilde{\mathscr{R}},$$

$$\widetilde{\mathscr{R}} := \left\{ \tilde{A} \in \mathscr{E}_1 \times \mathscr{E}_2 : \lim_{t \to 0} \sup_{(x_1, x_2) \in \tilde{A}} \left[1 - \widetilde{P}(t; x_1, x_2; \{(x_1, x_2)\}) \right] = 0 \right\}.$$

Concerning the total stability and conservativity of the q-pair of a coupling (or coupled) process, we have the following result.

Theorem 2.6.   The following assertions hold:

(1) A (equivalently, any) Markovian coupling is a jump process iff so are their marginals.

(2) A (equivalently, any) coupling q-pair is totally stable iff so are the marginals.

(3) [Y. H. Zhang, 1994]. A (equivalently, any) coupling q-pair is conservative iff so are the marginals.

Proof of parts (1) and (2). To obtain a feeling for the proof, we prove here the easier part of the theorem. This proof is taken from Chen (1994b).

(a) First, we consider the *jump condition*. Let $P_k(t, x_k, \mathrm{d}y_k)$ and $\widetilde{P}(t; x_1, x_2; \mathrm{d}y_1, \mathrm{d}y_2)$ be the marginal and coupled Markov processes, respectively. By the marginality for processes, we have

$$\widetilde{P}(t; x_1, x_2; \{x_1\} \times \{x_2\})$$
$$\geqslant \widetilde{P}(t; x_1, x_2; \{x_1\} \times E_2) - \widetilde{P}(t; x_1, x_2; E_1 \times (E_2 \setminus \{x_2\}))$$
$$\geqslant \widetilde{P}(t; x_1, x_2; \{x_1\} \times E_2) - 1 + \widetilde{P}(t; x_1, x_2; E_1 \times \{x_2\})$$
$$= P_1(t, x_1, \{x_1\}) - 1 + P_2(t, x_2, \{x_2\}).$$

If both of the marginals are jump processes, then $\underline{\lim}_{t \to 0} \widetilde{P}(t; x_1, x_2; \{x_1\} \times \{x_2\}) \geqslant 1$. Thus, a Markovian coupling $\widetilde{P}(t)$ must be a jump process.
  Conversely, since

$$\widetilde{P}(t; x_1, x_2; \{x_1\} \times \{x_2\}) \leqslant \widetilde{P}(t; x_1, x_2; \{x_1\} \times E_2) = P_1(t, x_1, \{x_1\}),$$

if $\widetilde{P}(t)$ is a jump process, then $\underline{\lim}_{t \to 0} P_1(t, x_1, \{x_1\}) \geqslant 1$, and so $P_1(t)$ is also a jump process. Symmetrically, so is $P_2(t)$.
  (b) Next, we consider the *equivalence of total stability*. Assume that all the processes concerned are jump processes. Denote by $(q_k(x_k), q_k(x_k, dy_k))$ the marginal q-pairs on $(E_k, \mathscr{R}_k)$, where

$$\mathscr{R}_k = \left\{ A \in \mathscr{E}_k : \lim_{t \to 0} \sup_{x \in A} \left[1 - P_k(l, x, \{x\}) \right] = 0 \right\}, \qquad k = 1, 2.$$

Denote by $(\tilde{q}(x_1, x_2), \tilde{q}(x_1, x_2; dy_1, dy_2))$ a coupling q-pair on $(E_1 \times E_2, \widetilde{\mathscr{R}})$. We need to show that $\tilde{q}(\tilde{x}) < \infty$ for all $\tilde{x} \in E_1 \times E_2$ iff $q_1(x_1) \vee q_2(x_2) < \infty$ for all $x_1 \in E_1$ and $x_2 \in E_2$. Clearly, it suffices to show that

$$q_1(x_1) \vee q_2(x_2) \leqslant \tilde{q}(x_1, x_2) \leqslant q_1(x_1) + q_2(x_2).$$

Note that we cannot use either the conservativity or uniqueness of the processes at this step. But the last assertion follows from (a) and the first part of (2.3) immediately. □

  Due to Theorem 2.6, *from now on, assume that all coupling operators considered below are conservative.* Then we have

$$\tilde{q}(x_1, x_2) = \lim_{t \to 0} \frac{1 - \widetilde{P}(t; x_1, x_2; \{x_1\} \times \{x_2\})}{t}, \qquad (x_1, x_2) \in E_1 \times F_2,$$

$$\tilde{q}(x_1, x_2; \tilde{A}) = \lim_{t \to 0} \frac{\widetilde{P}(t; x_1, x_2; \tilde{A})}{t}, \qquad (x_1, x_2) \notin \tilde{A} \in \mathscr{E}_1 \times \mathscr{E}_2.$$

Note that in the second line, the original set $\widetilde{\mathscr{R}}$ is replaced by $\mathscr{E}_1 \times \mathscr{E}_2$. Define

$$\Omega_1 f(x_1) = \int_{E_1} q_1(x_1, dy_1)[f(y_1) - f(x_1)], \qquad f \in b\mathscr{E}_1.$$

Similarly, we can define Ω_2. Corresponding to a coupling process $\widetilde{P}(t)$, we also have an operator $\widetilde{\Omega}$. Now, since the marginal q-pairs and the coupling

q-pairs are all conservative, it is not difficult to prove that (MP) implies the following:

$$\widetilde{\Omega}f(x_1, x_2) = \Omega_1 f(x_1), \qquad f \in {}_b\mathscr{E}_1,$$
$$\widetilde{\Omega}f(x_1, x_2) = \Omega_2 f(x_2), \qquad f \in {}_b\mathscr{E}_2, \ x_k \in E_k, \ k = 1, 2. \qquad \text{(MO)}$$

Again, on the left-hand side, f is regarded as a bivariate function. Refer to Chen (1986a) or Chen (1992a, Chapter 5). Here, "MO" means the marginality for operators.

Definition 2.7. Any operator $\widetilde{\Omega}$ satisfying (MO) is called a *coupling operator*.

Do there exist any coupling operators?

Examples of coupling operators for jump processes

The simplest example to answer the above question is the following.

Example 2.8 (Independent coupling $\widetilde{\Omega}_0$).

$$\widetilde{\Omega}_0 f(x_1, x_2) = [\Omega_1 f(\cdot, x_2)](x_1) + [\Omega_2 f(x_1, \cdot)](x_2), \qquad x_k \in E_k, \ k = 1, 2.$$

This coupling is trivial, but it does show that a coupling operator always exists.

To simplify our notation, in what follows, instead of writing down a coupling operator, we will use tables. For instance, a conservative q-pair can be expressed as follows:

$$x \to dy \setminus \{x\} \qquad \text{at rate} \qquad q(x, dy).$$

In particular, in the discrete case, a conservative Q-matrix can be expressed as

$$i \to j \neq i \qquad \text{at rate} \qquad q_{ij}.$$

Example 2.9 (Classical coupling $\widetilde{\Omega}_c$). Take $E_1 = E_2 = E$ and let $\Omega_1 = \Omega_2 = \Omega$. If $x_1 \neq x_2$, then take

$$
\begin{aligned}
(x_1, x_2) \quad &\to \quad (y_1, x_2) \quad \text{at rate} \ q(x_1, dy_1) \\
&\to \quad (x_1, y_2) \quad \text{at rate} \ q(x_2, dy_2).
\end{aligned}
$$

Otherwise,

$$(x, x) \to (y, y) \qquad \text{at rate} \qquad q(x, dy).$$

Each coupling has its own character. The classical coupling means that the marginals evolve independently until they meet. Then they move together. A nice way to interpret this coupling is to use a Chinese idiom: fall in love at first sight. That is, a boy and a girl had independent paths of their lives before the first time they met each other. Once they meet, they are in love at once and will have the same path of their lives forever. When the marginal Q-matrices are the same, all couplings considered below will have the property listed in the last line, and hence we will omit the last line in what follows.

Example 2.10 (Basic coupling $\widetilde{\Omega}_b$). For $x_1, x_2 \in E$, take

$$
\begin{aligned}
(x_1, x_2) \quad &\to \quad (y, y) &&\text{at rate} && \big[q_1(x_1, \cdot) \wedge q_2(x_2, \cdot)\big](dy) \\
&\to \quad (y_1, x_2) &&\text{at rate} && \big[q_1(x_1, \cdot) - q_2(x_2, \cdot)\big]^{+}(dy_1) \\
&\to \quad (x_1, y_2) &&\text{at rate} && \big[q_2(x_2, \cdot) - q_1(x_1, \cdot)\big]^{+}(dy_2).
\end{aligned}
$$

The basic coupling means that the components jump to the same place at the greatest possible rate. This explains where the term $q_1(x_1, dy_1) \wedge q_2(x_2, dy_2)$ comes from, which is the biggest one to guarantee the marginality. This term is the key of the coupling. Note that whenever we have a term $A \wedge B$, we should have the other two terms $(A - B)^{+}$ and $(B - A)^{+}$ automatically, again, due to the marginality. Thus, in what follows, we will write down the term $A \wedge B$ only for simplicity.

Example 2.11 (Coupling of marching soldiers $\widetilde{\Omega}_m$). Assume that E is an addition group. Take

$$
(x_1, x_2) \to (x_1 + y, x_2 + y) \qquad \text{at rate} \qquad q_1(x_1, x_1 + dy) \wedge q_2(x_2, x_2 + dy).
$$

The word "marching" is a Chinese name, which is the command to soldiers to start marching. Thus, this coupling means that at each step, the components maintain the same length of jumps at the biggest possible rate.

In the time-discrete case, the classical coupling and the basic coupling are due to W. Doeblin (1938) (which was the first paper to study the convergence rate by coupling) and L.N. Wasserstein (1969), respectively. The coupling of marching soldiers is due to Chen (1986b). The original purpose of the last coupling is mainly to preserve the order.

Let us now consider a birth–death process with regular Q-matrix:

$$
q_{i,i+1} = b_i, \quad i \geqslant 0; \qquad q_{i,i-1} = a_i, \quad i \geqslant 1.
$$

Then for two copies of the process starting from i_1 and i_2, respectively, we have the following two examples taken from (Chen, 1990).

Example 2.12 (Modified coupling of marching soldiers $\widetilde{\Omega}_{cm}$). Take $\widetilde{\Omega}_{cm} = \widetilde{\Omega}_c$ if $|i_1 - i_2| \leqslant 1$ and $\widetilde{\Omega}_{cm} = \widetilde{\Omega}_m$ if $|i_1 - i_2| \geqslant 2$.

Example 2.13 (Coupling by inner reflection $\widetilde{\Omega}_{ir}$). Again, take $\widetilde{\Omega}_{ir} = \widetilde{\Omega}_c$ if $|i_1 - i_2| \leqslant 1$. For $i_2 \geqslant i_1 + 2$, take

$$
\begin{aligned}
(i_1, i_2) \quad &\to \quad (i_1 + 1, i_2 - 1) &&\text{at rate} && b_{i_1} \wedge a_{i_2} \\
&\to \quad (i_1 - 1, i_2) &&\text{at rate} && a_{i_1} \\
&\to \quad (i_1, i_2 + 1) &&\text{at rate} && b_{i_2}.
\end{aligned}
$$

By exchanging i_1 and i_2, we can get the expression of $\widetilde{\Omega}_{ir}$ for the case that $i_1 \geqslant i_2$.

This coupling lets the components move to the closed place (not necessarily the same place as required by the basic coupling) at the biggest possible rate.

From these examples one sees that there are many choices of a coupling operator $\widetilde{\Omega}$. Indeed, there are infinitely many choices! Thus, in order to use the coupling technique, a basic problem we should study is the regularity (nonexplosive problem) of coupling operators, for which, fortunately, we have a complete answer [Chen (1986a) or Chen (1992a, Chapter 5)]. The following result can be regarded as a fundamental theorem for couplings of jump processes.

Theorem 2.14 (Chen, 1986a).

    (1) If a coupling operator is nonexplosive, then so are its marginals.

    (2) If the marginals are both nonexplosive, then so is every coupling operator.

    (3) In the nonexplosive case, (MP) and (MO) are equivalent.

Clearly, Theorem 2.14 simplifies greatly our study of couplings for general jump processes, since the marginality (MP) of a coupling process is reduced to the rather simpler marginality (MO) of the corresponding operators. The hard but most important part of the theorem is the second assertion, since there are infinitely many coupling operators having no unified expression.

Markovian couplings for diffusions

We now turn to study the couplings for diffusion processes in $\mathbb{R}^d$ with second-order differential operator

$$L = \frac{1}{2} \sum_{i,j=1}^d a_{ij}(x) \frac{\partial^2}{\partial x_i \partial x_j} + \sum_{i=1}^d b_i(x) \frac{\partial}{\partial x_i}.$$

For simplicity, we write $L \sim (a(x), b(x))$. Given two diffusions with operators

$$L_k \sim (a_k(x), b_k(x)), \qquad k = 1, 2,$$

respectively, an elliptic (may be degenerate) operator $\widetilde{L}$ on the product space $\mathbb{R}^d \times \mathbb{R}^d$ is called a *coupling* of L_1 and L_2 if it satisfies the following *marginality*:

$$\widetilde{L}f(x,y) = L_1 f(x) \qquad (\text{respectively, } \widetilde{L}f(x,y) = L_2 f(y)),$$
$$f \in C_b^2(\mathbb{R}^d), \quad x \neq y.$$
(MO)

Again, on the left-hand side, f is regarded as a bivariate function. From this, it is clear that the coefficients of any coupling operator $\widetilde{L}$ should be of the form

$$a(x,y) = \begin{pmatrix} a_1(x) & c(x,y) \\ c(x,y)^* & a_2(y) \end{pmatrix}, \qquad b(x,y) = \begin{pmatrix} b_1(x) \\ b_2(y) \end{pmatrix},$$

where the matrix $c(x, y)^*$ is the conjugate of $c(x, y)$. This condition and the nonnegative definite property of $a(x, y)$ constitute the *marginality* in the context of diffusions. Obviously, the only freedom is the choice of $c(x, y)$.

As an analogue of jump processes, we have the following examples.

Example 2.15 (Classical coupling). $c(x, y) \equiv 0$ for all $x \neq y$.

Example 2.16 (Coupling of marching soldiers [Chen and S.F. Li 1989]**).** Let $a_k(x) = \sigma_k(x)\sigma_k(x)^*$, $k = 1,\ 2$. Take $c(x, y) = \sigma_1(x)\sigma_2(y)^*$.

The two choices given in the next example are due to T. Lindvall and L.C.G. Rogers (1986), Chen and S.F. Li (1989), respectively.

Example 2.17 (Coupling by reflection). Let $L_1 = L_2$ and $a(x) = \sigma(x)\sigma(x)^*$. We have two choices:

$$c(x, y) = \sigma(x)\left[\sigma(y)^* - 2\frac{\sigma(y)^{-1}\bar{u}\bar{u}^*}{|\sigma(y)^{-1}\bar{u}|^2}\right], \quad \det\sigma(y) \neq 0, \qquad x \neq y,$$

$$c(x, y) = \sigma(x)\left[I - 2\bar{u}\bar{u}^*\right]\sigma(y)^*, \qquad x \neq y,$$

where $\bar{u} = (x - y)/|x - y|$.

This coupling was generalized to Riemannian manifolds by W.S. Kendall (1986) and M. Cranston (1991).

In the case that $x = y$, the first and the third couplings here are defined to be the same as the second one.

In probabilistic language, suppose that the original process is given by the stochastic differential equation

$$dX_t = \sqrt{2}\,\sigma(X_t)dB_t + b(X_t)dt,$$

where (B_t) is a Brownian motion. We want to construct a new process (X_t'),

$$dX_t' = \sqrt{2}\,\sigma'(X_t)dB_t' + b'(X_t)dt,$$

on the same probability space, having the same distribution as that of (X_t). Then, what we need is only to choose a suitable Brownian motion (B_t'). Corresponding to the above three examples, we have

(1) *Classical coupling*: B_t' is a new Brownian motion, independent of B_t.
(2) *Coupling of marching soldiers*: $B_t' = B_t$.
(3) *Coupling by reflection*: $B_t' = [I - 2\bar{u}\bar{u}^*](X_t, X_t')B_t$, where $\bar{u}$ is given in Example 2.17.

It is important to remark that in the constructions, we need only consider the time $t < T$, where T is the *coupling time*,

$$T = \inf\{t \geq 0 : X_t = X_t'\},$$

since $X_t = X_t'$ for all $t \geq T$. This avoids the degeneration of the coupling operators.

Before moving further, let us mention a conjecture:

Conjecture 2.18. The fundamental theorem (Theorem 2.14) holds for diffusions.

The following facts strongly support the conjecture.

(a) A well known sufficient condition says that the operator L_k $(k = 1, 2)$ is well posed if there exists a function φ_k such that $\lim_{|x| \to \infty} \varphi_k(x) = \infty$ and $L_k \varphi_k \leqslant c \varphi_k$ for some constant c. Then the conclusion holds for all coupling operators, simply taking

$$\tilde{\varphi}(x_1, x_2) = \varphi_1(x_1) + \varphi_2(x_2).$$

(b) Let $\tau_{n,k}$ be the first time of leaving the cube with side length n of the kth process $(k = 1, 2)$ and let $\tilde{\tau}_n$ be the first time of leaving the product cube of coupled process. Then we have

$$\tau_{n,1} \vee \tau_{n,2} \leqslant \tilde{\tau}_n \leqslant \tau_{n,1} + \tau_{n,2}.$$

Moreover, a process, the kth one for instance, is well posed iff

$$\lim_{n \to \infty} \mathbb{P}_k[\tau_{n,k} < t] = 0.$$

Having studied the Markovian couplings for Markov jump processes and diffusions, it is natural to study the Lévy processes.

Open Problem 2.19. What should be the representation of Markovian coupling operators for Lévy processes?

2.2   Optimality with respect to distances

Since there are infinitely many Markovian couplings, we asked ourselves several times in the past years, does there exist an optimal one? Now another question arises: What is the optimality we are talking about? We now explain how we obtained a reasonable notion for optimal Markovian couplings. The first time we touched this problem was in Chen and S.F. Li (1989). It was proved there for Brownian motion that coupling by reflection is optimal with respect to the total variation, and moreover, for different probability metrics, the effective couplings can be different. The second time, in Chen (1990), it was proved that for birth–death processes, we have an order as follows:

$$\widetilde{\Omega}_{ir} \succ \widetilde{\Omega}_b \succ \widetilde{\Omega}_c \succ \widetilde{\Omega}_{cm} \succ \widetilde{\Omega}_m,$$

where $A \succ B$ means that A is better than B in some sense. However, only in 1992 it did become clear to the author how to optimize couplings.

To explain our optimal couplings, we need more preparation. As was mentioned several times in previous publications [Chen (1989a; 1989b; 1992a) and Chen and S. F. Li (1989)], it should be helpful to keep in mind the relation between couplings and the probability metrics. It will be clear soon that this is actually one of the key ideas of the study. As far as we know, there are more than 16 different probability distances, including the total variation and the Lévy–Prohorov distance for weak convergence. But we often are concerned with another distance. We now explain our understanding of how to introduce this distance.

As we know, in probability theory, we usually consider the types of convergence for real random variables on a probability space shown in Figure 2.1.

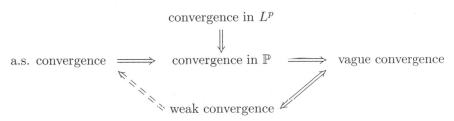

Figure 2.1 Typical types of convergence in probability theory

L^p-convergence, a.s. convergence, and convergence in $\mathbb{P}$ all depend on the reference frame, our probability space $(\Omega, \mathscr{F}, \mathbb{P})$. But vague (weak) convergence does not. By a result of Skorohod [cf. N. Ikeda and S. Watanabe (1988, p. 9 Theorem 2.7)], if P_n converges weakly to P, then we can choose a suitable reference frame $(\Omega, \mathscr{F}, \mathbb{P})$ such that $\xi_n \sim P_n$, $\xi \sim P$, and $\xi_n \to \xi$ a.s., where $\xi \sim P$ means that ξ has distribution P. Thus, all the types of convergence listed in Figure 2.1 are intrinsically the same, except L^p-convergence. In other words, if we want to find another intrinsic metric on the space of all probabilities, we should consider an analogue of L^p-convergence.

Let ξ_1, ξ_2: $(\Omega, \mathscr{F}, \mathbb{P}) \to (E, \rho, \mathscr{E})$. The usual L^p-metric is defined by

$$\|\xi_1 - \xi_2\|_p = \left\{\mathbb{E}\big[\rho(\xi_1, \xi_2)^p\big]\right\}^{1/p}.$$

Suppose that $\xi_i \sim P_i$, $i = 1, 2$, and $(\xi_1, \xi_2) \sim \widetilde{P}$. Then

$$\|\xi_1 - \xi_2\|_p = \left\{\int \rho(x_1, x_2)^p \widetilde{P}(\mathrm{d}x_1, \mathrm{d}x_2)\right\}^{1/p}.$$

Certainly, $\widetilde{P}$ is a coupling of P_1 and P_2. However, if we ignore our reference frame $(\Omega, \mathscr{F}, \mathbb{P})$, then there are many choices of $\widetilde{P}$ for given P_1 and P_2. Thus, the intrinsic metric should be defined as follows:

$$W_p(P_1, P_2) = \inf_{\widetilde{P}} \left\{\int \rho(x_1, x_2)^p \widetilde{P}(\mathrm{d}x_1, \mathrm{d}x_2)\right\}^{1/p}, \qquad p \geqslant 1,$$

where $\widetilde{P}$ varies over all couplings of P_1 and P_2.

Definition 2.20. The metric defined above is called the W_p-*distance* or pth *Wasserstein distance*. Briefly, we write $W = W_1$.

From the probabilistic point of view, the W_p-metrics have an intrinsic property that makes them more suitable for certain applications. For example, if (E, ρ) is the Euclidean space for P_2 obtained from P_1 by a translation, then $W_p(P_1, P_2)$ is just the length of the translation vector.

In general, it is quite hard to compute the W_p-distance exactly. Here are the main known results.

Theorem 2.21 (S.S. Vallender, 1973). Let P_k be a probability on the real line with distribution function $F_k(x)$, $k = 1, 2$. Then

$$W(P_1, P_2) = \int_{-\infty}^{+\infty} |F_1(x) - F_2(x)| \mathrm{d}x.$$

Theorem 2.22 (D.C. Dowson and B.V. Landau (1982), C.R. Givens and R.M. Shortt (1984), I. Olkin and R. Pukelsheim (1982)). Let P_k be the normal distribution on $(\mathbb{R}^d, \mathscr{B}(\mathbb{R}^d))$ $(d \geqslant 1)$ with mean value m_k and covariance matrix M_k, $k = 1, 2$. Then

$$W_2(P_1, P_2) = \big[|m_1 - m_2|^2 + \operatorname{Trace} M_1 + \operatorname{Trace} M_2$$
$$- 2 \operatorname{Trace}\big(\sqrt{M_1} M_2 \sqrt{M_1} \big)^{1/2} \big]^{1/2},$$

where $\operatorname{Trace} M$ denotes the trace of M.

Theorem 2.23 (R.L. Dobrushin, 1970).   (1) For bounded ρ, W is equivalent to the Lévy-Prohorov distance.

    (2) For discrete distance ρ, $W = \| \cdot \|_{\mathrm{Var}} / 2$.

Fortunately, in most cases, what we need is only certain estimates of an upper bound. Clearly, any coupling provides an upper bound of $W(P_1, P_2)$. Thus, it is very natural to introduce the following notion.

Definition 2.24. A coupling $\overline{P}$ of P_1 and P_2 is called ρ-*optimal* if

$$\int \rho(x_1, x_2) \overline{P}(\mathrm{d}x_1, \mathrm{d}x_2) = W(P_1, P_2).$$

Now, it is natural to define the optimal coupling for time-discrete Markov processes without restriction to the Markovian class. In the special case of ρ being the discrete metric (or equivalently, restricted to the total variation), it is just the *maximal coupling*, introduced by D. Griffeath (1978). However, the maximal couplings constructed in the quoted paper are usually non-Markovian. Even though the maximal couplings as well as other non-Markovian couplings now constitute an important part of the theory and have been widely studied in the literature (cf. T. Lindvall (1992), J.G. Propp and D.B. Wilson (1996), H. Thorrison (2000), and references therein), they are difficult to handle, especially when we come to the time-continuous situation. Moreover, it will be clear soon that in the context of diffusions, in dealing with the optimal Markovian coupling in terms of their operators, the discrete metric will lose its meaning. Thus, our optimal Markovian couplings are essentially different from the maximal ones. It should also be pointed out that the sharp estimates introduced in Chapter 1 were obtained from the exponential rate in the W-metric with respect to some much more refined metric ρ rather than the discrete one. Replacing P_k and $\widetilde{P}$ with $P_k(t)$ and $\widetilde{P}(t)$, respectively, and then going to the operators, it is not difficult to arrive at the following notion [cf. Chen (1994b; 1994a) for details].

Definition 2.25. A coupling operator $\overline{\Omega}$ is called ρ-*optimal* if

$$\overline{\Omega}\,\rho(x_1, x_2) = \inf_{\widetilde{\Omega}} \widetilde{\Omega}\,\rho(x_1, x_2) \qquad \text{for all } x_1 \neq x_2,$$

where $\widetilde{\Omega}$ varies over all coupling operators.

To see that the notion is useful, let us introduce one more coupling.

Example 2.26 (Coupling by reflection $\widetilde{\Omega}_r$). Given a birth–death process with birth rates b_i and death rates a_i, this coupling evolves in the following way. If $i_2 = i_1 + 1$, then

$$
\begin{aligned}
(i_1, i_2) \quad &\rightarrow \quad (i_1 - 1, i_2 + 1) \quad &&\text{at rate} \quad a_{i_1} \wedge b_{i_2} \\
&\rightarrow \quad (i_1 + 1, i_2) \quad &&\text{at rate} \quad b_{i_1} \\
&\rightarrow \quad (i_1, i_2 - 1) \quad &&\text{at rate} \quad a_{i_2}.
\end{aligned}
$$

If $i_2 \geqslant i_1 + 2$, then

$$
\begin{aligned}
(i_1, i_2) \quad &\rightarrow \quad (i_1 - 1, i_2 + 1) \quad &&\text{at rate} \quad a_{i_1} \wedge b_{i_2} \\
&\rightarrow \quad (i_1 + 1, i_2 - 1) \quad &&\text{at rate} \quad b_{i_1} \wedge a_{i_2}.
\end{aligned}
$$

By symmetry, we can write down the rates for the other case that $i_1 > i_2$.

Intuitively, the reflection in the outside direction is quite strange, since it separates the components by distance 2 but not by 1. For this reason, even though the coupling came to our attention years ago, we never believed that it could be better than the coupling by inner reflection. But the next result changed our mind.

Theorem 2.27 (Chen, 1994a). For birth–death processes, the coupling by reflection is ρ-optimal for any translation-invariant metric ρ on $\mathbb{Z}_+$ having the property that

$$u_k := \rho(0, k + 1) - \rho(0, k), \quad k \geqslant 0$$

is nonincreasing in k.

To see that the optimal coupling depends heavily on the metric ρ, note that the above metric ρ can be rewritten as

$$\rho(i,j) = \sum_{k<|i-j|} u_k$$

for some positive nonincreasing sequence (u_k). In this way, for any positive sequence (u_k), we can introduce another metric as follows:

$$\tilde{\rho}(i,j) = \left| \sum_{k<i} u_k - \sum_{k<j} u_k \right|.$$

Because $(u_k > 0)$ is arbitrary, this class of metrics is still quite large. Now, among the couplings listed above, which are $\tilde{\rho}$-optimal?

Theorem 2.28 (Chen, 1994a). For birth–death processes, every coupling mentioned above except the trivial (independent) one is $\tilde{\rho}$-optimal.

This result is again quite surprising, far from our probabilistic intuition. Thus, our optimality does produce some unexpected results.

We are now ready to study the optimal couplings for diffusion processes.

Definition 2.29. Given $\rho \in C^2(\mathbb{R}^d \times \mathbb{R}^d \setminus \{(x,x) : x \in \mathbb{R}^d\})$, a coupling operator $\overline{L}$ is called ρ-*optimal* if

$$\overline{L}\rho(x,y) = \inf_{\widetilde{L}} \widetilde{L}\rho(x,y), \qquad x \neq y,$$

where $\widetilde{L}$ varies over all coupling operators.

For the underlying Euclidean distance $|\cdot|$ in $\mathbb{R}^d$, we introduce a family of distances as follows:

$$\rho(x,y) = f(|x-y|), \quad \text{where } f(0) = 0,\ f' > 0,\ \text{and } f'' \leqslant 0. \qquad (2.4)$$

In order to make ρ a distance, the first two conditions of f are necessary and the third condition guarantees the triangle inequality. For this class of distance, as mentioned in the paper quoted below, the existence of ρ-optimal coupling for diffusion is not a serious problem. Here we introduce only some explicit constructions.

Theorem 2.30 (Chen, 1994a). Let $\rho(x,y)=f(|x-y|)$ for some $f\in C^2(\mathbb{R}_+;\mathbb{R}_+)$ satisfying (2.4). Then the ρ-optimal solution $c(x,y)$ is given as follows:

(1) If $d = 1$, then $c(x,y) = -\sqrt{a_1(x)a_2(y)}$, and moreover,

$$\overline{L}f(|x-y|) = \frac{1}{2}\left(\sqrt{a_1(x)} + \sqrt{a_2(y)}\right)^2 f''(|x-y|)$$
$$+ \frac{(x-y)(b_1(x) - b_2(y))}{|x-y|}f'(|x-y|).$$

Next, suppose that $a_k = \sigma_k^2$ ($k = 1, 2$) is nondegenerate and write

$$c(x, y) = \sigma_1(x)H^*(x, y)\sigma_2(y).$$

(2) If $f''(r) < 0$ for all $r > 0$, then $H(x, y) = U(\gamma)^{-1}\big[U(\gamma)U(\gamma)^*\big]^{1/2}$, where

$$\gamma = 1 - \frac{|x-y|f''(|x-y|)}{f'(|x-y|)}, \qquad U(\gamma) = \sigma_1(x)(I - \gamma\bar{u}\bar{u}^*)\sigma_2(y).$$

(3) If $f(r) = r$, then $H(x, y)$ is a solution to the equation

$$U(1)H = \big(U(1)U(1)^*\big)^{1/2}.$$

In particular, if $a_k(x) = \varphi_k(x)\sigma^2$ for some positive function φ_k ($k = 1, 2$), where σ is independent of x and $\det \sigma > 0$, then

(4) $H(x, y) = I - 2\sigma^{-1}\bar{u}\bar{u}^*\sigma^{-1}/|\sigma^{-1}\bar{u}|^2$ if $\rho(x, y) = |x - y|$. Moreover,

$$\overline{L}f(|x - y|) = \frac{1}{2|x - y|}\Big\{\big(\sqrt{\varphi_1(x)} - \sqrt{\varphi_2(y)}\,\big)^2\big[\operatorname{Trace}\sigma^2 - |\sigma\bar{u}|^2\big]$$
$$+ 2\langle x - y, b_1(x) - b_2(y)\rangle\Big\}.$$

(5) H is the same as in the last assertion if $\rho(x, y) = |x - y|$ is replaced by $\rho(x, y) = f(|\sigma^{-1}(x - y)|)$. Furthermore,

$$\overline{L}\rho(x, y) = \frac{1}{2}\big(\sqrt{\varphi_1(x)} + \sqrt{\varphi_2(y)}\,\big)^2 f''(|\sigma^{-1}(x - y)|)$$
$$+ \Big\{(d - 1)\big(\sqrt{\varphi_1(x)} - \sqrt{\varphi_2(y)}\,\big)^2$$
$$+ 2\langle \sigma^{-1}(x - y), \sigma^{-1}(b_1(x) - b_2(y))\rangle\Big\}$$
$$\times \frac{f'(|\sigma^{-1}(x - y)|)}{2|\sigma^{-1}(x - y)|}.$$

2.3   Optimality with respect to closed functions

As an extension of the optimal couplings with respect to distances, we can consider the optimal couplings with respect to a more general, nonnegative, closed (=lower semicontinuous) function φ.

Definition 2.31. Given a metric space $(E, \rho, \mathscr{E})$, let φ be a nonnegative, closed function on $(E, \rho, \mathscr{E})$. A coupling is called a φ-*optimal* (Markovian) coupling if in the definitions given in the last section, the distance function ρ is replaced by φ.

Here are some typical examples of φ.

Example 2.32. (1) φ is a distance of the form $f \circ \rho$ for some f having the properties $f(0) = 0$, $f' > 0$, and $f'' \leqslant 0$.

(2) φ is the discrete distance: $\varphi(x, y) = 1$ iff $x \neq y$; otherwise, $\varphi(x, y) = 0$.

(3) Let E be endowed with a measurable semiorder "$\prec$" and set $F = \{(x, y) : x \prec y\}$. Then F is a closed set. Take $\varphi = I_{F^c}$.

Before moving further, let us recall the definition of stochastic comparability.

Definition 2.33. Let $\mathscr{M}$ be the set of bounded monotone functions f: $x \prec y \Longrightarrow f(x) \leqslant f(y)$.

(1) We write $\mu_1 \prec \mu_2$ if $\mu_1(f) \leqslant \mu_2(f)$ for all $f \in \mathscr{M}$.

(2) Let P_1 and P_2 be transition probabilities. We write $P_1 \prec P_2$ if $P_1(f)(x_1) \leqslant P_2(f)(x_2)$ for all $x_1 \prec x_2$ and $f \in \mathscr{M}$.

(3) Let $P_1(t)$ and $P_2(t)$ be transition semigroups. We write $P_1(t) \prec P_2(t)$ if $P_1(t)(f)(x_1) \leqslant P_2(t)(f)(x_2)$ for all $t \geqslant 0$, $x_1 \prec x_2$, and $f \in \mathscr{M}$.

Here is a famous result about stochastic comparability.

Theorem 2.34 (V. Strassen, 1965). For a Polish space, $\mu_1 \prec \mu_2$ iff there exists a coupling measure $\bar{\mu}$ such that $\bar{\mu}(F^c) = 0$.

Usually, in practice, it is not easy to compare two measures directly. For this reason, one introduces stochastic comparability for processes. First, one constructs two processes with stationary distributions μ_1 and μ_2. Then the stochastic comparability of the two measures can be reduced to that of the processes. The advantage for the latter comparison comes from the intuition of the stochastic dynamics. One can even see the answer from the coefficients of the operators. See Examples 2.44–2.46 below.

A general result for φ-optimal coupling is the following.

Theorem 2.35 (S.Y. Zhang, 2000a). Let $(E, \rho, \mathscr{E})$ be Polish and $\varphi \geqslant 0$ be a closed function.

(1) Given $P_k(x_k, dy_k)$, $k = 1, 2$, there exists a transition probability $\overline{P}(x_1, x_2; dy_1, dy_2)$ such that $\overline{P}\varphi(x_1, x_2) = \inf_{\widetilde{P}^{(x_1, x_2)}} \widetilde{P}^{(x_1, x_2)}\varphi(x_1, x_2)$, where for fixed (x_1, x_2), $\widetilde{P}^{(x_1, x_2)}$ varies over all couplings of $P_1(x_1, dy_1)$ and $P_2(x_2, dy_2)$.

(2) Given operators Ω_k of regular jump processes, $k = 1, 2$, there exists a coupling operator $\overline{\Omega}$ of jump process such that $\overline{\Omega}\varphi = \inf_{\widetilde{\Omega}} \widetilde{\Omega}\varphi$, where $\widetilde{\Omega}$ varies over all coupling operators of Ω_1 and Ω_2.

According to Theorem 2.35 (1), Strassen's theorem can be restated as follows: the I_{F^c}-optimal Markovian coupling satisfies $\bar{\mu}(F^c) = 0$. This shows that Theorem 2.35 (1) is an extension of Strassen's theorem. Even though the proof of Theorem 2.35 is quite technical, the main root is still clear. Consider first finite state spaces. Then the conclusion follows from an existence theorem of linear programming regarding the marginality as a constraint. Next, pass to the general Polish space by using a tightness argument (a generalized Prohorov theorem) plus an approximation of φ by bounded Lipschitz functions.

Concerning stochastic comparability, we have the following result.

Theorem 2.36 (Chen (1992a, Chapter 5), Zhang (2000b)). For jump processes on a Polish space, under a mild assumption, $P_1(t) \prec P_2(t)$ iff

$$\Omega_1 I_B(x_1) \leqslant \Omega_2 I_B(x_2),$$

for all $x_1 \prec x_2$ and B with $I_B \in \mathscr{M}$.

Here we mention an additional result, which provides us the optimal solutions within the class of order-preserving couplings.

Theorem 2.37 (T. Lindvall, 1999). Again, let Δ denote the diagonals.

(1) Let $\mu_1 \prec \mu_2$. Then

$$\inf_{\tilde{\mu}(F^c)=0} \tilde{\mu}(\Delta^c) = \frac{1}{2}\|\mu_1 - \mu_2\|_{\text{Var}}.$$

(2) Let P_1 and P_2 be transition probabilities that satisfy $P_1 \prec P_2$. Then

$$\inf_{\tilde{P}(x_1,x_2;F^c)=0} \tilde{P}(x_1,x_2;\Delta^c) = \frac{1}{2}\|P_1(x_1,\cdot) - P_2(x_2,\cdot)\|_{\text{Var}}$$

for all $x_1 \prec x_2$.

In fact, the left-hand sides of the formulas in Theorem 2.37 can be replaced, respectively, by the I_{F^c}-optimal coupling given in Theorem 2.35 (1).

For order-preserving Markovian coupling for diffusions, refer to F.Y. Wang and M.P. Xu (1997).

Open Problem 2.38. Let $\varphi \in C^2(\mathbb{R}^{2d} \setminus \Delta)$. Prove the existence of φ-optimal Markovian couplings for diffusions under some reasonable hypotheses.

Open Problem 2.39. Construct φ-optimal Markovian couplings.

2.4 Applications of coupling methods

It should be helpful for readers, especially newcomers, to see some applications of couplings. Of course, the applications discussed below cannot be complete, and additional applications will be presented in Chapters 3, 5, and 9. One may refer to T.M. Liggett (1985), T. Lindvall (1992), and H. Thorrison (2000) for much more information. The coupling method is now a powerful tool in statistics, called "copulas" (cf. R.B. Nelssen (1999)). It is also an active research topic in PDE and related fields, named "optimal transportation" (cf. S.T. Rachev and L. Ruschendorf (1998), L. Ambrosio et al. (2003), C. Villani (2003)).

Spectral gap; exponential L^2-convergence

We introduce two general results, due to Chen and F.Y. Wang (1993b) [see also Chen (1994a)], on the estimation of the first nontrivial eigenvalue (spectral gap) by couplings.

Definition 2.40. Let L be an operator of a Markov process $(X_t)_{t\geqslant 0}$. We say that a function f is in the weak domain of L, denoted by $\mathscr{D}_w(L)$, if f satisfies the forward Kolmogorov equation

$$\mathbb{E}^x f(X_t) = f(x) + \int_0^t \mathbb{E}^x Lf(X_s)\mathrm{d}s,$$

or equivalently, if

$$f(X_t) - \int_0^t Lf(X_s)\mathrm{d}s$$

is a $\mathbb{P}^x$-martingale with respect to the natural flow of σ-algebras $\{\mathscr{F}_t\}_{t\geqslant 0}$, where $\mathscr{F}_t = \sigma\{X_s : s \leqslant t\}$.

Definition 2.41. We say that g is an eigenfuction of L corresponding to λ in the weak sense if g satisfies the eigenequation $Lg = -\lambda g$ pointwise.

Note that the eigenfunction defined above may not belong to $L^2(\pi)$, where π is the stationary distribution of $(X_t)_{t\geqslant 0}$. In the reversible case, all of the eigenvalues are nonnegative and all of the eigenfunctions are real.

The next two results remain true in the irreversible case (where λ and g are often complex), provided λ is replaced by $|\lambda|$.

Theorem 2.42. Let (E, ρ) be a metric space and let $\{X_t\}_{t\geqslant 0}$ be a reversible Markov process with operator L. Denote by g the eigenfunction corresponding to $\lambda \neq 0$ in the weak sense. Next, let (X_t, Y_t) be a coupled process, starting from (x, y), with coupling operator $\widetilde{L}$, and let $\gamma : E \times E \to [0, \infty)$ satisfy $\gamma(x,y) = 0$ iff $x = y$. Suppose that

(1) $g \in \mathscr{D}_w(L)$,
(2) $\gamma \in \mathscr{D}_w(\widetilde{L})$,
(3) $\widetilde{L}\gamma(x, y) \leqslant -\alpha\gamma(x, y)$ for all $x \neq y$ and some constant $\alpha \geqslant 0$,
(4) g is Lipschitz with respect to γ in the sense that

$$c_{g,\gamma} := \sup_{y \neq x} \gamma(y, x)^{-1}|g(y) - g(x)| < \infty.$$

Then we have $\lambda \geqslant \alpha$.

Proof. Without loss of generality, assume that $\alpha > 0$. Otherwise, the conclusion is trivial. By conditions (2), (3) and Lemma A.6, we have

$$\widetilde{\mathbb{E}}^{x,y}\gamma(X_t, Y_t) \leqslant \gamma(x, y)e^{-\alpha t}, \qquad t \geqslant 0.$$

Next, by condition (1) and the definition of g,

$$g(X_t) - \int_0^t Lg(X_s)\mathrm{d}s = g(X_t) + \lambda \int_0^t g(X_s)\mathrm{d}s$$

is a $\mathbb{P}^x$-martingale with respect to the natural flow of σ-algebras $\{\mathscr{F}_t\}_{t\geqslant0}$. In particular,

$$g(x) = \mathbb{E}^x\left[g(X_t) + \lambda \int_0^t g(X_s)\mathrm{d}s\right].$$

Because of the coupling property,

$$\mathbb{E}^x\left[g(X_t) + \lambda \int_0^t g(X_s)\mathrm{d}s\right] = \widetilde{\mathbb{E}}^{x,y}\left[g(X_t) + \lambda \int_0^t g(X_s)\mathrm{d}s\right].$$

Thus, we obtain

$$g(x) - g(y) = \widetilde{\mathbb{E}}^{x,y}\left[g(X_t) - g(Y_t) + \lambda \int_0^t [g(X_s) - g(Y_s)]\mathrm{d}s\right].$$

Therefore

$$|g(x) - g(y)| \leqslant \widetilde{\mathbb{E}}^{x,y}\big|g(X_t) - g(Y_t)\big| + \lambda\widetilde{\mathbb{E}}^{x,y}\int_0^t |g(X_s) - g(Y_s)|\mathrm{d}s$$

$$\leqslant c_{g,\gamma}\widetilde{\mathbb{E}}^{x,y}\gamma(X_t, Y_t) + \lambda c_{g,\gamma}\widetilde{\mathbb{E}}^{x,y}\int_0^{t\wedge T}\gamma(X_s, Y_s)\mathrm{d}s$$

$$\leqslant c_{g,\gamma}\gamma(x,y)e^{-\alpha t} + \lambda c_{g,\gamma}\gamma(x,y)\int_0^t e^{-\alpha s}\mathrm{d}s.$$

Noting that g is not a constant, since $\lambda \neq 0$, we have $c_{g,\gamma} \neq 0$. Dividing both sides by $\gamma(x,y)$ and choosing a sequence (x_n, y_n) such that

$$|g(y_n) - g(x_n)|/\gamma(y_n, x_n) \to c_{g,\gamma},$$

we obtain

$$1 \leqslant e^{-\alpha t} + \lambda \int_0^t e^{-\alpha s}\mathrm{d}s = e^{-\alpha t} + \lambda(1 - e^{-\alpha t})/\alpha$$

for all t. This implies that $\lambda \geqslant \alpha$ as required. $\square$

One may compare this probabilistic proof with the analytic one sketched in Section 1.2.

When γ is a distance, $\widetilde{\mathbb{E}}^{x,y}\gamma(X_t, Y_t)$ is nothing but the Wasserstein metric $W = W_1$ with respect to γ of the distributions at time t. The above proof shows that W_1 can be used to study the Poincaré inequality (i.e., λ_1). Noting that W_2 is stronger than W_1 for a fixed underframe distance ρ, it is natural to study the stronger logarithmic Sobolev inequality in terms of W_2 with respect to the Euclidean distance, for instance. To deal with the inequalities themselves, it is helpful but not necessary to go to the dynamics, since they are mainly concerned with measures. Next, it was discovered in the 1990s that in many cases, for two given probability densities, the optimal coupling for W_2 exists uniquely, and the mass of the coupling measure is concentrated on the set $\{(x, T(x)) : x \in \mathbb{R}^d\}$. Moreover, the optimal transport T can be expressed by $T = \nabla \Psi$ for some convex function Ψ that solves a nonlinear Monge–Ampère equation. It turns out that this transportation solution provides a new way to prove a class of logarithmic Sobolev (or even more general) inequalities in $\mathbb{R}^d$. This explains roughly the interaction between probability distances (couplings) and PDE. Considerable progress has been made recently in this field, as shown in the last two books mentioned in the first paragraph of this section.

Condition (3) in Theorem 2.42 is essential. The other conditions can often be relaxed or avoided by using a localizing procedure. Define the *coupling time* $T = \inf\{t \geqslant 0 : X_t = Y_t\}$. The next, weaker, result is useful. It has a different meaning, as will be explained in Section 5.6. Indeed, the condition "$\sup_{x \neq y} \widetilde{\mathbb{E}}^{x,y} T < \infty$" used in the next theorem is closely related to the strong ergodicity of the process rather than $\lambda_1 > 0$.

Theorem 2.43. Let $\{X_t\}_{t \geqslant 0}$, L, λ, and g be the same as in the last theorem. Suppose that

(1) $g \in \mathscr{D}_w(L)$,
(2) $\sup_{x \neq y} |g(x) - g(y)| < \infty$.

Then for every coupling $\widetilde{\mathbb{P}}^{x,y}$, we have $\lambda \geqslant \left(\sup_{x \neq y} \widetilde{\mathbb{E}}^{x,y} T\right)^{-1}$.

Proof. Set $f(x, y) = g(x) - g(y)$. By the martingale formulation as in the last proof, we have

$$f(x,y) = \widetilde{\mathbb{E}}^{x,y} f\big(X_{t \wedge T}, Y_{t \wedge T}\big) - \widetilde{\mathbb{E}}^{x,y} \int_0^{t \wedge T} \widetilde{L} f\big(X_s, Y_s\big) \mathrm{d}s$$

$$= \widetilde{\mathbb{E}}^{x,y} f\big(X_{t \wedge T}, Y_{t \wedge T}\big) + \lambda \widetilde{\mathbb{E}}^{x,y} \int_0^{t \wedge T} f\big(X_s, X_s\big) \mathrm{d}s.$$

Hence

$$|g(x) - g(y)| \leqslant \widetilde{\mathbb{E}}^{x,y} \big|g\big(X_{t \wedge T}\big) - g\big(Y_{t \wedge T}\big)\big| + \lambda \widetilde{\mathbb{E}}^{x,y} \int_0^{t \wedge T} \big|g\big(X_s\big) - g\big(Y_s\big)\big| \mathrm{d}s.$$

Assume $\sup_{x \neq y} \widetilde{\mathbb{E}}^{x,y} T < \infty$, and so $\widetilde{\mathbb{P}}^{x,y}[T < \infty] = 1$. Letting $t \uparrow \infty$, we obtain

$$|g(x) - g(y)| \leqslant \lambda \widetilde{\mathbb{E}}^{x,y} \int_0^T \big|g\big(X_s\big) - g\big(Y_s\big)\big| \mathrm{d}s.$$

Choose x_n and y_n such that

$$\lim_{n\to\infty} |g(x_n) - g(y_n)| = \sup_{x,y} |g(x) - g(y)|.$$

Without loss of generality, assume that $\sup_{x,y} |g(x) - g(y)| = 1$. Then

$$1 \leqslant \lambda \lim_{n\to\infty} \widetilde{\mathbb{E}}^{x_n, y_n}(T).$$

Therefore $1 \leqslant \lambda \sup_{x \neq y} \widetilde{\mathbb{E}}^{x,y} T$. $\square$

For the remainder of this section, we emphasize the main ideas by using some simple examples. In particular, from now on, the metric is taken to be $\rho(x,y) = |x - y|$. That is, $f(r) = r$. In view of Theorem 2.30, this metric may not be optimal, since $f'' = 0$. Thus, in practice, additional work is often needed to figure out an effective metric ρ. The details will be discussed in the next chapter. Additional discrete examples are included in Appendix B.

To conclude this subsection, let us consider the Ornstein–Uhlenbeck process in $\mathbb{R}^d$. By Theorem 2.30 (4), we have $\overline{L}\rho(x,y) \leqslant -\rho(x,y)$, and so

$$\mathbb{E}^{x,y} \rho(X_t, Y_t) \leqslant \rho(x,y)e^{-t}. \tag{2.5}$$

By using Theorem 2.42 with the help of a localizing procedure, this gives us $\lambda_1 \geqslant 1$, which is indeed exact!

Ergodicity

Coupling methods are often used to study the ergodicity of Markov processes. For instance, for an Ornstein–Uhlenbeck process, from (2.5), it follows that

$$W(P(t,x,\cdot), \pi) \leqslant C(x)e^{-t}, \qquad t \geqslant 0, \tag{2.6}$$

where π is the stationary distribution of the process. The estimate (2.6) simply means that the process is exponentially ergodic with respect to W.

Recall that $T = \inf\{t \geqslant 0 : X_t = Y_t\}$. Starting from time T, we can adopt the coupling of marching soldiers so that the two components will move together. Then we have

$$\|P(t,x,\cdot) - P(t,y,\cdot)\|_{\mathrm{Var}} \leqslant 2\,\widetilde{\mathbb{E}}^{x,y} I_{[X_t \neq Y_t]} = 2\,\widetilde{\mathbb{P}}^{x,y}[T > t]. \tag{2.7}$$

Thus, if $\widetilde{\mathbb{P}}^{x,y}[T > t] \to 0$ as $t \to \infty$, then the existence of a stationary distribution plus (2.7) gives us the ergodicity with respect to the total variation. See T. Lindvall (1992) for details and references on this topic. Actually, for Brownian motion, as pointed out in Chen and S.F. Li (1989), coupling by reflection provides a sharp estimate for the total variation. We will come back to this topic in Chapter 5.

Gradient estimate

Recall that for every suitable function f, we have

$$f(x) - f(y) = \widetilde{\mathbb{E}}^{x,y}\big[f(X_{t\wedge T}) - f(Y_{t\wedge T})\big] - \widetilde{\mathbb{E}}^{x,y}\int_0^{t\wedge T}\big[Lf(X_s) - Lf(Y_s)\big]\mathrm{d}s.$$

Thus, if f is L-harmonic, i.e., $Lf = 0$, then we have

$$f(x) - f(y) = \widetilde{\mathbb{E}}^{x,y}\big[f(X_{t\wedge T}) - f(Y_{t\wedge T})\big].$$

Hence

$$|f(x) - f(y)| \leqslant 2\,\|f\|_\infty\,\widetilde{\mathbb{P}}^{x,y}[T > t].$$

Letting $t \to \infty$, we obtain

$$|f(x) - f(y)| \leqslant 2\,\|f\|_\infty\,\widetilde{\mathbb{P}}^{x,y}[T = \infty].$$

Now, if f is bounded and $\widetilde{\mathbb{P}}^{x,y}[T = \infty] = 0$, then $f =$ constant. Otherwise, if $\widetilde{\mathbb{P}}^{x,y}[T = \infty] \leqslant$ constant$\cdot \rho(x,y)$, then we get

$$\|\nabla f\|_\infty \leqslant \text{constant}\,\cdot\,\|f\|_\infty,$$

which is the gradient estimate we are looking for [cf. M. Cranston (1991; 1992) and F.Y. Wang (1994a; 1994b)]. For Brownian motion in $\mathbb{R}^d$, the optimal coupling gives us $\widetilde{\mathbb{P}}^{x,y}[T < \infty] = 1$, and so $f =$ constant. We have thus proved a well-known result: every bounded harmonic function should be constant.

Comparison results

The stochastic order occupies a crucial position in the study of probability theory, since the usual order relation is a fundamental structure in mathematics.

The coupling method provides a natural way to study the order-preserving property (i.e., stochastic comparability). Refer to Chen (1992a, Chapter 5) for a study on jump processes. Here is an example for diffusions.

Example 2.44. Consider two diffusions in $\mathbb{R}$ with

$$a_1(x) = a_2(x) = a(x), \qquad b_1(x) \leqslant b_2(x). \tag{2.8}$$

Then we have $P_1(t) \prec P_2(t)$.

The conclusion was proved in N. Ikeda and S. Watanabe (1988, Section 6.1), using stochastic differential equations. The same proof with a slight modification works if we adopt the coupling of marching soldiers.

A criterion for order preservation for multidimensional diffusion processes was presented in Chen and F.Y. Wang (1993a), from which we see that condition (2.8) is not only sufficient but also necessary. A related topic, the preservation of positive correlations for diffusions, was also solved in the same paper, as mentioned at the beginning of this chapter.

To illustrate an application of the study, let us introduce a simple example.

Example 2.45. Let μ^λ be the Poisson measure on $\mathbb{Z}_+$ with parameter λ:

$$\mu^\lambda(k) = \frac{\lambda^k}{k!} e^{-\lambda}, \qquad k \geqslant 0.$$

Then we have $\mu^\lambda \prec \mu^{\lambda'}$ whenever $\lambda \leqslant \lambda'$.

In some publications, one proves such a result by constructing a coupling measure $\tilde{\mu}$ such that $\tilde{\mu}\{(x, y) : x \prec y\} = 1$. Of course, such a proof is lengthy. So we now introduce a very short proof based on the coupling argument.

Consider a birth–death process with rate

$$a(k) \equiv 1, \qquad b^\lambda(k) = \frac{\mu^\lambda(k+1)}{\mu^\lambda(k)} = \frac{\lambda}{k+1} \uparrow \quad \text{as} \quad \lambda \uparrow.$$

Denote by $P^\lambda(t)$ the corresponding process. It should be clear that

$$P^\lambda(t) \prec P^{\lambda'}(t) \quad \text{whenever} \quad \lambda \leqslant \lambda'$$

[cf. Chen (1992a, Theorem 5.26; Theorem 5.41 in the 2nd edition)]. Then, by the ergodic theorem,

$$\mu^\lambda(f) = \lim_{t \to \infty} P^\lambda(t)f \leqslant \lim_{t \to \infty} P^{\lambda'}(t)f = \mu^{\lambda'}(f)$$

for all $f \in \mathcal{M}$. Clearly, the technique using stochastic processes [goes back to R. Holley (1974)] provides an intrinsic insight into order preservation for probability measures.

We now return to the FKG inequality mentioned at the beginning of this chapter. Clearly, the inequality is meaningful in the higher-dimensional space $\mathbb{R}^d$ with respect to the ordinary partial ordering. The inequality for a Markov semigroup $P(t)$ becomes

$$P(t)(fg) \geqslant P(t)fP(t)g, \qquad t \geqslant 0, \ f, g \in \mathcal{M}.$$

The study of the FKG inequality in terms of semigroups is exactly the same as above. Choose a Markov process having the given measure as a stationary distribution. Then, study the inequality for the dynamics. Finally, passing to the limit as $t \to \infty$, we return to (2.1).

An aspect of the applications of coupling methods is to compare a rather complicated process with a simpler one. To provide an impression, we introduce an example that was used by Chen and Y.G. Lu (1990) in the study of large deviations for Markov chains.

Example 2.46. Consider a single birth Q-matrix $Q = (q_{ij})$, which means that

$$q_{i,i+1} > 0 \qquad \text{and} \quad q_{ij} = 0 \quad \text{for all} \quad j > i + 1,$$

and a birth–death Q-matrix $\overline{Q} = (\bar{q}_{ij})$ with $\bar{q}_{i,i-1} = \sum_{j<i} q_{ij}$. If $\bar{q}_{i,i+1} \geqslant q_{i,i+1}$ for all $i \geqslant 0$. Then $P(t) \prec \overline{P}(t)$.

The conclusion can be easily deduced by the following coupling:

$$
\begin{aligned}
(i_1, i_2) \;\; &\to\;\; (i_1 - k,\, i_2 - 1) \quad & \text{at rate} \quad & q_{i_1,i_1-k} \wedge q_{i_2,i_2-k} \\
&\to\;\; (i_1 - k,\, i_2) & \text{at rate} \quad & (q_{i_1,i_1-k} - q_{i_2,i_2-k})^+ \\
&\to\;\; (i_1,\, i_2 - 1) & \text{at rate} \quad & (q_{i_2,i_2-k} - q_{i_1,i_1-k})^+ \\
&\to\;\; (i_1 + 1,\, i_2 + 1) & \text{at rate} \quad & q_{i_1,i_1+1} \wedge \bar{q}_{i_2,i_2+1} \\
&\to\;\; (i_1 + 1,\, i_2) & \text{at rate} \quad & (q_{i_1,i_1+1} - \bar{q}_{i_2,i_2+1})^+ \\
&\to\;\; (i_1,\, i_2 + 1) & \text{at rate} \quad & (\bar{q}_{i_2,i_2+1} - q_{i_1,i_1+1})^+,
\end{aligned}
$$

where we have used the convention $q_{ij} = 0$ if $j < 0$. Refer to Chen (1992a, Theorem 8.24) for details. This example illustrates the flexibility in the application of couplings.

The details of this chapter, except for diffusions, are included in Chapter 5 of the second edition of Chen (1992a).

Finally, we mention that the coupling methods are also powerful for time-inhomogeneous Markov processes, not touched on in this book. In fact, the fundamental theorem 2.14 is valid for Markov jump processes valued in Polish spaces [cf. J.L. Zheng (1993)]. For estimation of convergence rate, refer to A.I. Zeifman (1995), B.L. Granovsky and A.I. Zeifman (1997).

Chapter 3

New Variational Formulas for the First Eigenvalue

This chapter is devoted to the proofs of the main variational formulas introduced in Section 1.2. Two quick proofs for the discrete case are given in Section 3.2. Then, three sections are used to explain the ideas in detail for the proof in the geometric case. In Section 3.6, we compare the coupling methods with other techniques. The last two sections are more technical. In Section 3.7 we show that the new variational formulas are indeed complete in dimension one. The results for the first Dirichlet eigenvalue are similar and are presented in Section 3.8 for the discrete case.

Let us begin with some background on this topic.

3.1 Background

As mentioned in the first chapter, since the spectral theory is central to many branches of mathematics and the first nontrivial eigenvalue is the leading term of the spectrum, it should not be surprising that the study of λ_1 has a very wide range of applications. Here we mention two fashionable applications only.

Phase transitions

In the study of interacting particle systems, a physical model is described by a Markov process with semigroup $\{P_t\}_{t\geqslant 0}$ (depending on temperature $1/\beta$) having stationary distribution π. Let $L^2(\pi)$ be the usual real L^2-space with norm $\|\cdot\|$.

Figure 3.1 shows that at higher temperatures (small β), the corresponding semigroup $\{P_t\}_{t\geqslant 0}$ is exponentially ergodic in the L^2-sense:

$$\|P_t f - \pi(f)\| \leqslant \|f - \pi(f)\| e^{-\lambda_1 t},$$

where $\pi(f) = \int f \mathrm{d}\pi$, with the largest rate λ_1, and when the temperature goes to the critical value $1/\beta_c$, the rate will go to zero. This provides a way to des-

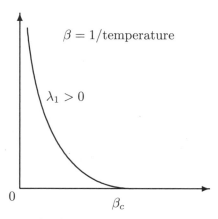

Figure 3.1   The first eigenvalue and phase transitions

cribe the phase transitions, and this is now an active research field. Further remarks are given at the end of Chapter 9. The next application we would like to mention is to Monte Carlo Markov chains.

Monte Carlo Markov chains (MCMC)

Consider a function with several local minima. The usual algorithms go at each step to a place that decreases the value of the function. The problem is that one may fall into a local trap (Figure 3.2).

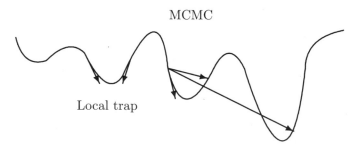

Figure 3.2   The first eigenvalue and random algorithm

The MCMC algorithm avoids this by allowing the possibility of visiting other places, not only toward a local minimum. The random algorithm consists of two steps.

- Construct a distribution according to the local minima, staying at a lower place with higher probability, in terms of the Gibbs principle.

- Construct a Markov chain with the stationary distribution given above, and with a fast convergence rate (i.e., λ_1).

The idea is a great one, since it reduces some NP problems to the P problems in computer science. The effectiveness of a random algorithm is determined by λ_1 of the Markov chain. Refer to M.R. Jerrum and A.J. Sinclair (1989), and A.J. Sinclair (1993) for further information.

Here is a practical example, called the traveling salesman problem: find the shortest closed path (without loops) among 144 cities in China. For a computer with speed of computing 100^9 paths per second, a brute-force search would require

$$\frac{143!}{10^9 \times 365 \times 24 \times 60 \times 60} \approx 100^{111}$$

years for the computation. This is a typical NP problem. However, by using MCMC, it can be done quickly, as in L.S. Kang et al. (1994). The resulting best path has length 30,421 kilometers, only about 40 kilometers different from the best known length: 30,380 kilometers.

3.2 Partial proof in the discrete case

In the section, we introduce a short proof for the lower bound of λ_1 in the discrete case. Even though the proof is very elementary, it does illustrate a good use of the Cauchy–Schwarz inequality. Recall that

$$\mu_0 = 1, \qquad \mu_i = \frac{b_0 \cdots b_{i-1}}{a_1 \cdots a_i}, \qquad i \geqslant 1.$$

For an infinite matrix, we need the assumption

$$\sum_{k=0}^{\infty} \frac{1}{b_k \mu_k} \sum_{i=0}^{k} \mu_i = \infty \qquad \text{and} \qquad Z = \sum_{i=0}^{\infty} \mu_i < \infty. \tag{3.1}$$

Let

$$\pi_i = \mu_i/Z, \qquad L^2(\pi) - \{f : \pi(f^2) < \infty\},$$

where $\pi(f) = \sum_{i=0}^{\infty} \pi_i f_i$. The first eigenvalue is defined by the *classical variational formula* as follows:

$$\lambda_1 = \inf \{D(f) : \pi(f) = 0, \pi(f^2) = 1\}, \tag{3.2}$$

where $D(f) = \sum_{i=0}^{\infty} \pi_i b_i (f_{i+1} - f_i)^2$. We are now going to prove the variational formula for the lower bounds (cf. Theorem 1.5):

$$\lambda_1 \geqslant \sup_{w \in \widehat{\mathscr{W}}} \inf_{i \geqslant 0} I_i(w)^{-1} \geqslant \sup_{w \in \mathscr{W}} \inf_{i \geqslant 0} I_i(\bar{w})^{-1}, \tag{3.3}$$

where

$$\mathscr{W} = \{w : w_0 = 0,\ w_i \uparrow\uparrow\}, \qquad \widehat{\mathscr{W}} = \{w : w_i \uparrow\uparrow,\ \pi(w) \geqslant 0\},$$
$$\bar{w}_i = w_i - \pi(w), \qquad i \geqslant 0,$$

$$I_i(w) = \frac{1}{\mu_i b_i (w_{i+1} - w_i)} \sum_{j=i+1}^{\infty} \mu_j w_j, \qquad i \geqslant 0, \quad w \in \widehat{\mathscr{W}},$$

and "$\uparrow\uparrow$" means strictly increasing.

Analytic proof of (3.3)

Clearly, it suffices to prove the first estimate in (3.3), since $\{\bar{w} : w \in \mathscr{W}\} \subset \widehat{\mathscr{W}}$.

(a) First, we prove that $I_i(w) > 0$ for each $w \in \widehat{\mathscr{W}}$ and all $i \geqslant 1$. Equivalently, $\sum_{j=i+1}^{\infty} \mu_j w_j > 0$ for all $i \geqslant 0$. Otherwise, let i_0 satisfy $\sum_{j=i_0+1}^{\infty} \mu_j w_j \leqslant 0$. Then, since w_j is strictly increasing, it follows that $w_{i_0} < 0$, and furthermore,

$$0 \leqslant \sum_{j=0}^{\infty} \mu_j w_j = \sum_{j=0}^{i_0} \mu_j w_j + \sum_{j=i_0+1}^{\infty} \mu_j w_j \leqslant \sum_{j=0}^{i_0} \mu_j w_j \leqslant w_{i_0} \sum_{j=0}^{i_0} \mu_j < 0.$$

This is a contradiction.

(b) For each $i \geqslant 0$, define a bond $e_i := \langle i, i+1 \rangle$. Next, for each pair i, j ($i < j$), let γ_{ij} be the path (only one) consisting of the bonds $e_i, e_{i+1}, \ldots, e_{j-1}$. Given $w \in \widehat{\mathscr{W}}$, choose a positive weight function $(w(e))$ on the bonds, $w(e_i) = w_{i+1} - w_i$, and define the length of the path γ_{ij} to be $|\gamma_{ij}|_w = \sum_{e \in \gamma_{ij}} w(e)$. Set

$$J(w)(e) = \frac{1}{a(e)w(e)} \sum_{\{i,j\}:\ \gamma_{ij} \ni e} |\gamma_{ij}|_w \pi_i \pi_j,$$

where $a(e_i) = \pi_i b_i$. At the same time, we write $f(e_i) = f_{i+1} - f_i$.

(c) As a good application of the Cauchy–Schwarz inequality, we have

$$(f_i - f_j)^2 = \left(\sum_{e \in \gamma_{ij}} f(e) \right)^2 = \left(\sum_{e \in \gamma_{ij}} \frac{f(e)}{\sqrt{w(e)}} \cdot \sqrt{w(e)} \right)^2 \leqslant \left(\sum_{e \in \gamma_{ij}} \frac{f(e)^2}{w(e)} \right) |\gamma_{ij}|_w.$$

Thus, for every f with $\pi(f) = 0$ and $\pi(f^2) = 1$, we obtain

$$1 = \frac{1}{2} \sum_{i,j} \pi_i \pi_j (f_i - f_j)^2 = \sum_{\{i,j\}} \pi_i \pi_j \left(\sum_{e \in \gamma_{ij}} f(e) \right)^2$$

$$\leqslant \sum_{\{i,j\}} \pi_i \pi_j \left(\sum_{e \in \gamma_{ij}} \frac{f(e)^2}{w(e)} \right) |\gamma_{ij}|_w$$

$$= \sum_e a(e) f(e)^2 \frac{1}{a(e)w(e)} \sum_{\{i,j\}:\, \gamma_{ij} \ni e} |\gamma_{ij}|_w \pi_i \pi_j$$

$$\leqslant D(f) \sup_e J(w)(e),$$

where $\{i,j\}$ denotes the unordered pair of i and j. The first equality follows by expanding the sum on the right-hand side of the equality; the last equality follows by exchanging the order of the sums. Clearly, the proof in this paragraph works for general Markov chains on a graph.

Note that for every $\ell > k$,

$$|\gamma_{k\ell}|_w = (w_{k+1} - w_k) + \cdots + (w_\ell - w_{\ell-1}) = w_\ell - w_k.$$

If a path $\gamma_{k\ell}$ ($k < \ell$) contains $e_i = \langle i, i+1 \rangle$, then $k \in \{0, 1, \dots, i\}$ and $\ell \in \{i+1, i+2, \dots\}$. Hence, once $\pi(w) \geqslant 0$, we have

$$\sum_{\{k,\ell\}:\, \gamma_{k\ell} \supset e_i} |\gamma_{k\ell}|_w \pi_k \pi_\ell$$

$$= \sum_{k=0}^{i} \sum_{\ell=i+1}^{\infty} \pi_k \pi_\ell (w_\ell - w_k) = \sum_{k=0}^{i} \pi_k \sum_{\ell=i+1}^{\infty} \pi_\ell w_\ell - \sum_{k=0}^{i} \pi_k w_k \sum_{\ell=i+1}^{\infty} \pi_\ell$$

$$= \sum_{\ell=i+1}^{\infty} \pi_\ell w_\ell - \left(\sum_{k=i+1}^{\infty} \pi_k \right) \sum_{\ell=i+1}^{\infty} \pi_\ell w_\ell - \sum_{k=0}^{i} \pi_k w_k \sum_{\ell=i+1}^{\infty} \pi_\ell$$

$$= \sum_{\ell=i+1}^{\infty} \pi_\ell w_\ell - \left(\sum_{k=i+1}^{\infty} \pi_k \right) \left(\sum_{\ell=0}^{\infty} \pi_\ell w_\ell \right) \leqslant \sum_{\ell=i+1}^{\infty} \pi_\ell w_\ell, \qquad i \geqslant 0.$$

Collecting the above two inequalities together whenever $\pi(w) \geqslant 0$, we have

$$1 \leqslant D(f) \sup_{i \geqslant 0} J(w)(e_i) \leqslant D(f) \sup_{i \geqslant 0} \frac{1}{a(e_i) w(e_i)} \sum_{j=i+1}^{\infty} \pi_j w_j$$

$$= D(f) \sup_{i \geqslant 0} \frac{1}{\pi_i b_i (w_{i+1} - w_i)} \sum_{j=i+1}^{\infty} \pi_j w_j$$

$$= D(f) \sup_{i \geqslant 0} \frac{1}{\mu_i b_i (w_{i+1} - w_i)} \sum_{j=i+1}^{\infty} \mu_j w_j.$$

Combining this with (a), it follows that $D(f) \geqslant \inf_{i \geqslant 0} I_i(w)^{-1}$. Now, by conditions $\pi(f) = 0$, $\pi(f^2) = 1$, and (3.2), we get $\lambda_1 \geqslant \inf_{i \geqslant 0} I_i(w)^{-1}$. Since $w \in \widehat{\mathscr{W}}$ is arbitrary, we obtain $\lambda_1 \geqslant \sup_{w \in \widehat{\mathscr{W}}} \inf_{i \geqslant 0} I_i(w)^{-1}$. That is what we required. $\qquad \square$

Coupling proof of (3.3)

We use the same notation $\widehat{\mathscr{W}}$ and $I(w)$ introduced below (3.3).

Fix $w \in \widehat{\mathscr{W}}$ and define

$$u_i = \frac{1}{b_i \mu_i} \sum_{j \geqslant i+1} \mu_j w_j, \qquad i \geqslant 0,$$

$$g_i = \sum_{k<i} u_k, \qquad i \geqslant 0,$$

$$\rho(i,j) = |g_i - g_j|, \qquad i, j \geqslant 0.$$

From part (a) in the last proof, it follows that $u_i > 0$ for all $i \geqslant 0$. Hence g is strictly increasing, and so ρ is a distance.

Next, we adopt the classical coupling. Because of the symmetry $\mu_i b_i = \mu_{i+1} a_{i+1}$ $(i \geqslant 0)$, we have

$$\begin{aligned}
\widetilde{\Omega}_c \rho(i, i+1) &= [\Omega \rho(\cdot, i+1)](i) + [\Omega \rho(i, \cdot)](i+1) \\
&= [\Omega(g_{i+1} - g_{\bullet})](i) + [\Omega(g_{\bullet} - g_i)](i+1) \\
&= \Omega g(i+1) - \Omega g(i) \\
&= b_{i+1} u_{i+1} - a_{i+1} u_i - b_i u_i + a_i u_{i-1} \\
&= \left[\frac{1}{\mu_{i+1}} \sum_{j \geqslant i+2} \mu_j w_j - \frac{a_{i+1}}{b_i \mu_i} \sum_{j \geqslant i+1} \mu_j w_j \right] \\
&\quad - \left[\frac{1}{\mu_i} \sum_{j \geqslant i+1} \mu_j w_j - \frac{a_i}{b_{i-1} \mu_{i-1}} \sum_{j \geqslant i} \mu_j w_j \right] \\
&= -w_{i+1} + w_i \\
&= -\frac{(w_{i+1} - w_i) b_i \mu_i}{\sum_{j \geqslant i+1} \mu_j w_j} \cdot \frac{\sum_{j \geqslant i+1} \mu_j w_j}{b_i \mu_i} \\
&= -I_i(w)^{-1} \rho(i, i+1) \\
&\leqslant -\left[\inf_{k \geqslant 0} I_k(w)^{-1} \right] \rho(i, i+1), \qquad i \geqslant 1.
\end{aligned}$$

On the other hand, since $\sum_j \mu_j w_j \geqslant 0$, we have

$$\begin{aligned}
\widetilde{\Omega}_c \rho(0, 1) &= \Omega g(1) - \Omega g(0) = b_1 u_1 - a_1 u_0 - b_0 u_0 \\
&= \frac{1}{\mu_1} \sum_{j \geqslant 2} \mu_j w_j - \frac{a_1 + b_0}{b_0 \mu_0} \sum_{j \geqslant 1} \mu_j w_j \\
&= -w_1 - \frac{1}{\mu_0} \sum_{j \geqslant 1} \mu_j w_j \leqslant -(w_1 - w_0) \\
&\leqslant -\left[\inf_{i \geqslant 0} I_i(w)^{-1} \right] \rho(0, 1).
\end{aligned}$$

Collecting these two estimates together, we obtain

$$\begin{aligned}
\widetilde{\Omega}_c \rho(i, j) &= \widetilde{\Omega}_c \rho(i, i+1) + \cdots + \widetilde{\Omega}_c \rho(j-1, j) \\
&\leqslant -\left[\inf_{i \geqslant 0} I_i(w)^{-1} \right] \rho(i, j), \qquad i < j.
\end{aligned}$$

This proves the key condition (1.23), from which the conclusion (3.3) follows by a localizing procedure. Refer to Theorem 2.42. □

Clearly, the key point in the last proof is the choice of the distance ρ, which is not obvious at all. We will explain this point in detail in Section 3.4. Actually, the two equalities in (3.3) all hold and so we have complete variational formulas for the lower bound of the first eigenvalue. Since the proof is more technical, we would like to delay it to Section 3.7.

In the next three sections, we explain the ideas of the proof in the geometric case for the variational formula (1.11).

3.3 The three steps of the proof in the geometric case

In this section, we explain the three steps to prove the variational formula (1.11).

Choosing a coupling

Let (B_t) be the standard Brownian motion (BM) in $\mathbb{R}^d$ and let (X_t) be the solution to the stochastic differential equation (SDE):

$$\mathrm{d}X_t = \sqrt{2}\,\mathrm{d}B_t, \qquad x_0 = x. \tag{3.4}$$

The process corresponds to the operator Δ (half of it corresponds to the BM). Certainly, we can define a process (Y_t) in the same way but with different starting point:

$$\mathrm{d}Y_t = \sqrt{2}\,\mathrm{d}B_t, \qquad y_0 = y. \tag{3.5}$$

Now, because the processes (X_t) and (Y_t) are defined on the same probability space, we obtain a coupling, that is, the *coupling of marching soldiers* (X_t, Y_t). However, in what follows, we will use another process (Y_t), which is defined by

$$\mathrm{d}Y_t = \sqrt{2}\,H(X_t, Y_t)\mathrm{d}B_t, \qquad y_0 = y, \tag{3.6}$$

where $H(x, y) = I - 2(x - y)(x - y)^*/|x - y|^2$. Note that $H(x, y)$ has no meaning when $x = y$, so the process (Y_t) given in (3.6) is meaningful only up to the *coupling time*

$$T := \inf\{t \geqslant 0 : X_t = Y_t\}.$$

Starting from the time T, we define $Y_t = X_t$. We have thus constructed a process (Y_t). Clearly, this (Y_t) strongly depends on (X_t). Of course, the solutions of (3.5) and (3.6) are different, but they do have the same distribution, due to the invariance of orthogonal transform of BM and the fact that $H(x, y)$ is a reflection matrix. The last couple (X_t, Y_t) is the *coupling by reflection* discussed in the last chapter.

Intuitively, the construction of (Y_t) can be completed in two steps: Let $y \neq x$.

- Transport X_t from x to y in parallel along the line (x, y).
- Make the mirror reflection of the transported image of X_t in the hyperplane that is perpendicular to the line (x, y) at y.

Then, the mirror image gives us the process (Y_t).

For the diffusion (X_t) on manifold M with generator Δ, a process (Y_t) can be constructed in a similar way. Roughly speaking, one simply replaces the phrase "the line (x, y)" in the above construction by "the unique shortest geodesic γ between x and y." Certainly, there are some technical details and geometric difficulty (the cutlocus for instance) in the construction; refer to W.S. Kendall (1986) and M. Cranston (1992). An account of this coupling is now contained in E.P. Hsu (2002).

The appearance of the coupling by reflection is a crucial step in the development of the coupling theory. For a long period, one knew mainly the classical coupling; it is successful (i.e., $\widetilde{\mathbb{P}}[T < \infty] = 1$) for BM in $\mathbb{R}^d$ iff $d = 1$ [cf. Chen and S.F. Li (1989)]. Thus, one may have an impression that a process having a successful coupling ought to be recurrent. But the coupling by reflection shows that the success can be much weaker than the recurrence, since this coupling is successful in any dimension [cf. T. Lindvall and L.C.G. Rogers (1986), Chen and S.F. Li (1989)]. The key point is that the strong dependence of (Y_t) on (X_t) enables us to reduce the higher-dimensional case to dimension one.

Computing the distance

Throughout this chapter, we consider a connected, complete Riemannian manifold M with Ricc $\geqslant K$ for some $K \in \mathbb{R}$. In most cases, we consider compact M only. Denote by ρ the Riemannian distance on M. For the distance of the coupled process (X_t, Y_t), the following formula was proved by W.S. Kendall (1986) and M. Cranston (1992):

$$\mathrm{d}\rho(X_t, Y_t) = 2\sqrt{2}\,\mathrm{d}B_t + \left[\int_{X_t}^{Y_t} \sum_{i=2}^{d} \left(|\nabla_U W^i|^2 - \langle R(W^i, U)U, W^i \rangle \right) \right] \mathrm{d}t$$

$$- \mathrm{d}L_t, \qquad t < T, \tag{3.7}$$

where W^i, $i = 2, \ldots, d$, are Jacobi fields along the unique shortest geodesic γ between X_t and Y_t, U is the unit tangent vector to γ, the integral in $[\cdots]$ is along γ, (B_t) is a BM in $\mathbb{R}$, and (L_t) is an increasing process with support contained in $\{t \geqslant 0 : (X_t, Y_t) \in \boldsymbol{C}\}$, $\boldsymbol{C} := \{(x, y) : x$ is the cutlocus of $y\}$. When $(X_t, Y_t) \in \boldsymbol{C}$, the coefficient of $\mathrm{d}t$ is taken to be 0.

The formula is a finer version of the deterministic situation. The second term on the right-hand side of (3.7) is more or less familiar, and comes from the second variation of arc length. The first and the last terms are new in the stochastic case. Since the measure of the cutlocus equals zero, the last term is not essential. Next, because the mean of the first term is zero, it will be ignored once we take the expectation, as we will see soon in the next step. However, the condition "$t < T$" is crucial to avoid the singularity at $t = T$. This is the main place for which the present proof is probabilistic.

To estimate $\rho(X_t, Y_t)$, we need only to deal with the second term on the right-hand side of (3.7). By comparing M with a manifold with constant sectional curvature, M. Cranston (1992) proved that when $K < 0$, this term is controlled by

$$2\sqrt{-K(d-1)}\,\tanh\left(\frac{\rho_t}{2}\sqrt{\frac{-K}{d-1}}\right), \qquad \rho_t := \rho(X_t, Y_t). \qquad (3.8)$$

It was then proved by Chen and F.Y. Wang (1993b) that the same conclusion remains true when $K > 0$, in which case, (3.8) can be rewritten as

$$-2\sqrt{K(d-1)}\,\tan\left(\frac{1}{2}\sqrt{\frac{K}{d-1}}\,\rho_t\right).$$

Set

$$\gamma(r) = 2\sqrt{-K(d-1)}\,\tanh\left(\frac{1}{2}\sqrt{\frac{-K}{d-1}}\,r\right).$$

Then we obtain

$$d\rho_t \leqslant 2\sqrt{2}\,dB_t + \gamma(\rho_t)dt - dL_t \leqslant 2\sqrt{2}\,dB_t + \gamma(\rho_t)dt, \qquad t < T. \qquad (3.9)$$

Equivalently,

$$\rho_{t\wedge T} - \rho_0 \leqslant 2\sqrt{2}\int_0^{t\wedge T} dB_s + \int_0^{t\wedge T} \gamma(\rho_s)ds.$$

Taking the expectation, we get

$$\widetilde{\mathbb{E}}^{x,y}\rho_{t\wedge T} \leqslant \rho_0 + \widetilde{\mathbb{E}}^{x,y}\int_0^{t\wedge T} \gamma(\rho_s)ds. \qquad (3.10)$$

In order to get an exponential rate, we need the condition

$$\gamma(r) \leqslant -\alpha\,r \qquad \text{for some } \alpha > 0. \qquad (3.11)$$

When $K > 0$, since $\tan\theta \geqslant \theta$ on $[0, \pi/2]$, we have $\alpha = K$. Under (3.11), we have

$$\widetilde{\mathbb{E}}^{x,y}\int_0^{t\wedge T} \gamma(\rho_s)ds \leqslant -\alpha\widetilde{\mathbb{E}}^{x,y}\int_0^{t\wedge T} \rho_s ds = -\alpha\widetilde{\mathbb{E}}^{x,y}\int_0^{t} \rho_{s\wedge T}ds$$

$$= -\alpha\int_0^{t} \widetilde{\mathbb{E}}^{x,y}\rho_{s\wedge T}ds,$$

since $\rho_{t\wedge T} = 0$ for all $t \geqslant T$. Combining this with (3.10), we obtain

$$\widetilde{\mathbb{E}}^{x,y}\rho_{t\wedge T} \leqslant \rho_0 e^{-\alpha t}.$$

Equivalently,

$$\widetilde{\mathbb{E}}^{x,y}\rho_t \leqslant \rho_0 e^{-\alpha t}, \qquad t \geqslant 0. \qquad (3.12)$$

This is the key estimate of our method.

Estimating λ_1

Let g be an eigenfunction of λ_1: $-\Delta g = \lambda_1 g$, $g \neq$ constant. Then $\mathbb{E}^x g(X_t) = g(x)e^{-\lambda_1 t}$ for all $t \geq 0$. This gives us a relation between λ_1, g, and the process (X_t). The same relation holds for (Y_t). Note that the coupling property gives us $\widetilde{\mathbb{E}}^{x,y} g(X_t) = \mathbb{E}^x g(X_t)$. By (3.12), we have

$$
\begin{aligned}
e^{-\lambda_1 t} |g(x) - g(y)| = |\mathbb{E}^x g(X_t) - \mathbb{E}^y g(Y_t)| &= \left| \widetilde{\mathbb{E}}^{x,y} \left[g(X_t) - g(Y_t) \right] \right| \\
&\leqslant L(g)\widetilde{\mathbb{E}}^{x,y} \rho_t \leqslant L(g)\rho_0 e^{-\alpha t} \\
&= L(g)\rho(x,y)e^{-\alpha t}, \qquad t \geqslant 0,
\end{aligned}
$$

where $L(g)$ is the Lipschitz constant of g. This gives us immediately $\lambda_1 \geqslant \alpha$, and hence our proof is complete.

The proof is very much the same as sketched in Section 1.2.

The last step of the proof is rather simple but may not be so easy to discover. This is indeed a characteristic of various applications of the coupling methods: once the idea is understood, the proof often becomes quite straightforward.

3.4   Two difficulties

Roughly speaking, we have explained half of the first version of the paper by Chen and F.Y. Wang (1993b). The problem is that the above arguments are still not enough to obtain the sharp estimates listed in Table 1.2. For instance, when $K > 0$, we get the lower bound $\alpha = K$ only, as mentioned right after (3.11). The best we can get (when $K > 0$) is $8/D^2$ rather than the sharp one π^2/D^2, where D is the diameter of the compact manifold M. Even for the bound $8/D^2$, we still need to estimate $\widetilde{\mathbb{E}}^{x,y} T$ (cf. Theorem 2.43), which we are not going to discuss here.

We now return to analyze the proof discussed in the last section. In the last step, we need the Lipschitz property of g. The noncompact case can often be reduced to the compact one [cf. Chen and Wang (1995)] and in the latter case, g is smooth and hence the Lipschitz property is automatic. Thus, in the whole proof, the key is the estimate (3.12), for which we require not only a good coupling but also a good distance. This is not surprising. Since the convergence rate is not a topological concept, it certainly depends heavily on the choice of a distance. There is no reason why the underlying Riemannian distance should always be a correct choice.

Optimal Markovian coupling

The first question relates to the effectiveness of the coupling used above. Is there an optimal choice? This problem is quite hard, as explained in Section 2.2. However, the goal for optimality becomes clear now, that is, choosing a coupling to make the rate α as big as possible, or in a slightly wider sense, to make

$$\widetilde{\mathbb{E}}^{x,y}\rho(X_t, Y_t)$$

as small as possible for all $t \geqslant 0$, and for every fixed pair (x, y) and fixed ρ. Because we are dealing with Markovian coupling, we can use the language of coupling operators, studied in the last chapter. Of course, one can translate the discussions here into stochastic differential equations. Note that under a mild assumption, the last statement is equivalent to that $\widetilde{L}\rho(x, y)$ is as small as possible for every pair (x, y), $x \neq y$. This leads to the definition of ρ-*optimal coupling operator* $\overline{L}$:

$$\overline{L}\rho(x, y) = \inf_{\widetilde{L}} \widetilde{L}\rho(x, y), \qquad x \neq y,$$

where $\widetilde{L}$ varies over all coupling operators (cf. Section 2.2).

Some constructions for the optimal Markovian couplings are presented in the last chapter. In particular, Theorem 2.30 (4) tells us that the coupling by reflection is already good enough even for the BM on manifolds. Furthermore, it suggests that we use $f \circ \rho$ instead of the original Riemannian distance ρ. The construction of a new distance is the second main difficulty of the study, and this is the content of the remainder of this section.

Modification of Riemannian distance

To illustrate the use of the above idea, assume that $K \geqslant 0$ and take $\bar{\rho} = \sin \frac{\pi\rho}{2D}$. Since $\pi \leqslant D$, $\bar{\rho}$ is a distance. To compute $d\bar{\rho}_t$, noting that $d\rho_t \leqslant 2\sqrt{2}\,dB_t$, apply Itô's formula plus a comparison argument,

$$d\bar{\rho}_t \leqslant \frac{\pi}{2D}\cos\frac{\pi\rho_t}{2D} \cdot 2\sqrt{2}\,dB_t - \frac{1}{2}\cdot\frac{\pi^2}{4D^2}\cdot\sin\frac{\pi\rho_t}{2D}\cdot 8dt, \qquad t < T.$$

The first term is a martingale, denoted by M_t. We then obtain

$$d\bar{\rho}_t \leqslant dM_t - \frac{\pi^2}{D^2}\bar{\rho}_t dt, \qquad t < T.$$

Repeating the proof given in the last section, we get

$$\widetilde{\mathbb{E}}^{x,y}\bar{\rho}_t \leqslant \bar{\rho}_0 \exp\left[-\frac{\pi^2}{D^2}t\right].$$

Thus, we obtain luckily $\lambda_1 \geqslant \pi^2/D^2$, which is optimal in the case of zero curvature. By using the sine function again with a slight modification (which comes from some controlling equations of (3.9) with constant coefficients), we can obtain the other two optimal lower bounds [i.e., (1.1) and (1.7)], as shown in the final version of Chen and F.Y. Wang (1993b, Theorem 1.8). Finally, it is interesting to remark that $2\theta/\pi \leqslant \sin\theta \leqslant \theta$ on $[0, \pi/2]$, and so the distances $\bar{\rho}$ and ρ used above are actually equivalent. However, the resulting rates are essentially different.

Redesignated distances

Is there any other choice of the distance? The question is again easy to state but not so easy to answer. Indeed, we did not know for a long time where we should start from. This problem becomes more serious when one goes to the noncompact situation. Intuitively, distances cannot be good if with respect to them the eigenfunction g is too far away from being Lipschitz. As usual, we are taught by simple examples. Consider the diffusion on the half-line $[0, \infty)$ with operator

$$L = a\, d^2/dx^2 - b\, d/dx$$

for some constants a, $b > 0$ and with the Neumann boundary condition at the origin. If one adopts the Euclidean distance, then it gives us nothing. So what distance should we take? Our goal is to look at the eigenfunction of $\lambda_1 = b^2/4$ in the weak sense (without loss of generality, set $a = 1$):

$$g(x) = (1 - bx/2)\exp[bx/2] \in L^1(\pi) \setminus L^2(\pi).$$

This suggests that we construct a new distance ρ from the leading part of g:

$$\rho(x, y) = |\exp[\gamma x] - \exp[\gamma y]|$$

for suitable $\gamma > 0$. Surprisingly, it gives us the exact estimate of λ_1 even though the eigenfunction g is still not Lipschitz with respect to this distance [cf. Chen and F.Y. Wang (1995)]. Furthermore, if g is strictly monotone (which is indeed the case in dimension one but the proof is rather technical, cf. Section 3.7 below), we can always take $|g(x) - g(y)|$ as the distance we require. This provides us a way to construct and to classify the distances according to different classes of elementary functions [cf. Chen (1996), Chen and F.Y. Wang (1997b)].

   However, there is still a serious difficulty in the construction of the new distance, since the eigenvalue λ_1 and its eigenfunctions g are either known or unknown simultaneously. To see this, consider another example on the half-line with operator $L = a(x)d^2/dx^2$. A beautiful estimate due to I.S. Kac and M.G. Krein (1958), S. Kotani and S. Watanabe (1982) says that

$$\frac{1}{4}\left(\sup_{x>0} x \int_x^\infty \frac{du}{a(u)}\right)^{-1} \leqslant \lambda_1 \leqslant \left(\sup_{x>0} x \int_x^\infty \frac{du}{a(u)}\right)^{-1}.$$

Now, in order to recover this estimate using our method, according to what was discussed above, we have to know some information about the eigenfunction g. Even in such a simple situation, it is still hopeless to try to solve g from $a(x)$ explicitly. What can we do now? Once again, we examine the eigenequation:

$$a(x)g'' = -\lambda_1 g \iff g'(s) = \int_s^\infty \frac{\lambda_1 g(u)}{a(u)}du \qquad (\text{since } g'(\infty) = 0)$$

$$\iff g(x) = g(0) + \int_0^x ds \int_s^\infty \frac{\lambda_1 g(u)}{a(u)}du. \qquad (3.13)$$

Since we are dealing with the ergodic case, we can regard ∞ as a Neumann boundary, and so $g'(\infty) = 0$. What we have done is just to rewrite the differential equation as the corresponding integral equation. Is the last equation helpful? The answer is affirmative. We now move step by step as follows:

- Regard $\lambda_1 g$ as a new function f.
- Regard the right-hand side of (3.13) as an approximation of the left-hand side g.
- Ignore the constant $g(0)$ on the right-hand side, since we are interested in the difference $g(x) - g(y)$ only.

In other words, these considerations suggest to us to take

$$\tilde{g}(x) = \int_0^x \mathrm{d}s \int_s^\infty \frac{f(u)}{a(u)} \mathrm{d}u \tag{3.14}$$

as an approximation of q (up to a constant) and then to take $\rho(x, y) = |g(x) - \tilde{g}(y)|$. The function f used above is called a *test function*. A slightly different explanation of the construction goes as follows. Even though the equation (3.13) cannot be solved explicitly, as usual we have a successive approximation procedure. Thus, one may regard (3.14) as the first step of the approximation and go further step by step. However, the further approximations are not completely necessary, since on the one hand it becomes too complicated and on the other hand it is not as effective as modifying the test function f directly.

Next, we consider the general operator on the half-line:

$$L = a(x)\mathrm{d}^2/\mathrm{d}x^2 + b(x)\mathrm{d}/\mathrm{d}x.$$

By standard ODE, it can be reduced to the above simple case. The approximation function now becomes (cf. Chen and Wang (1997b))

$$g(r) = \int_0^r e^{-C(s)} \mathrm{d}s \int_s^\infty \frac{f(u)e^{C(u)}}{a(u)} \mathrm{d}u, \qquad C(r) := \int_0^r \frac{b(u)}{a(u)} \mathrm{d}u. \tag{3.15}$$

We have thus obtained a general construction of the mimic eigenfunctions and furthermore of the required distances. It should be not surprising that the reconstruction of the distances is a powerful tool in many situations. This will be illustrated in the next section.

Optimizing the distances

Before moving further, let us mention that an optimizing method of the distance induced from (3.15) as well as some comparison methods is developed in Chen and F.Y. Wang (1995). In short, the condition "$\tilde{L}\rho(x, y) \leqslant -\alpha\rho(x, y)$ [which is equivalent to (3.12)] holds for all large enough $\rho(x, y)$" but not necessarily "for all $x \neq y$" is enough to guarantee a positive lower bound of λ_1.

3.5   The final step of the proof of the formula

Up to now, we have discussed only the construction of the mimic eigenfunctions g in the case of a half-line. But how do we go to the whole line and further to $\mathbb{R}^d$ and a manifold M? This seems quite difficult. However, the answer is still rather simple once the idea has been figured out. As we have seen from Section 3.3, the coupling methods reduce the higher-dimensional case to computing the distance of the coupled process, and then the distance itself consists of a process valued in the half-line $[0, \infty)$. We have thus returned to what was treated in the last section.

Recall that

$$\gamma(r) = 2\sqrt{-K(d-1)} \tanh \left(\frac{1}{2} \sqrt{\frac{-K}{d-1}} \, r \right)$$

and $\rho_t = \rho(X_t, Y_t)$. From (3.9), it is known that

$$\mathrm{d}\rho_t \leqslant 2\sqrt{2}\,\mathrm{d}B_t + \gamma(\rho_t)\mathrm{d}t, \qquad t < T. \qquad (3.16)$$

The one-dimensional diffusion operator corresponding to (3.16) with equality is

$$L_1 = 4\mathrm{d}^2/\mathrm{d}x^2 + \gamma(x)\mathrm{d}/\mathrm{d}x$$

on $[0, D]$ with absorbing boundary at 0 and reflecting boundary at D. This is indeed simpler than what we discussed in the last section $(a(x) \equiv 4)$. Redefine

$$C(r) = \exp\left[\frac{1}{4} \int_0^r \gamma(s)\mathrm{d}s \right].$$

Then the approximation function defined by (3.15) becomes

$$g(r) = \int_0^r C(s)^{-1}\mathrm{d}s \int_s^D C(u)f(u)\mathrm{d}u,$$

up to a constant factor. Now the same proof as given in Section 3.3 and the second subsection of Section 3.4 implies rather easily the formula (1.11).

To derive Corollary 1.2 from Theorem 1.1, one needs to deal with the double integral in (1.11). For this, the main tool is the FKG inequality (2.1).

It remains to explain the choice of the test functions used in Corollary 1.2, $f(r) = \sin \frac{\pi r}{2D}$ for instance. Recall that the main result obtained by Chen and F.Y. Wang (1993b) compares λ_1 of the original operator with the first eigenvalue $\lambda_0^{(1)}$ of the one-dimensional operator L_1 with Dirichlet boundary at 0 and Neumann boundary at D. Since in general, $\lambda_0^{(1)}$ is not computable explicitly, one uses a constant instead of the function γ. In the case of $K \geqslant 0$, since $\gamma(r) \leqslant 0$, it is natural to replace γ by the constant 0. Then the first eigenfunction for the new operator is $f(r) = \sin \frac{\pi r}{2D}$, which is just the test function used in Corollary 1.2 (1).

Similarly, when $K < 0$, since

$$\gamma(r) = 2\sqrt{-K(d-1)} \, \tanh\left[\frac{r}{2}\sqrt{\frac{-K}{d-1}}\right] \leqslant 2\sqrt{-K(d-1)} \, \tanh\left[\frac{D}{2}\sqrt{\frac{-K}{d-1}}\right] := \gamma_0,$$

it is natural to use the operator $L_2 = 4\mathrm{d}^2/\mathrm{d}x^2 + \gamma_0 \mathrm{d}/\mathrm{d}x$, for which the eigenfunction of the first eigenvalue

$$\lambda_0^{(2)} := -\frac{K(d-1)}{4} \, \tanh^2\left[\frac{D}{2}\sqrt{\frac{-K}{d-1}}\right] \operatorname{sech}^2\theta$$

of L_2 has the form $f(r) = \exp[-\gamma_0 r/8] \sinh\left(\gamma_0 \delta r/8\right)$, where $\delta = \sqrt{1 - 16\lambda_0^{(2)}/\gamma_0^2}$ and θ is a root of the equation $\theta = \theta_1 \tanh\theta$, which is the (decreasing) limit of θ_n:

$$\theta_1 = \gamma_0 D/8, \qquad \theta_n = \theta_1 \tanh\theta_{n-1}, \qquad n \geqslant 2.$$

This is the test function used in Chen (1994a, Theorem 6.6) to deduce the following result:

$$\lambda_1 \geqslant \lambda_0^{(2)}.$$

Noting that $\tanh x \sim 1$ as $x \to \infty$, $\gamma_0 \sim 2\sqrt{-K(d-1)}$ and $\theta \sim \theta_1 \sim \frac{D}{4}\sqrt{-K(d-1)}$ as $K \to -\infty$, the leading order of $\lambda_0^{(2)}$ grows as

$$\operatorname{sech}^2\theta = \cosh^{-2}\theta \sim \cosh^{1-d}\left[\frac{D}{2}\sqrt{\frac{-K}{d-1}}\right]$$

when $K \to -\infty$. This means that as $K \to -\infty$, the leading term of the lower bound provided by Corollary 1.2 (2) is quite good. Note that as $K \to -\infty$, $\delta \sim 1$, and so the above eigenfunction f grows as follows

$$\frac{1}{2}\left(1 - \exp\left[-\frac{r}{2}\sqrt{-K(d-1)}\right]\right) \sim \frac{1}{2}\left(1 - \cosh^{1-d}\left[\frac{r}{2}\sqrt{\frac{-K}{d-1}}\right]\right).$$

The right-hand side is close to zero for small $|K|$. The change to the test function used in Corollary 1.2 is to keep the balance of $K = 0$ and $K \to -\infty$.

We have thus completed the proof in the geometric case. Our proof is universal in the sense that it works for general Markov processes, as shown by Theorems 2.42 and 2.43. We also obtain variational formulas for noncompact manifolds, elliptic operators in $\mathbb{R}^d$ [Chen and F.Y. Wang (1997b)], and Markov chains (Chen, 1996). It is more difficult to derive the variational formulas for the elliptic operators and Markov chains due to the presence of infinite parameters in these cases. In contrast, there are only three parameters (d, D, and K) in the geometric case. In fact, formula (1.11) is a particular consequence of our general formula (which is complete in dimension one) for elliptic operators.

Finally, we mention that the same method is used by Y.Z. Wang (1999) and Y.H. Mao (2002d; 2002e) to show that for diffusion on a compact manifold, the rate of strongly ergodic convergence is bounded below by

$$4\left[\int_0^D C(s)^{-1}ds \int_s^D C(u)du\right]^{-1}. \tag{3.17}$$

Note that this lower bound coincides with (1.11) by setting the test function $f = 1$. We will come back to this topic in Chapter 5.

3.6   Comments on different methods

First, we would like to make some remarks on the Dirichlet eigenvalue (called the D-problem for short). Similarly, we have an N-problem (Neumann or closed eigenvalue problem). It is interesting to note that historically, most of the publications in this field are devoted to the D-problem rather than the N-problem. The main reason is that the D-problem is equivalent to the maximum principle. Let $B(p,n)$ be the ball centered at p with radius n. It is well known [go back to J. Barta (1937); refer to H. Berestycki, L. Nirenberg and S.R.S. Varadhan (1994), and references within] that

$$\lambda_1 \geqslant \sup_f \inf_{B(p,n)} (-Lf)/f, \tag{3.18}$$

where f varies over all $C^2(B(p,n))$-functions with $f|_{\partial B(p,n)} = 0$ and $f > 0$ on $B(p,n)$. In other words, we have a variational formula for the lower bound for the D-problem. Note that the maximum principle is a powerful tool in PDE. One should not be surprised that one can do a lot with the D-problem. However, this formula does not work for the N-problem. The reason is simply

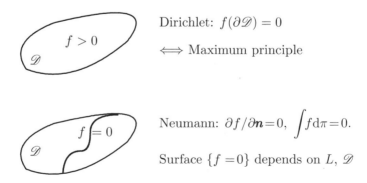

Dirichlet: $f(\partial\mathscr{D}) = 0$

$\Longleftrightarrow$ Maximum principle

Neumann: $\partial f/\partial \boldsymbol{n}=0$, $\int f d\pi=0$.

Surface $\{f=0\}$ depends on $L, \mathscr{D}$

Figure 3.3   Intrinsic difference of the methods

that the eigenfunction g in the Neumann case must cross zero and so is Lg, because the mean of g equals zero. See Figure 3.3. Hence, there is a singularity of $(-Lg)/g$ around the zero point, which causes serious difficulty when the eigenfunction g is replaced by its perturbation f. Traditionally, one transfers the N-problem to the D-problem, as will be studied in the next chapter. This explains why one often thinks that the N-problem is more difficult than the D-problem. It seems that the N-problem is also more difficult than the closed problem. For instance, for the Neumann eigenvalue λ_1 with convex boundary, the best known lower bound is Lichnerowicz's estimate obtained by J.F. Escobar (1990) in the case of $K > 0$, and up to now we have not seen in the literature a proof about "$\lambda_1 \geqslant \pi^2/D^2$" for general $K \geqslant 0$. The known estimates of λ_1 for the N-problem in the case of $K < 0$ are all less than the known estimates for the closed eigenvalue (refer to the books quoted at the beginning of Section 1.2). However, as we mentioned above, Theorem 1.1 and its corollaries are all suitable for the Neumann eigenvalue λ_1 with convex boundary. These discussions also show that the use of coupling enables us to avoid the singularity, just as mentioned above. The degeneracy of the coupled process appears at time T only, and before time T, the process is quite regular. This is somehow similar to the D-problem, for which the degeneracy appears at the boundary only. In other words, the coupling method plays an analogous role in our proof as the maximum principle played for the D-problem.

Geometric proof

We now recall Li–Yau's method (1980).

Let g be the eigenfunction corresponding to λ_1. By using a normalizing procedure, assume that $1 = \sup g > \inf g =: -k$. Here is the Li–Yau's key estimate:

$$|\nabla g| \leqslant \frac{2\lambda_1}{1+k}(1-g)(k+g).$$

That is often called *the method of gradient estimation*. To improve Li–Yau's estimate of λ_1, a key result is Zhong–Yang's estimate (1984):

$$|\nabla \theta|^2 \leqslant \lambda_1\big(1 + a_\varepsilon \psi(\theta)\big),$$

where

$$\theta = \arcsin(\text{a linear function of } g), \qquad a_\varepsilon = \frac{1-k}{(1+k)(1+\varepsilon)},$$

$$\psi(\theta) = \begin{cases} \dfrac{2[2\theta + \sin(2\theta)]/\pi - 2\sin\theta}{\cos^2\theta}, & \theta \in \left(-\dfrac{\pi}{2}, \dfrac{\pi}{2}\right), \\ 1, & \theta = \dfrac{\pi}{2}, \\ -1, & \theta = -\dfrac{\pi}{2}. \end{cases}$$

This estimate has been improved step by step by H.C. Yang, F. Jia, D. Zhao, et al. See also P. Kröger (1992; 1997), D. Bakry and Z.M. Qian (2000). All of the proofs are based on the maximum principle. From this, it should be clear that Zhong–Yang's proof cannot be simple, and is completely different from the probabilistic proof discussed in Sections 3.3–3.5.

Here is one more comment. As mentioned in the last section, the coupling method enables us to bound λ_1 in terms of the first mixed eigenvalue $\lambda_0^{(1)}$ of the operator $L_1 = 4d^2/dr^2 + \gamma(r)d/dr$ (with boundary conditions $f(0) = 0$ and $f'(D) = 0$). Since γ is odd, $\gamma(-r) = -\gamma(r)$, the first mixed eigenvalue $\lambda_0^{(1)}$ of L_1 on $(0, D)$ coincides with the first nontrivial Neumann eigenvalue $\lambda_1^{(1)}$ of L_1 on $(-D, D)$. Hence, $\lambda_1^{(1)}$ bounds λ_1 from below. This is the main result presented in Theorem 2 and its remark of P. Kröger (1992), and Theorem 14 of D. Bakry and Z.M. Qian (2000). Nevertheless, to obtain formula (1.11), as explained in the last two sections, one more step is required: expressing $\lambda_0^{(1)}$ as the right-hand side of (1.11) (see also Theorem 6.1 (2)).

No doubt, our method should be useful for complex manifolds. However, much work is expected to be done.

Open Problem 3.1. Study the first eigenvalue for complex manifolds by couplings.

3.7   Proof in the discrete case (continued)

The main purpose of this section is to prove that each sign of the equalities in (3.3) holds. We use the notation given in Section 3.2 and restate the result as follows.

Theorem 3.2. We have

$$\lambda_1 = \sup_{w \in \widehat{\mathscr{W}}} \inf_{i \geqslant 0} I_i(w)^{-1} = \sup_{w \in \mathscr{W}} \inf_{i \geqslant 0} I_i(\bar{w})^{-1}. \tag{3.19}$$

To prove these two equalities, we need some properties of the corresponding eigenfunction, and so the proof is rather technical.

Proposition 3.3. Let $\lambda > 0$ and $g \not\equiv 0$ be a solution to the equation $\Omega g = -\lambda g$. Then $g_0 \neq 0$ and

$$\pi_n b_n(g_{n+1} - g_n) = -\lambda \sum_{i=0}^{n} \pi_i g_i, \qquad n \geqslant 0. \tag{3.20}$$

Proof. The formula (3.20) follows from

$$-\lambda \sum_{i=0}^{n} \pi_i g_i = \sum_{i=0}^{n} \pi_i \Omega g(i) = \sum_{i=0}^{n} \left[\pi_i a_i (g_{i-1} - g_i) + \pi_i b_i (g_{i+1} - g_i) \right]$$

$$= \sum_{i=0}^{n} \left[- \pi_i a_i (g_i - g_{i-1}) + \pi_{i+1} a_{i+1} (g_{i+1} - g_i) \right]$$

$$= -\pi_0 a_0 (g_0 - g_{-1}) + \pi_{n+1} a_{n+1} (g_{n+1} - g_n)$$

$$= \pi_n b_n (g_{n+1} - g_n).$$

Here the additional term g_{-1} can be ignored, since $a_0 = 0$.

If $g_0 = 0$, then by induction and (3.20), it follows that $g_i \equiv 0$. This is a contradiction. □

Proposition 3.4. Let $\lambda_1 > 0$ and let g be a solution to the equation $\Omega g = -\lambda_1 g$ with $g_0 < 0$. Then g_i is strictly increasing.

Proof. Since $g_0 < 0$, by (3.20), we have $g_1 > g_0$. If g_i were not strictly increasing, then there would exist an $n \geqslant 1$ such that

$$g_0 < g_1 < \cdots < g_{n-1} < g_n \geqslant g_{n+1}. \tag{3.21}$$

We are going to prove that this is impossible.

By (3.20), we have

$$g_k < (\text{respectively } =) g_{k+1} \iff \sum_{i=0}^{k} \pi_i g_i < (\text{respectively } =) 0. \tag{3.22}$$

Define $\tilde{g}_n = -\sum_{i=0}^{n-1} \pi_i g_i / \pi_n$ and $\tilde{g}_i = g_i I_{[i<n]} + \tilde{g}_n I_{[i \geqslant n]}$. Then, from (3.20)–(3.22), it follows that

$$\sum_{i \leqslant n-1} \pi_i g_i + \pi_n \tilde{g}_n = 0, \tag{3.23}$$

$$g_n \geqslant \tilde{g}_n = [\pi_{n-1} b_{n-1}(g_n - g_{n-1})]/(\lambda_1 \pi_n) = [a_n(g_n - g_{n-1})]/\lambda_1 > 0. \tag{3.24}$$

Define $\bar{g}_i = g_i I_{\{i<n\}} + g_n I_{\{i \geqslant n\}}$. Then we have

$$\sum_i \pi_i \bar{g}_i^2 = \sum_{i \leqslant n-1} \pi_i g_i^2 + g_n^2 \sum_{i \geqslant n} \pi_i,$$

$$\sum_i \pi_i \bar{g}_i = \sum_{i \leqslant n-1} \pi_i g_i + g_n \sum_{i \geqslant n} \pi_i = g_n \sum_{i \geqslant n} \pi_i - \pi_n \tilde{g}_n \quad (\text{by (3.23)}).$$

Hence

$$\sum_i \pi_i \bar{g}_i^2 - \left(\sum_i \pi_i \bar{g}_i \right)^2 = \sum_{i \leqslant n-1} \pi_i g_i^2 + g_n^2 \sum_{i \geqslant n} \pi_i - \left(g_n \sum_{i \geqslant n} \pi_i - \pi_n \tilde{g}_n \right)^2. \tag{3.25}$$

Next,

$$-\sum_i \pi_i (\bar{g} \Omega \bar{g})(i) = \lambda_1 \sum_{i \leqslant n-1} \pi_i g_i^2 + \pi_n a_n g_n (g_n - g_{n-1})$$

$$= \lambda_1 \sum_{i \leqslant n-1} \pi_i g_i^2 + \lambda_1 \pi_n g_n \tilde{g}_n \quad (\text{by (3.24)}). \tag{3.26}$$

We now prove that

$$\pi_n g_n \tilde{g}_n < g_n^2 \sum_{i \geqslant n} \pi_i - \left(g_n \sum_{i \geqslant n} \pi_i - \pi_n \tilde{g}_n \right)^2. \tag{3.27}$$

By (3.24), $g_n > 0$. Thus, (3.27) is equivalent to

$$\pi_n \frac{\tilde{g}_n}{g_n} < \sum_{i \geqslant n} \pi_i - \left(\sum_{i \geqslant n} \pi_i - \pi_n \tilde{g}_n / g_n \right)^2.$$

That is,

$$\left(\sum_{i \geqslant n} \pi_i - \pi_n \frac{\tilde{g}_n}{g_n} \right)^2 < \sum_{i \geqslant n} \pi_i - \pi_n \frac{\tilde{g}_n}{g_n}.$$

This clearly holds, since $0 < \tilde{g}_n \leqslant g_n$,

$$0 < \sum_{i \geqslant n} \pi_i - \pi_n \tilde{g}_n / g_n = \sum_{i \geqslant n+1} \pi_i + \pi_n (1 - \tilde{g}_n / g_n) < 1.$$

We have thus proved (3.27). Collecting (3.25)–(3.27), it follows that

$$\lambda_1 \leqslant \frac{-\sum_i \pi_i \big(\bar{g} \Omega \bar{g} \big)(i)}{\sum_i \pi_i \bar{g}_i^2 - \big(\sum_i \pi_i \bar{g}_i \big)^2}$$

$$= \frac{\lambda_1 \sum_{i \leqslant n-1} \pi_i g_i^2 + \lambda_1 \pi_n g_n \tilde{g}_n}{\sum_{i \leqslant n-1} \pi_i g_i^2 + g_n^2 \sum_{i \geqslant n} \pi_i - \big(g_n \sum_{i \geqslant n} \pi_i - \pi_n \tilde{g}_n \big)^2} < \lambda_1,$$

which is a contradiction.   □

Proposition 3.5. Let $\lambda_1 > 0$ and let g be the function given by Proposition 3.4. Then $g \in L^1(\pi)$ and $\pi(g) = 0$.

Proof. By Proposition 3.4, we can define a positive sequence $u_i = g_{i+1} - g_i$, $i \geqslant 0$. From the eigenequation $\Omega g = -\lambda_1 g$, it follows that

$$b_i u_i - a_i u_{i-1} = -\lambda_1 g_i \quad (a_0 := 0), \qquad i \geqslant 0. \tag{3.28}$$

Replacing i with $i+1$, we obtain another equation. Taking the difference of these two equations, we get

$$R_i(u) := (a_{i+1} u_i - b_{i+1} u_{i+1} - a_i u_{i-1} + b_i u_i)/u_i = \lambda_1 > 0, \qquad i \geqslant 0.$$

By Propositions 3.3 and 3.4, if $g_i \leqslant 0$ for all i, then $\mu_n b_n u_n$ is increasing. Otherwise, if there is some $g_{i_0} > 0$, then $g_i > 0$ for all $i \geqslant i_0$ and hence $\mu_n b_n u_n$ is strictly decreasing for large n. Therefore, there is a limit $c := \lim_{n \to \infty} \mu_n b_n u_n \geqslant 0$. Set $u_{-1} = 0$. Define

$$w_i = a_i u_{i-1} - b_i u_i + c/(Z - \mu_0) = \lambda_1 g_i + c/(Z - \mu_0) \quad \text{(by (3.28))}.$$

Then

$$(w_{i+1} - w_i)/u_i = R_i(u) = \lambda_1 > 0, \qquad i \geqslant 0. \tag{3.29}$$

It follows that w_i is strictly increasing. On the other hand, we have

$$
\begin{aligned}
\sum_{j\geqslant i+1} \mu_j w_j &= \sum_{j\geqslant i+1} [\mu_j a_j u_{j-1} - \mu_j b_j u_j + c\mu_j/(Z - \mu_0)] \\
&= \sum_{j\geqslant i+1} [\mu_{j-1} b_{j-1} u_{j-1} - \mu_j b_j u_j + c\mu_j/(Z - \mu_0)] \\
&= \sum_{j\geqslant i+1} [\mu_{j-1} b_{j-1} u_{j-1} - \mu_j b_j u_j] + \frac{c}{Z - \mu_0} \sum_{j\geqslant i+1} \mu_j \qquad (3.30) \\
&= b_i \mu_i u_i - c + \frac{c}{Z - \mu_0} \sum_{j\geqslant i+1} \mu_j \\
&= b_i \mu_i u_i - \frac{c}{Z - \mu_0} \sum_{1\leqslant j\leqslant i} \mu_j \leqslant b_i \mu_i u_i, \qquad i \geqslant 0.
\end{aligned}
$$

In particular, $\sum_{j\geqslant 1} \mu_j w_j = \mu_0 b_0 u_0 \in (0, \infty)$ and so $w \in L^1(\pi)$.
Next, because $w_0 = -b_0 u_0 + c/(Z - \mu_0)$, we see that

$$
\sum_j \mu_j w_j = w_0 + \sum_{j\geqslant 1} \mu_j w_j = c/(Z - \mu_0) \geqslant 0.
$$

This fact plus $w_i \uparrow\uparrow$ implies that $\sum_{j\geqslant i+1} \mu_j w_j > 0$ for all $i \geqslant 0$, as proved in part (a) of the analytic proof of (3.3) (given in Section 3.2).
Collecting the above facts, we obtain

$$
\begin{aligned}
I_i(w)^{-1} &= b_i \mu_i (w_{i\,|\,1} - w_i) \Big/ \sum_{j\geqslant i+1} \mu_j w_j \\
&= b_i \mu_i R_i(u) u_i \Big/ \left[b_i \mu_i u_i - \frac{c}{Z - \mu_0} \sum_{1\leqslant j\leqslant i} \mu_j \right] \qquad (3.31) \\
&= \lambda_1 \left[1 - \frac{c}{(Z - \mu_0) b_i \mu_i u_i} \sum_{1\leqslant j\leqslant i} \mu_j \right]^{-1} \\
&\geqslant \lambda_1, \qquad i \geqslant 0.
\end{aligned}
$$

Thus, $\inf_{i\geqslant 0} I_i(w)^{-1} \geqslant \lambda_1$. Combining this with (3.3), we get $\inf_{i\geqslant 0} I_i(w)^{-1} = \lambda_1$.

Finally, we claim that $c = 0$ and so $\pi(g) = 0$ by the definition of w and (3.30). Otherwise, $c > 0$. Because $\lim_{n\to\infty} \mu_n b_n u_n = c$ and $\lim_{n\to\infty} \sum_{1\leqslant j\leqslant n} \mu_j = Z - \mu_0$, we would have $\lim_{i\to\infty} I_i(w)^{-1} = \infty$ by (3.31). Then, there would exist a k such that $\inf_{i\geqslant 0} I_i(w)^{-1} = \min_{i\leqslant k} I_i(w)^{-1} > \lambda_1$ by (3.31). This is a contradiction. $\square$

Having these preparations at hand, the proof of Theorem 3.2 is quite easy.

Proof of Theorem 3.2. Since $\inf_{i\geqslant 0} I_i(\bar{w})^{-1} \geqslant 0$, by (3.3), the equalities of (3.19) become trivial when $\lambda_1 = 0$.

We now assume that $\lambda_1 > 0$. By (3.3), we have $\lambda_1 \geqslant \sup_{w \in \widehat{\mathscr{W}}} \inf_{i \geqslant 0} I_i(w)^{-1}$. Combining this with Proposition 3.5 and its proof (by setting $w = \lambda_1 g$), we see that the sign of the last equality holds:

$$\lambda_1 = \sup_{w \in \widehat{\mathscr{W}}} \inf_{i \geqslant 0} I_i(w)^{-1}.$$

One can replace the right-hand side by $\sup_{w \in \mathscr{W}} \inf_{i \geqslant 0} I_i(\bar{w})^{-1}$, since $I_i(\bar{w})$ is invariant under the transform $w_i \to \alpha w_i + \beta$ for all $\alpha > 0$.   □

3.8   The first Dirichlet eigenvalue

We now turn to study the first Dirichlet eigenvalue. This is a more traditional topic than the Neumann one, as explained in Section 3.6, and will be studied subsequently. Here we consider Markov chains only. The results given in this section are quite similar to those of the last section.

Fix a point, say $0 \in E$. Then the first Dirichlet eigenvalue is defined by

$$\lambda_0 = \inf\{D(f) : f(0) = 0 \text{ and } \pi(f^2) = 1\}.$$

For each $i \in E$, choose a path γ_i from 0 to i (without a loop). Again, choose a positive weight function $\{w(e)\}$ on the edges and define $|\gamma_i|_w = \sum_{e \in \gamma_i} w(e)$,

$$I(w)(e) = \frac{1}{a(e)w(e)} \sum_{i \neq 0: \gamma_i \ni e} |\gamma_i|_w \pi_i.$$

Theorem 3.6. We have $\lambda_0 \geqslant \sup_w \inf_e I(w)(e)^{-1}$.

Proof.

$$1 = \sum_{i \neq 0} \pi_i f_i^2 = \sum_{i \neq 0} \pi_i (f_i - f_0)^2 = \sum_{i \neq 0} \pi_i \left(\sum_{e \in \gamma_i} f(e) \right)^2$$

$$\leqslant \sum_{i \neq 0} \pi_i \sum_{e \in \gamma_i} \frac{f(e)^2}{w(e)} |\gamma_i|_w = \sum_e a(e) f(e)^2 I(w)(e)$$

$$\leqslant D(f) \sup_e I(w)(e).   □$$

We now consider birth–death processes. Let $E = \{0, 1, 2, \ldots, N\}, N \leqslant \infty$, $q_{i,i+1} = b_i > 0 \ (0 \leqslant i \leqslant N-1)$, $q_{i,i-1} = a_i > 0 \ (1 \leqslant i \leqslant N)$, and $q_{ij} = 0$ for other $i \neq j$. Define

$$\mu_0 = 1, \quad \mu_n = \frac{b_0 \cdots b_{n-1}}{a_1 \cdots a_n}, \qquad 1 \leqslant n \leqslant N,$$

$$Z = \sum_{n=0}^{N} \mu_n, \qquad \pi_n = \mu_n / Z, \qquad \mathscr{W} = \{w : w_0 = 0, w_i \uparrow\uparrow\},$$

$$I_i(w) = \frac{1}{b_i \mu_i (w_{i+1} - w_i)} \sum_{j=i+1}^{N} \mu_j w_j, \qquad 0 \leqslant i \leqslant N-1, \qquad w \in \mathscr{W}.$$

For general $b_0 \geqslant 0$, define

$$\tilde{\mu}_1 = 1, \quad \tilde{\mu}_n = \frac{b_1 \cdots b_{n-1}}{a_2 \cdots a_n}, \quad 2 \leqslant n \leqslant N,$$

$$I_i(w) = \frac{1}{a_{i+1}\tilde{\mu}_{i+1}(w_{i+1} - w_i)} \sum_{j=i+1}^{N} \tilde{\mu}_j w_j, \quad 0 \leqslant i \leqslant N-1, \quad w \in \mathcal{W},$$

$$\tilde{\pi}_i = \tilde{\mu}_i \Big/ \sum_{1 \leqslant j \leqslant N} \tilde{\mu}_j, \quad 1 \leqslant i \leqslant N,$$

$$\tilde{D}(f) = \sum_{1 \leqslant i \leqslant N-1} \tilde{\pi}_i b_i (f_{i+1} - f_i)^2 + \tilde{\pi}_1 a_1 f_1^2.$$

Even though part (1) below is a particular case of part (2), it is kept for convenience in applications.

Theorem 3.7.   (1) If $b_0 > 0$, then we have

$$\lambda_0 = \sup_{w \in \mathcal{W}} \inf_{0 \leqslant i \leqslant N-1} I_i(w)^{-1}.$$

(2) For general $b_0 \geqslant 0$, the conclusion in part (1) remains true if we redefine $\lambda_0 = \inf \left\{ \tilde{D}(f) : f_0 = 0, \tilde{\pi}(f^2) = 1 \right\}$.

Proof. Until the last step of the proof, assume that $b_0 > 0$.

(a) Again, let e_i be the edge $\langle i, i+1 \rangle$. For each $i \geqslant 1$, there is a path γ_i consisting of $e_0, e_1, \ldots, e_{i-1}$. Take $w(e_i) = w_{i+1} - w_i$. Then

$$\sum_{k:\, \gamma_k \ni e_i} |\gamma_k|_w \pi_k = \sum_{k=i+1}^{N} (w_k - w_0)\pi_k = \sum_{k=i+1}^{N} \pi_k w_k.$$

Now the inequality "$\lambda_0 \geqslant \cdots$" in part (1) follows from Theorem 3.6.

(b) The remainder of the proof is similar to the proof of Theorem 3.2 given in Section 3.7. However, we still present the details here for completeness. Let $\lambda_0 > 0$ and $g \not\equiv 0$ with $g_0 = 0$ be a solution to the equation $\Omega g(i) = -\lambda_0 g_i$, $1 \leqslant i \leqslant N$. Here, we adopt the convention that $a_0 = 0$ and $b_N = 0$. The key to prove the equality in part (1) is to show the strict monotonicity of (g_i). Once this is done, without less of generality, assume that $g_i \uparrow\uparrow$. Then we have

$$I_i(g) = \frac{1}{a_{i+1}\mu_{i+1}(g_{i+1} - g_i)} \sum_{j=i+1}^{N} \mu_j g_j \equiv \frac{1}{\lambda_0} \tag{3.32}$$

for all $0 \leqslant i \leqslant N-1$, and hence the required assertion follows.

(c) To see that (3.32) holds, first, we show that

$$-\lambda_0 \sum_{i=1}^{n} \pi_i g_i = \pi_{n+1} a_{n+1}(g_{n+1} - g_n) - \pi_1 a_1 g_1, \quad 1 \leqslant n \leqslant N. \tag{3.33}$$

Here we use the convention $a_{N+1} = 0$, provided $N < \infty$. The proof is easy:

$$-\lambda_0 \sum_{i=1}^{n} \pi_i g_i = \sum_{i=1}^{n} \pi_i \Omega g(i) = \sum_{i=1}^{n} \left[\pi_i a_i (g_{i-1} - g_i) + \pi_i b_i (g_{i+1} - g_i) \right]$$

$$= \sum_{i=1}^{n} \left[-\pi_i a_i (g_i - g_{i-1}) + \pi_{i+1} a_{i+1} (g_{i+1} - g_i) \right]$$

$$= \pi_{n+1} a_{n+1} (g_{n+1} - g_n) - \pi_1 a_1 g_1.$$

Let $u_i = g_{i+1} - g_i$, $0 \leqslant i \leqslant N - 1$. Even though it is not necessary, for specificity we set $u_N = 1$ when $N < \infty$. By the eigenequation, we have

$$b_i u_i - a_i u_{i-1} = -\lambda_0 g_i, \qquad 1 \leqslant i \leqslant N.$$

Then

$$R_i(u) := (a_{i+1} u_i - b_{i+1} u_{i+1} - a_i u_{i-1} + b_i u_i)/u_i = \lambda_0 > 0, \qquad 1 \leqslant i \leqslant N - 1.$$

By (3.33) and the assumption $g_i \uparrow\uparrow$, $g_0 = 0$, it follows that

$$0 \leqslant \mu_{n+1} a_{n+1} u_n = \mu_1 a_1 g_1 - \lambda_0 \sum_{i=1}^{n} \mu_i g_i \leqslant \mu_1 a_1 g_1, \qquad 1 \leqslant n \leqslant N - 1.$$

Thus, $\mu_{n+1} a_{n+1} u_n$ is decreasing in n and

$$0 \leqslant c := \lim_{n \to N} \mu_{n+1} a_{n+1} u_n \leqslant \mu_1 a_1 g_1.$$

Note that $c = 0$ when $N < \infty$. Next, let

$$w_i = a_i u_{i-1} - b_i u_i + c/(Z - \mu_0) = \lambda_0 g_i + c/(Z - \mu_0) > 0, \qquad 1 \leqslant i \leqslant N.$$

Then

$$(w_{i+1} - w_i)/u_i = R_i(u) = \lambda_0 > 0, \qquad 1 \leqslant i \leqslant N - 1.$$

This implies that $w_i \uparrow\uparrow$. Therefore

$$\sum_{j=i+1}^{N} \mu_j w_j = \sum_{j=i+1}^{N} (\mu_j a_j u_{j-1} - \mu_j b_j u_j) + \frac{c}{Z - \mu_0} \sum_{j=i+1}^{N} \mu_j$$

$$= \sum_{j=i+1}^{N} (\mu_j a_j u_{j-1} - \mu_{j+1} a_{j+1} u_j) + \frac{c}{Z - \mu_0} \sum_{j=i+1}^{N} \mu_j$$

$$= \mu_{i+1} a_{i+1} u_i - c + \frac{c}{Z - \mu_0} \sum_{j=i+1}^{N} \mu_j$$

$$= \mu_{i+1} a_{i+1} u_i - \frac{c}{Z - \mu_0} \sum_{1 \leqslant j \leqslant i} \mu_j, \qquad 0 \leqslant i \leqslant N - 1.$$

Define additionally $w_0 = 0$. Since $w_1 > 0$, it is clear that $w \in \mathcal{W}$. We have

$$I_i(w)^{-1} = \mu_{i+1}a_{i+1}(w_{i+1} - w_i) \bigg/ \sum_{j=i+1}^{N} \mu_j w_j$$

$$= \mu_{i+1}a_{i+1}R_i(u)u_i \bigg/ \left[\mu_{i+1}a_{i+1}u_i - \frac{c}{Z - \mu_0} \sum_{1 \leqslant j \leqslant i} \mu_j \right]$$

$$= \lambda_0 \left[1 - \frac{c}{(Z - \mu_0)\mu_{i+1}a_{i+1}u_i} \sum_{1 \leqslant j \leqslant i} \mu_j \right]^{-1}$$

$$\geqslant \lambda_0, \qquad 1 \leqslant i \leqslant N - 1,$$

$$I_0(w)^{-1} = \mu_1 a_1 w_1 \bigg/ \sum_{j=1}^{N} \mu_j w_j - \mu_1 a_1 \left(\lambda_0 u_0 + \frac{c}{Z - \mu_0} \right) \bigg/ a_1 \mu_1 u_0$$

$$- \lambda_0 + \frac{c}{(Z - \mu_0)u_0} \geqslant \lambda_0.$$

Collecting these two estimates, we get

$$\sup_{\tilde{w} \in \mathcal{W}} \inf_{0 \leqslant i \leqslant N-1} I_i(\tilde{w})^{-1} \geqslant \inf_{0 \leqslant i \leqslant N-1} I_i(w)^{-1} \geqslant \lambda_0.$$

Combining this with proof (a), we know that $\inf_{0 \leqslant i \leqslant N-1} I_i(w)^{-1} = \lambda_0$. When $N < \infty$, we have $c = 0$ and so $w_i = \lambda_0 g_i$. Hence (3.32) holds. We now show that when $N = \infty$, we still have $c = 0$ and so (3.32) also holds. Otherwise, since $\mu_{i+1}a_{i+1}u_i$ is decreasing in i, we have $\inf_{i \geqslant 1} I_i(w)^{-1} = I_1(w)^{-1}$. From this, we must have a contradiction with $\inf_{i \geqslant 0} I_i(w)^{-1} = \lambda_0$, provided $c > 0$.

We have thus completed the proof of (3.32) under the assumption that $g_i \uparrow\uparrow$.

(d) We now prove the strict monotonicity of the eigenfunction (g_i) of λ_0. By (3.33), we have $g_1 \neq 0$. Otherwise, by induction, we would have $g_i \equiv 0$ for all $i \geqslant 1$. Thus, we may assume that $g_1 > 0$. Suppose that there is an n with $1 \leqslant n \leqslant N - 1$ such that

$$0 = g_0 < g_1 < \cdots < g_{n-1} < g_n \geqslant g_{n+1}.$$

Define $\bar{g}_i = g_i I_{[i<n]} + g_n I_{[i \geqslant n]}$. Then we have

$$\sum_i \pi_i \bar{g}_i^2 = \sum_{i \leqslant n-1} \pi_i g_i^2 + g_n^2 \sum_{i=n}^{N} \pi_i,$$

$$-\sum_i \pi_i (\bar{g}\Omega\bar{g})(i) = \lambda_0 \sum_{i \leqslant n-1} \pi_i g_i^2 + \pi_n a_n g_n(g_n - g_{n-1}).$$

Note that

$$\lambda_0 g_n = -\Omega g(n) = b_n(g_n - g_{n+1}) + a_n(g_n - g_{n-1}) \geqslant a_n(g_n - g_{n-1}).$$

We have

$$\pi_n a_n g_n(g_n - g_{n-1}) \leqslant \lambda_0 \pi_n g_n^2 < \lambda_0 g_n^2 \sum_{i=n}^{N} \pi_i.$$

Therefore

$$\lambda_0 \leqslant \frac{-\sum_i \pi_i (\bar{g} \Omega \bar{g})(i)}{\sum_i \pi_i \bar{g}_i^2} = \frac{\lambda_0 \sum_{i \leqslant n-1} \pi_i g_i^2 + \pi_n a_n g_n(g_n - g_{n-1})}{\sum_{i \leqslant n-1} \pi_i g_i^2 + g_n^2 \sum_{i=n}^{N} \pi_i} < \lambda_0,$$

which is a contradiction.

(e) As for the second assertion of the theorem, simply note that in the above proofs (a)–(d), we make no use of π_0 (recall that $g_0 = 0$) and b_0. Moreover, the original $I_i(w)$ is homogeneous in (μ_i). Actually, when $b_0 > 0$,

$$\lambda_0 = \inf_{f_0=0, f \neq 0} \sum_{0 \leqslant i \leqslant N-1} \pi_i b_i (f_{i+1} - f_i)^2 \bigg/ \sum_{0 \leqslant i \leqslant N} \pi_i f_i^2$$

$$= \inf_{f \neq 0} \left[\sum_{1 \leqslant i \leqslant N-1} \pi_i b_i (f_{i+1} - f_i)^2 + \pi_1 a_1 f_1^2 \right] \bigg/ \sum_{1 \leqslant i \leqslant N} \pi_i f_i^2$$

$$= \inf_{f \neq 0} \left[\sum_{1 \leqslant i \leqslant N-1} \mu_i b_i (f_{i+1} - f_i)^2 + \mu_1 a_1 f_1^2 \right] \bigg/ \sum_{1 \leqslant i \leqslant N} \mu_i f_i^2$$

$$= \inf_{f \neq 0} \left[\sum_{1 \leqslant i \leqslant N-1} \tilde{\mu}_i b_i (f_{i+1} - f_i)^2 + \tilde{\mu}_1 a_1 f_1^2 \right] \bigg/ \sum_{1 \leqslant i \leqslant N} \tilde{\mu}_i f_i^2.$$

Thus, we are studying the process with Dirichlet form $\tilde{D}(f)$ on the state space $\{1, 2, \ldots, N\}$ and with killing rate a_1. No role is played by b_0.   □

Chapter 4

Generalized Cheeger's Method

From the previous chapters, we have seen an application of a probabilistic method to a problem in Riemannian geometry. This chapter goes in the opposite direction. We use Cheeger's method, which comes from Riemannian geometry, to study some probabilistic problems. We begin with a review of Cheeger's method in geometry (Section 4.1). We then move to a generalization (Section 4.2) and present our new results (Section 4.3). In particular, we examine Cheeger's splitting technique and prove an existence criterion for the spectral gap (Section 4.4). In Sections 4.5–4.8, we sketch the proofs of the main theorems. Applications to birth–death processes are collected in the last section (Section 4.9).

4.1 Cheeger's method

Let us recall Cheeger's inequality in geometry.

Again, let M be a connected compact Riemannian manifold. We consider the first nontrivial eigenvalue λ_1 of the Laplacian Δ. We will also study the first Dirichlet eigenvalue, denoted by λ_0. Here is the geometric result.

Theorem 4.1 (Cheeger's inequality, 1970). We have

$$k \geqslant \lambda_1 \geqslant \frac{1}{4}k^2,$$

where *Cheeger's constant k* is defined by

$$k = \inf_{\substack{\partial M_1 = S = \partial M_2 \\ M_1 \cup S \cup M_2 = M}} \frac{\mathrm{Area}(S)}{\mathrm{Vol}(M_1) \wedge \mathrm{Vol}(M_2)},$$

where S varies over all hypersurfaces dividing M into two parts having the same boundary S.

As usual, $\mathrm{Vol}(M)$ and $\mathrm{Area}(S)$ denote the Riemannian volume of M and the area of S, respectively.

The key ideas in establishing this inequality are the following:

- *Splitting technique.* $\lambda_1 \geqslant \inf_B[\lambda_0(B) \vee \lambda_0(B^c)]$. That is, we split the space into two parts B and B^c, and then estimate the first eigenvalue λ_1 in terms of $\lambda_0(B)$ and $\lambda_0(B^c)$, where $\lambda_0(B)$ is the first (local) Dirichlet eigenvalue, to be defined later.

- Estimate $\lambda_0(B)$ in terms of another *Cheeger's constant*

$$h = \inf_{M_1 \subset M, \, \partial M_1 \cap \partial M = \emptyset} \frac{\mathrm{Area}(\partial M_1)}{\mathrm{Vol}(M_1)},$$

where M_1 varies over all subdomains of M.

The last constant is closely related to the *isoperimetric inequality*:

$$\frac{\mathrm{Area}(\partial A)}{\mathrm{Vol}(A)^{(d-1)/d}} \geqslant \frac{\mathrm{Area}(\mathbb{S}^{d-1})}{\mathrm{Vol}(\mathbb{B}^d)^{(d-1)/d}},$$

where $\mathbb{B}^d$ and $\mathbb{S}^{d-1}$ denote the unit ball and unit sphere in $\mathbb{R}^d$, respectively. It is standard that $\mathrm{Area}(\mathbb{S}^{d-1}) = 2\pi^{d/2}/\Gamma(d/2)$ and $\mathrm{Vol}(\mathbb{B}^d) = \mathrm{Area}(\mathbb{S}^{d-1})/d$. The ratio on the right-hand side is called the *isoperimetric constant*. It was observed first by J. Cheeger (1970) that the proof of the classical isoperimetric inequality can be also used to study the first eigenvalue λ_1. Certainly, one can replace Lebesgue measure with others. The isoperimetric inequality with respect to Gaussian measure was studied by P. Lévy (1919) [see Lévy (1951, Chapter IV)] and extended by M. Gromov (1980; 1999), S.G. Bobkov (1996; 1997), S.G. Bobkov and F. Götze (1999a), D. Bakry and M. Ledoux (1996) even in the infinite-dimensional setting. A broad account of the results relating to this type of inequality in Euclidean space or on manifolds was presented in I. Chavel (2001). Mainly, these publications concern differential operators. However, in this chapter, we are going in another direction, that of studying integral operators.

4.2   A generalization

Let $(E, \mathscr{E}, \pi)$ be a probability space satisfying $\{(x, x) : x \in E\} \in \mathscr{E} \times \mathscr{E}$. Denote by $L^p(\pi)$ the usual real L^p-space with norm $\|\cdot\|_p$. Write $\|\cdot\| = \|\cdot\|_2$ for simplicity. In this chapter, we consider mainly a *symmetric form* $(D, \mathscr{D}(D))$ (not necessarily a Dirichlet form) on $L^2(\pi)$:

$$D(f) = \frac{1}{2} \int_{E \times E} J(\mathrm{d}x, \mathrm{d}y)[f(y) - f(x)]^2,$$
$$\mathscr{D}(D) = \{f \in L^2(\pi) : D(f) < \infty\},$$
(4.1)

where $J \geqslant 0$ is a symmetric measure, having no charge on the diagonal set $\{(x, x) : x \in E\}$. A typical example is as follows. For a q-pair $(q(x), q(x, \mathrm{d}y))$, reversible with respect to π (i.e., $\pi(\mathrm{d}x)q(x, \mathrm{d}y) = \pi(\mathrm{d}y)q(y, \mathrm{d}x)$), we simply take

$$J(\mathrm{d}x, \mathrm{d}y) = \pi(\mathrm{d}x)q(x, \mathrm{d}y).$$

More especially, for a Q-matrix $Q = (q_{ij})$, reversible with respect to $(\pi_i > 0)$ (i.e., $\pi_i q_{ij} = \pi_j q_{ji}$ for all i, j), we take $J_{ij} = \pi_i q_{ij}$ $(j \neq i)$ as the density of the symmetric measure $J(\mathrm{d}x, \mathrm{d}y)$ with respect to the counting measure.

Naturally, define

$$\lambda_1 = \inf\{D(f) : \pi(f) = 0, \|f\| = 1\}, \qquad \pi(f) := \int f \mathrm{d}\pi.$$

We call λ_1 the *spectral gap* of $(D, \mathscr{D}(D))$.

For bounded jump processes, the fundamental known result is due to G.F. Lawler and A.D. Sokal (1988), stated in Theorem 1.6. As mentioned in Section 1.3, the last result has a very wide range of applications and has been collected in several books. The main shortcoming is the restriction on bounded operators.

On the other hand, for differential operators, there are many publications on the logarithmic Sobolev inequality (1.26). Refer to D. Bakry and M. Emery (1985), D. Bakry (1992), L. Gross (1993), S. Aida and I. Shigekawa (1994), F.Y. Wang (1997), M. Ledoux (1999; 2000), A. Guionnet and B. Zegarlinski (2003), and references within. However, the known results for integral operators are still rather limited. Here is a general result for Markov chains.

Theorem 4.2 (P. Diaconis and L. Saloff-Coste, 1996). Let the Q-matrix $Q = (q_{ij})$ be reversible with respect to $(\pi_i > 0)$ and satisfy $\sum_j |q_{ij}| = 1$. Define

$$\mathrm{Ent}(f) = \sum_i \pi_i f_i \log f_i - \sum_i \pi_i f_i \log \sum_i \pi_i f_i, \qquad f \geqslant 0,$$

and $D(f) = \frac{1}{2} \sum_{i,j} \pi_i q_{ij}(f_j - f_i)^2$. Then the optimal constant σ in the logarithmic Sobolev inequality

$$\mathrm{Ent}\,(f^2) \leqslant \frac{2}{\sigma} D(f)$$

satisfies

$$\sigma \geqslant \frac{2(1 - 2\pi_*)\lambda_1}{\log[1/\pi_* - 1]}, \qquad \pi_* := \min_i \pi_i.$$

This result is very good, since it can be sharp. Clearly, it works only for a finite state space.

For the Nash inequality (1.27), our knowledge is more or less at the same level.

4.3   New results

To avoid unboundedness, our goal is to use a renormalizing procedure. Choose a nonnegative symmetric function r such that

$$J^{(1)}(\mathrm{d}x, E)/\pi(\mathrm{d}x) \leqslant 1, \qquad \pi\text{-a.e.,} \tag{4.2}$$

where

$$J^{(\alpha)}(\mathrm{d}x, \mathrm{d}y) = I_{\{r(x,y)^{\alpha} > 0\}} \frac{J(\mathrm{d}x, \mathrm{d}y)}{r(x,y)^{\alpha}}, \quad \alpha \in [0,1], \qquad J^{(0)} := J.$$

Then, corresponding to each inequality, define a new Cheeger's constant, as listed in Table 1.3. Finally, one of our main results can be stated as follows (Theorem 1.7).

Theorem 4.3. If $k^{(1/2)} > 0$, then the corresponding inequality holds.

Even though the result is very simple to state, its proof is completed in four papers: Chen and F.Y. Wang (1998), Chen (1999b; 2000b), and F.Y. Wang (2001). Of course, many more results were proved in these papers. For instance, here is a lower estimate of λ_1.

Theorem 4.4.

$$\lambda_1 \geqslant \frac{k^{(1/2)^2}}{1 + \sqrt{1 - k^{(1)^2}}}. \tag{4.3}$$

Surprisingly, this estimate can be sharp, which is rather unusual in using Cheeger's approach.

The proof of Theorem 4.4 is delayed to Section 4.5. In parallel, the lower estimates for the logarithmic Sobolev and the Nash inequalities are presented in Sections 4.6 and 4.8, respectively. Some upper estimates are given in Section 4.7.

The main advantage of Cheeger's approach is that it works in a very general setting. Here is an example.

Corollary 4.5. Let $(E, \mathscr{E}, \pi)$ be a probability space and let $j(x, y) \geqslant 0$ be a symmetric function satisfying $j(x, x) = 0$ and

$$j(x) := \int_E j(x, y)\pi(\mathrm{d}y) < \infty, \qquad x \in E.$$

Then, for the symmetric form generated by $J(\mathrm{d}x, \mathrm{d}y) = j(x, y)\pi(\mathrm{d}x)\pi(\mathrm{d}y)$, we have

$$\lambda_1 \geqslant \frac{1}{8} \inf_{x \neq y} \frac{j(x, y)^2}{j(x) \vee j(y)}.$$

Proof. Note that

$$
\begin{aligned}
k^{(\alpha)} &= \inf_{\pi(A)\in(0,1/2]} \frac{1}{\pi(A)} \int_{A\times A^c} \frac{j(x,y)}{[j(x)\vee j(y)]^\alpha} \pi(\mathrm{d}x)\pi(\mathrm{d}y) \\
&\geqslant \inf_{x\neq y} \frac{j(x,y)}{[j(x)\vee j(y)]^\alpha} \inf_{\pi(A)\in(0,1/2]} \pi(A^c) \\
&\geqslant \frac{1}{2} \inf_{x\neq y} \frac{j(x,y)}{[j(x)\vee j(y)]^\alpha}.
\end{aligned}
$$

The conclusion now follows immediately from (4.3). $\quad\square$

The generalized Cheeger's approach and the isoperimetric method have also been applied to the L^p-setup for jump processes by F. Wang and Y.H. Zhang (2003), Y.H. Mao (2001a; 2001b), to diffusions by F.Y. Wang (2000a; 2004b), and to the weaker Poincaré inequalities (which will be discussed in Chapter 7) by M. Röckner and F.Y. Wang (2001).

4.4 Splitting technique and existence criterion

Recall the definitions

$$
\lambda_1 = \inf\{D(f) : f \in \mathscr{D}(D), \pi(f) = 0, \pi(f^2) = 1\},
$$
$$
\lambda_0(A) = \inf\{D(f) : f \in \mathscr{D}(D), f|_{A^c} = 0, \pi(f^2) = 1\}.
$$

As mentioned in the last section, the reduction of the Neumann case to the Dirichlet one is based on Cheeger's splitting technique. Here is the result proved by G.F. Lawler and A.D. Sokal (1988) for bounded operators:

$$
\lambda_1 \geqslant \inf_B \{\lambda_0(B) \vee \lambda_0(B^c)\}.
$$

However, we are unable to extend this result to unbounded symmetric forms. Instead, a weaker version is as follows:

$$
\lambda_1 \geqslant \inf_{\pi(B)\subset(0,1/2]} \lambda_0(B).
$$

More precisely, we have the following result.

Theorem 4.6 (Chen and F.Y. Wang (1998), Chen (2000c)). For the symmetric form (4.1) or general Dirichlet form $(D, \mathscr{D}(D))$, we have

$$
\inf_{\pi(A)\in(0,1/2]} \lambda_0(A) \leqslant \lambda_1 \leqslant \inf_{\pi(A)\in(0,1)} \min\left\{\frac{\lambda_0(A)}{\pi(A^c)}, \frac{\lambda_0(A^c)}{\pi(A)}\right\}
$$
$$
\leqslant 2 \inf_{\pi(A)\in(0,1/2]} \lambda_0(A). \tag{4.4}
$$

Proof. (a) Let $f \in \mathscr{D}(D)$ be such that $f|_{A^c} = 0$ and $\pi(f^2) = 1$. Then

$$\pi(f^2) - \pi(f)^2 = 1 - \pi(fI_A)^2 \geqslant 1 - \pi(f^2)\pi(A)$$
$$= 1 - \pi(A) = \pi(A^c).$$

Hence

$$\lambda_1 \leqslant \frac{D(f)}{\pi(f^2) - \pi(f)^2} \leqslant \frac{D(f)}{\pi(A^c)},$$

which implies that $\lambda_1 \leqslant \lambda_0(A)/\pi(A^c)$. Furthermore,

$$\lambda_1 \leqslant \inf_{\pi(A)\in(0,1)} \min\left\{\frac{\lambda_0(A)}{\pi(A^c)}, \frac{\lambda_0(A^c)}{\pi(A)}\right\}$$
$$= \inf_{\pi(A)\in(0,1/2]} \min\left\{\frac{\lambda_0(A)}{\pi(A^c)}, \frac{\lambda_0(A^c)}{\pi(A)}\right\}$$
$$\leqslant \inf_{\pi(A)\in(0,1/2]} \lambda_0(A)/\pi(A^c)$$
$$\leqslant 2 \inf_{\pi(A)\in(0,1/2]} \lambda_0(A).$$

This part of the proof works for general $D(f)$.

(b) Next, for $\varepsilon > 0$, choose f_ε such that $\pi(f_\varepsilon) = 0$, $\pi(f_\varepsilon^2) = 1$, and $\lambda_1 + \varepsilon \geqslant D(f_\varepsilon)$. Choose c_ε such that $\pi(f_\varepsilon < c_\varepsilon) \leqslant 1/2$ and $\pi(f_\varepsilon > c_\varepsilon) \leqslant 1/2$. Set $f_\varepsilon^\pm = (f_\varepsilon - c_\varepsilon)^\pm$, and define $B_\varepsilon^\pm = \{f_\varepsilon^\pm > 0\}$. Recall that

$$D(f) = \frac{1}{2}\int J(\mathrm{d}x, \mathrm{d}y)[f(y) - f(x)]^2.$$

We have

$$\lambda_1 + \varepsilon \geqslant D(f_\varepsilon) = D(f_\varepsilon - c_\varepsilon)$$
$$= \frac{1}{2}\int J(\mathrm{d}x, \mathrm{d}y)\left[\left|f_\varepsilon^+(y) - f_\varepsilon^+(x)\right| + \left|f_\varepsilon^-(y) - f_\varepsilon^-(x)\right|\right]^2.$$

Therefore,

$$\lambda_1 + \varepsilon \geqslant \frac{1}{2}\int J(\mathrm{d}x,\mathrm{d}y)\left(f_\varepsilon^+(y) - f_\varepsilon^+(x)\right)^2 + \frac{1}{2}\int J(\mathrm{d}x,\mathrm{d}y)\left(f_\varepsilon^-(y) - f_\varepsilon^-(x)\right)^2$$
$$\geqslant \lambda_0(B_\varepsilon^+)\pi\left(\left(f_\varepsilon^+\right)^2\right) + \lambda_0(B_\varepsilon^-)\pi\left(\left(f_\varepsilon^-\right)^2\right)$$
$$\geqslant \inf_{\pi(B)\in(0,1/2]} \lambda_0(B)\,\pi\left(\left(f_\varepsilon^+\right)^2 + \left(f_\varepsilon^-\right)^2\right)$$
$$= (1 + c_\varepsilon^2) \inf_{\pi(B)\in(0,1/2]} \lambda_0(B)$$
$$\geqslant \inf_{\pi(B)\in(0,1/2]} \lambda_0(B).$$

Because ε is arbitrary, the proof is done for the symmetric form (4.1).

(c) Finally, for the general Dirichlet form, since

$$D(f) = \lim_{t \downarrow 0} \frac{1}{2t} \int_{E \times E} \pi(\mathrm{d}x) P_t(x, \mathrm{d}y)[f(y) - f(x)]^2, \qquad (4.5)$$

the proof needs a little modification only. $\quad\square$

Theorem 4.6 presents a traditional way of studying λ_1, that is, estimating λ_1 in terms of the Dirichlet eigenvalue λ_0. The latter one is usually easier to handle. As an illustration, here we consider jump processes. Let $(q(x), q(x, \cdot))$ be a totally stable q-pair, $q(x, E) \leqslant q(x) < \infty$ for all $x \in E$, and symmetrizable with respect to a measure μ:

$$\int_A q(x, B)\mu(\mathrm{d}x) = \int_B q(x, A)\mu(\mathrm{d}x), \qquad A, B \in \mathscr{E}.$$

Equivalently, the corresponding operator Ω,

$$\Omega f(x) = \int_E q(x, \mathrm{d}y)[f(y) - f(x)] - r(x)f(x),$$

where $r(x) = q(x) - q(x, E)$ for $x \in E$, is symmetric in $L^2(\mu)$. Set $E_n = \{x \in E : q(x) \leqslant n\}$. Define

$$D(f) = \frac{1}{2} \int_{E \times E} \mu(\mathrm{d}x)q(x, \mathrm{d}y)[f(y) - f(x)]^2 + \int \mu(\mathrm{d}x)r(x)f(x)^2$$

and $\|f\|_D^2 = D(f) + \|f\|^2$. Next, set

$$\mathscr{D}_0 = \{f \in L^2(\mu) : f \text{ vanishes out of some } E_n\}$$

and let $\mathscr{D}(D)$ be the completion of $\mathscr{D}_0$ with respect to $\|\cdot\|_D$. Roughly speaking, the form $(D, \mathscr{D}(D))$ corresponds to the minimal jump process. Define

$$\lambda_0 = \inf\{D(f) : f \in \mathscr{D}(D), \|f\| = 1\}.$$

Then we have the following variational formula for the lower bound of λ_0.

Theorem 4.7 (Chen, 2000d). $\lambda_0 \geqslant \sup_{0 < g \in \mathscr{E}} \mu\text{-ess inf } (-\Omega g)/g.$

Applications to the Neumann eigenvalue

We need the following result. Let E be a locally compact separable metric space with Borel σ-algebra $\mathscr{E}$, μ an everywhere dense Radon measure on E, and $(D, \mathscr{D}(D))$ a Dirichlet form on $L^2(\mu) = L^2(E; \mu)$.

The next result is due to V.G. Maz'ya (1973) (cf. Maz'ya (1985) and references within) in a particular case, where the use of capacity was begun. The general case is due to Z. Vondraček (1996). Its proof was simplified by M. Fukushima and T. Uemura (2003).

Theorem 4.8. For a regular transient Dirichlet form on $A \in \mathscr{E}$,

$$(4\Theta(A))^{-1} \leqslant \lambda_0(A) \leqslant \Theta(A)^{-1},$$

where $\Theta(A) = \sup_{\text{compact } K \subset A} \pi(K)/\text{Cap}(K)$,

$$\text{Cap}(K) = \inf \{ D(f) : f \in \mathscr{D}(D) \cap C_0(E), \ f|_K \geqslant 1 \},$$

and $C_0(E)$ is the set of continuous functions with compact support.

Combining the above two results, we obtain the following one.

Theorem 4.9. Let $\mu(E) < \infty$. Then for a regular irreducible Dirichlet form, we have

$$\inf_{\substack{\text{open } A: \, \pi(A) \in (0,1/2]}} (4\Theta(A))^{-1} \leqslant \lambda_1 \leqslant 2 \inf_{\substack{\text{open } A: \, \pi(A) \in (0,1/2]}} \Theta(A)^{-1}.$$

In particular,
$$\lambda_1 > 0 \quad \text{iff} \quad \sup_{\substack{\text{open } A: \, \pi(A) \in (0,1/2]}} \Theta(A) < \infty.$$

We will return to this capacitary method in Chapter 7.

We now study the existence criterion in a different way. For compact state spaces $(E, \mathscr{E})$, it is often true that $\lambda_1 > 0$. Thus, we need only consider the noncompact case. The idea is to use Cheeger's splitting technique. Split the space E into two parts A and A^c. Mainly, there are two boundary conditions: the Dirichlet and the Neumann boundary conditions, which are absorbing and reflecting at the boundary, respectively. See Figure 4.1. The corresponding eigenvalue problems are denoted by (D) and (N), respectively.

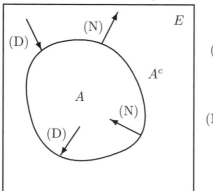

(D): Dirichlet boundary

(N): Neumann boundary

Figure 4.1   Four choices of boundary condition

Next, let A be compact for a moment. Then on A^c, one should consider the problem (D). Otherwise, since A^c is noncompact, the solution to problem (N) is unknown, and it is indeed what we are also interested in. On A, we can use either of the boundary conditions. However, it is better to use the Neumann one, since the corresponding λ_1 is closer to the original λ_1 when A becomes larger. In other words, we want to describe the original λ_1 in terms of the local $\lambda_1(A)$ and $\lambda_0(A^c)$.

We now state our criterion informally, which is easier to remember.

Criterion (Informal description [Chen and F.Y. Wang, 1998]). $\lambda_1 > 0$ iff there exists a compact A such that $\lambda_0(A^c) > 0$, where

$$\lambda_0(A^c) = \inf\{D(f, f): \ f|_A = 0, \pi(f^2) = 1\}.$$

To state the precise result, define

$$\lambda_1(B) = \inf\left\{D_B(f) : \pi(f) = 0, \pi\!\left(f^2\right)/\pi(B) = 1\right\},$$

where

$$D_B(f) = \frac{1}{2}\int_{B \times B} J(\mathrm{d}x, \mathrm{d}y)(f(y) - f(x))^2.$$

Theorem 4.10 (Criterion [Chen and F.Y. Wang, 1998]). Let $A \subset B$ satisfy $0 < \pi(A)$, $\pi(B) < 1$. Then

$$\frac{\lambda_0(A^c)}{\pi(A)} \geqslant \lambda_1 \geqslant \frac{\lambda_1(B)[\lambda_0(A^c)\pi(B) - 2M_A\pi(B^c)]}{2\lambda_1(B) + \pi(B)^2[\lambda_0(A^c) + 2M_A]}, \tag{4.6}$$

where $M_A = \mathrm{ess\,sup}_{A,\,\pi}J(\mathrm{d}x, A^c)/\pi(\mathrm{d}x)$, $\mathrm{ess\,sup}_{A,\,\pi}$ denoting the essential supremum over the set A with respect to the measure π.

As mentioned before, usually $\lambda_1(B) > 0$ for all compact B. Hence the result means, as stated in the heuristic description, that $\lambda_1 > 0$ iff $\lambda_0(A^c) > 0$ for some compact A, because we can first fix such an A and then make B large enough so that the right-hand side of (4.6) becomes positive. The reason we have to use two sets A and B rather than a single A is that the operator is not local, and there may exist an interaction with a very long range. The final choice of regions and the boundary conditions are shown in Figure 4.2.

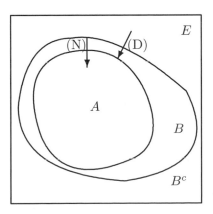

Figure 4.2 Nonlocal, need intersection

Proof of Theorem 4.10. Let f satisfy $\pi(f) = 0$ and $\pi(f^2) = 1$. Our aim is to bound $D(f)$ in terms of $\lambda_0(A^c)$ and $\lambda_1(B)$.

(a) First, we use $\lambda_1(B)$:

$$D(f) \geqslant D_B(fI_B) \geqslant \lambda_1(B)\pi(B)^{-1}\left[\pi\left(f^2I_B\right) - \pi(B)^{-1}\pi\left(fI_B\right)^2\right]$$
$$= \lambda_1(B)\pi(B)^{-1}\left[\pi\left(f^2I_B\right) - \pi(B)^{-1}\pi\left(fI_{B^c}\right)^2\right]. \tag{4.7}$$

Here in the last step, we have used $\pi(f) = 0$.

(b) Next, we use $\lambda_0(A^c)$. We need the following elementary inequality:

$$|(fI_{A^c})(x)-(fI_{A^c})(y)| \leqslant |f(x)-f(y)| + I_{A\times A^c \cup A^c \times A}(x,y)|(fI_A)(x)-(fI_A)(y)|.$$

Then

$$\lambda_0(A^c)\pi\left(f^2I_{A^c}\right) \leqslant D(fI_{A^c}) = \frac{1}{2}\int J(\mathrm{d}x,\mathrm{d}y)\left[(fI_{A^c})(y) - (fI_{A^c})(x)\right]^2$$
$$\leqslant 2D(f) + 2\int_{A\times A^c} J(\mathrm{d}x,\mathrm{d}y)\left[(fI_A)(y) - (fI_A)(x)\right]^2$$
$$\leqslant 2D(f) + 2M_A\pi\left(f^2I_A\right). \tag{4.8}$$

(c) Estimating the right-hand sides of (4.7) and (4.8) in terms of $\gamma := \pi\left(f^2I_B\right)$, we obtain two inequalities $D(f) \geqslant c_1\gamma + c_2$ and $D(f) \geqslant -c_3\gamma + c_4$ for some constants $c_1, c_3 > 0$. Hence

$$D(f) \geqslant \inf_{\gamma \in [0,1]} \max\{c_1\gamma + c_2,\ -c_3\gamma + c_4\}.$$

Clearly, the infimum is achieved at γ_0, which is the intersection of the two lines Γ_1 and Γ_2 in $\{\cdots\}$. See Figure 4.3. Then the required lower bound of λ_1 is given by $c_1\gamma_0 + c_2$. $\quad\square$

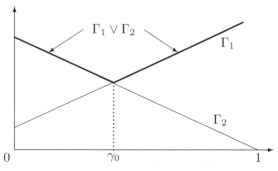

Figure 4.3   Intersection of Γ_1 and Γ_2

4.5 Proof of Theorem 4.4

The proof of Theorem 4.4 is based on Cheeger's splitting idea, that is, to estimate λ_1 in terms of λ_0 for a more general symmetric form

$$D(f) = \frac{1}{2} \int_{E \times E} J(\mathrm{d}x, \mathrm{d}y)[f(y) - f(x)]^2 + \int_E K(\mathrm{d}x)f(x)^2, \qquad (4.9)$$

where K is a nonnegative measure on $(E, \mathscr{E})$. Studying λ_0 is meaningful, since $D(1) \neq 0$ whenever $K \neq 0$. It is called the *Dirichlet eigenvalue of* $(D, \mathscr{D}(D))$. Thus, in what follows, when dealing with λ_0 (respectively, λ_1), we consider only the symmetric form given by (4.9) (respectively, (4.1)). Instead of (4.2), we now require that

$$[J^{(1)}(\mathrm{d}x, E) + K^{(1)}(\mathrm{d}x)]/\pi(\mathrm{d}x) \leqslant 1, \qquad \pi\text{-a.s}, \qquad (4.10)$$

where $J^{(\alpha)}$ is the same as before and

$$K^{(\alpha)}(\mathrm{d}x) = I_{\{s(x)^\alpha > 0\}} \frac{K(\mathrm{d}x)}{s(x)^\alpha}$$

for some nonnegative function $s(x)$. Corresponding to $(J^{(\alpha)}, K^{(\alpha)})$, we have a symmetric form $D^{(\alpha)}$, defined by (4.9). Next, define

$$h^{(\alpha)} = \inf_{\pi(A) > 0} \frac{J^{(\alpha)}(A \times A^c) + K^{(\alpha)}(A)}{\pi(A)}.$$

Theorem 4.11 (Chen and F.Y. Wang, 1998). For the symmetric form given by (4.9), under (4.10), we have

$$\lambda_0 \geqslant \frac{h^{(1/2)2}}{1 + \sqrt{1 - h^{(1)2}}}.$$

Proof. (a) First, we express $h^{(\alpha)}$ by the following functional form

$$h^{(\alpha)} = \inf\left\{\frac{1}{2} \int J^{(\alpha)}(\mathrm{d}x, \mathrm{d}y)|f(x) - f(y)| + K^{(\alpha)}(f) : f \geqslant 0, \pi(f) = 1\right\}.$$

By setting $f = I_A/\pi(A)$, one returns to the original set form of $h^{(\alpha)}$. For the reverse assertion, simply consider the set $A_\gamma = \{f > \gamma\}$ for $\gamma \geqslant 0$. The proof is also not difficult:

$$\int_{\{f(x) > f(y)\}} J^{(\alpha)}(\mathrm{d}x, \mathrm{d}y)[f(x) - f(y)] + K^{(\alpha)}(f)$$

$$= \int_0^\infty \mathrm{d}\gamma\left\{J^{(\alpha)}(\{f(x) > \gamma \geqslant f(y)\}) + K^{(\alpha)}(\{f > \gamma\})\right\}$$

$$\text{(coarea formula)}$$

$$= \int_0^\infty \left[J^{(\alpha)} \left(A_\gamma \times A_\gamma^c \right) + K^{(\alpha)} (A_\gamma) \right] d\gamma$$

$$\geqslant h^{(\alpha)} \int_0^\infty \pi (A_\gamma) d\gamma = h^{(\alpha)} \pi(f).$$

The appearance of K makes the notation heavier. To avoid this, one can enlarge the state space to $E^* = E \cup \{\infty\}$. At the same time, extend f to a function f^* on E^*: $f^* = f I_E$. Regarding K as a killing measure on E^*, the form $D(f, g)$ can be extended to the product space $E^* \times E^*$ but can be expressed by a symmetric measure J^* only:

$$J^{*(\alpha)}(C) = \begin{cases} J^{(\alpha)}(C), & C \in \mathscr{E} \times \mathscr{E}, \\ K^{(\alpha)}(A), & C = A \times \{\infty\} \text{ or } \{\infty\} \times A, \ A \in \mathscr{E}, \\ 0, & C = \{\infty\} \times \{\infty\}. \end{cases}$$

Then, we have $J^{*(\alpha)}(dx, dy) = J^{*(\alpha)}(dy, dx)$ and

$$\int J^{(\alpha)}(dx, E) f(x)^2 + K^{(\alpha)}(f^2) = \int_{E^*} J^{*(\alpha)}(dx, E^*) f^*(x)^2,$$

$$D^{(\alpha)}(f, f) = \frac{1}{2} \int_{E^* \times E^*} J^{*(\alpha)}(dx, dy) \left(f^*(y) - f^*(x) \right)^2,$$

$$\frac{1}{2} \int J^{(\alpha)}(dx, dy) |f(y) - f(x)| + \int K^{(\alpha)}(dx) |f(x)|$$

$$= \frac{1}{2} \int_{E^* \times E^*} J^{*(\alpha)}(dx, dy) |f^*(y) - f^*(x)|,$$

$$h^{(\alpha)} = \inf \left\{ \frac{1}{2} \int_{E^* \times E^*} J^{*(\alpha)}(dx, dy) |f^*(x) - f^*(y)| : f \geqslant 0, \ \pi(f) = 1 \right\}.$$

(b) Take f with $\pi(f^2) = 1$, by (a), the Cauchy–Schwarz inequality, and condition (4.10), we have

$$h^{(1)2} \leqslant \left\{ \frac{1}{2} \int_{E^* \times E^*} J^{*(1)}(dx, dy) |f^*(y)^2 - f^*(x)^2| \right\}^2$$

$$\leqslant \frac{1}{2} D^{(1)}(f) \int_{E^* \times E^*} J^{*(1)}(dx, dy) \left[f^*(y) + f^*(x) \right]^2$$

$$= \frac{1}{2} D^{(1)}(f) \left\{ 2 \int_{E^* \times E^*} J^{*(1)}(dx, dy) \left[f^*(y)^2 + f^*(x)^2 \right] \right.$$

$$\left. - \int_{E^* \times E^*} J^{*(1)}(dx, dy) \left[f^*(y) - f^*(x) \right]^2 \right\}$$

$$\leqslant D^{(1)}(f) \left[2 - D^{(1)}(f) \right]. \tag{4.11}$$

Solving this quadratic inequality in $D^{(1)}(f)$, one obtains

$$D^{(1)}(f) \geqslant 1 - \sqrt{1 - h^{(1)2}}.$$

(c) Repeating the above proof but by a more careful use of the Cauchy–Schwarz inequality, we obtain

$$h^{(1/2)^2} \leqslant \left\{ \frac{1}{2} \int_{E^* \times E^*} J^{*(1/2)}(dx, dy) \big| f^*(y)^2 - f^*(x)^2 \big| \right\}^2$$

$$= \left\{ \frac{1}{2} \int_{E^* \times E^*} J^*(dx, dy) |f^*(y) - f^*(x)| \cdot I_{\{r(x,y)>0\}} \frac{|f^*(y) + f^*(x)|}{\sqrt{r(x,y)}} \right\}^2$$

$$\leqslant \frac{1}{2} D(f) \int_{E^* \times E^*} J^{*(1)}(dx, dy) \big[f^*(y) + f^*(x) \big]^2$$

$$\leqslant D(f) \big[2 - D^{(1)}(f) \big].$$

From this and (b), the required assertion follows. □

Proof of Theorem 4.4. (a) For any $B \subset E$ with $\pi(B) > 0$, define a local form as follows:

$$\widetilde{D}_B^{(\alpha)}(f) = \frac{1}{2} \int_{B \times B} J^{(\alpha)}(dx, dy)[f(y) - f(x)]^2 + \int_B J^{(\alpha)}(dx, B^c) f(x)^2.$$

Obviously, $\widetilde{D}_B^{(\alpha)}(f) = \widetilde{D}_B^{(\alpha)}(f I_B)$. Moreover,

$$\lambda_0(B) := \inf\{D(f) : f|_{B^c} = 0, \|f\| = 1\} = \inf \big\{ \widetilde{D}_B(f) : \pi(f^2 I_B) = 1 \big\}.$$

Let $\pi^B = \pi(\cdot \cap B)/\pi(B)$ and set

$$h_B^{(\alpha)} = \inf_{A \subset B, \pi(A) > 0} \frac{J^{(\alpha)}(A \times (B \setminus A)) + J^{(\alpha)}(A \times B^c)}{\pi(A)} = \inf_{A \subset B, \pi(A) > 0} \frac{J^{(\alpha)}(A \times A^c)}{\pi(A)}.$$

Applying Theorem 4.11 to the local form on $L^2(B, \mathscr{E} \cap B, \pi^B)$ generated by $J^B = \pi(B)^{-1} J|_{B \times B}$ and $K^B = J(\cdot, B^c)|_B$, we obtain

$$\lambda_0(B) \geqslant \frac{h_B^{(1/2)^2}}{1 + \sqrt{1 - h_B^{(1)^2}}}.$$

(b) Noting that $\inf_{\pi(B) \leqslant 1/2} h_B^{(\alpha)} = k^{(\alpha)}$, by Theorem 4.6, we get

$$\lambda_1 \geqslant \inf_{\pi(B) \leqslant 1/2} \frac{h_B^{(1/2)^2}}{1 + \sqrt{1 - h_B^{(1)^2}}} \geqslant \inf_{\pi(B) \leqslant 1/2} \frac{\inf_{\pi(B) \leqslant 1/2} h_B^{(1/2)^2}}{1 + \sqrt{1 - h_B^{(1)^2}}}$$

$$\geqslant \frac{\inf_{\pi(B) \leqslant 1/2} h_B^{(1/2)^2}}{1 + \sqrt{1 - \inf_{\pi(B) \leqslant 1/2} h_B^{(1)^2}}} = \frac{k^{(1/2)^2}}{1 + \sqrt{1 - k^{(1)^2}}}.$$

We obtain the required conclusion. □

4.6   Logarithmic Sobolev inequality

Theorem 4.12 (Chen, 2000b). Denote by σ the optimal constant in the logarithmic Sobolev inequality:

$$\mathrm{Ent}\left(f^2\right) \leqslant \frac{2}{\sigma}D(f).$$

We have

$$2\kappa \geqslant \sigma \geqslant \frac{2\lambda_1\kappa^{(1/2)}}{\sqrt{\lambda_1\left(2 - \lambda_1^{(1)}\right) + 3\kappa^{(1/2)}}} \geqslant \frac{1}{8}\kappa^{(1/2)^2},$$

where

$$\kappa^{(\alpha)} = \inf_{\pi(A)\in(0,1)} \frac{J^{(\alpha)}(A \times A^c)}{-\pi(A)\log\pi(A)}, \qquad \kappa = \kappa^{(0)},$$

and $\lambda_1^{(\alpha)} = \inf\{D^{(\alpha)}(f) : \pi(f) = 0, \|f\| = 1\}.$

Proof. The proof is partially due to F.Y. Wang. To get the upper bound, simply apply the inequality to the test function $f = I_A/\sqrt{\pi(A)}$, $\pi(A) \in (0,1)$. To prove the lower bound, let $\pi(f) = 0$ and $\|f\| = 1$.

(a) Set $\varepsilon = \sqrt{2 - \lambda_1^{(1)}}\big/\big[2\kappa^{(1/2)}\big]$ and $E(f) = \pi\left(f^2 \log f^2\right)$. We claim that

$$E(f) \leqslant 2\varepsilon\sqrt{D(f)} + 1. \tag{4.12}$$

Actually, one shows first that

$$I := \frac{1}{2}\int J^{(1/2)}(\mathrm{d}x, \mathrm{d}y)\big|f(y)^2 - f(x)^2\big| \leqslant \sqrt{\left(2 - \lambda_1^{(1)}\right)D(f)}. \tag{4.13}$$

The proof is standard, as used before (cf. the proof (c) of Theorem 4.11). Next, set $A_t = \{f^2 > t\}$ and prove that

$$I \geqslant \kappa^{(1/2)}[E(f) - 1]. \tag{4.14}$$

The proof goes as follows. Note that $h_t := \pi(A_t) \leqslant 1 \wedge t^{-1}$. We have

$$I = \int_0^\infty J^{(1/2)}(A_t \times A_t^c)\mathrm{d}t \geqslant \kappa^{(1/2)} \int_0^\infty (-h_t \log h_t)\mathrm{d}t$$

$$\geqslant \kappa^{(1/2)} \int_0^\infty h_t \log t\,\mathrm{d}t = \kappa^{(1/2)} \int_0^\infty h_t(\log t + 1)\mathrm{d}t - \kappa^{(1/2)}$$

$$= \kappa^{(1/2)} \int \mathrm{d}\pi \int_0^{f^2} (\log t + 1)\mathrm{d}t - \kappa^{(1/2)}$$

$$= \kappa^{(1/2)}[E(f) - 1].$$

Combining (4.13) with (4.14), we get (4.12).

(b) By (4.12), on the one hand, we have

$$E(f) \leqslant 2\varepsilon\sqrt{D(f)} + 1 \leqslant \gamma\varepsilon D(f) + \varepsilon/\gamma + 1,$$

where $\gamma > 0$ is a constant to be specified below. On the other hand, by [Bakry (1992); Theorem 3.6 and Proposition 3.9], the inequality

$$\pi\left(f^2 \log f^2\right) \leqslant C_1 D(f) + C_2, \qquad \pi(f) = 0, \quad \|f\| = 1, \tag{4.15}$$

implies that

$$\sigma \geqslant 2/[C_1 + (C_2 + 2)\lambda_1^{-1}]. \tag{4.16}$$

In other words, if $\lambda_1 > 0$, then the weaker inequality (4.15) is indeed equivalent to the original logarithmic Sobolev inequality. We will prove this fact soon. Combining these facts, it follows that

$$\sigma \geqslant \frac{2}{\varepsilon\gamma + [\varepsilon/\gamma + 3]/\lambda_1}.$$

Maximizing the right-hand side with respect to γ, we get

$$\sigma \geqslant \frac{2\lambda_1\kappa^{(1/2)}}{\sqrt{(2 - \lambda_1^{(1)})\lambda_1 + 3\kappa^{(1/2)}}}. \tag{4.17}$$

Next, applying Theorem 4.4 to $J^{(1)}$, we have $k^{(1/2)} = k^{(1)}$ and hence $\lambda_1^{(1)} \geqslant 1 - \sqrt{1 - k^{(1)^2}}$. Combining this with Theorem 4.4 and noting that $k^{(1/2)} \geqslant (\log 2)\kappa^{(1/2)}$, it follows that the right-hand side of (4.17) is bounded below by

$$\frac{2(\log 2)^2\kappa^{(1/2)^2}}{(\log 2 + 3)\left[1 + \sqrt{1 - k^{(1)^2}}\right]} \geqslant \frac{1}{8}\kappa^{(1/2)^2}. \qquad \square$$

We now introduce a more powerful result; its proof is quite technical and is omitted here.

Theorem 4.13 (Chen, 2000b). Define

$$\xi^{(\delta)} = \inf_{\pi(A)>0} \frac{J^{(1/2)}(A \times A^c) + \delta\pi(A)}{\pi(A)\sqrt{1 - \log\pi(A)}},$$

$$A(\delta) = \frac{(2 + \delta)(\lambda_1 + \delta)}{(\xi^{(\delta)})^2}, \qquad \delta > 0.$$

Then we have

$$2\kappa \geqslant \sigma \geqslant \frac{2\lambda_1}{1 + 16\inf_{\delta>0} A(\delta)}.$$

Proof of (4.16). Let $\bar{f} = f - \pi(f)$. Then the assertion follows by Lemma 4.14 below, (4.15), and the Poincaré inequality:

$$\mathrm{Ent}\,(f^2) \leqslant \mathrm{Ent}\,(\bar{f}^2) + 2\|\bar{f}\|^2 \leqslant C_1 D(f) + (2+C_2)\|\bar{f}\|^2 = \left[C_1 + \frac{2+C_2}{\lambda_1}\right]D(f). \quad \square$$

The following result was proved by J.D. Deuschel and D.W. Stroock (1989, page 247), and goes back to O.S. Rothaus (1985, Lemma 9).

Lemma 4.14. We have

$$\sup_{c \in \mathbb{R}} \mathrm{Ent}\,((f+c)^2) \leqslant \mathrm{Ent}\,(f^2) + 2\pi\,(f^2), \qquad \pi(f) = 0.$$

Proof. Obviously, the assertion is equivalent to

$$\sup_{c \in \mathbb{R}} \mathrm{Ent}\,((f+c)^2) \leqslant \mathrm{Ent}\,((f - \pi(f))^2) + 2\pi\,((f - \pi(f))^2).$$

Without loss of generality, assume that $\|f - \pi(f)\| = 1$ and set $h = f - \pi(f)$. Then we have $\pi(h) = 0$ and $\|h\| = 1$. Using $h + 1/t$ instead of $f + c$, it suffices to show that

$$\int (1+th)^2 \log \frac{(1+th)^2}{1+t^2}\mathrm{d}\pi \leqslant t^2 \int h^2 \log h^2 \mathrm{d}\pi + 2t^2, \qquad t \in \mathbb{R}.$$

Define

$$h_\delta(t) = \int (1+th)^2 \log \frac{(1+th)^2 + \delta}{1+t^2}\mathrm{d}\pi - t^2 \int h^2 \log h^2 \mathrm{d}\pi.$$

Then

$$h'_\delta(t) = 2\int (1+th)h\log[(1+th)^2 + \delta]\mathrm{d}\pi + 2\int \frac{(1+th)^3 h}{(1+th)^2 + \delta}\mathrm{d}\pi$$
$$- 2t\left[1 + \log(1+t^2) + \int h^2 \log h^2 \mathrm{d}\pi\right],$$

$$h''_\delta(t) = 2\int h^2 \log \frac{(1+th)^2 + \delta}{(1+t^2)h^2}\mathrm{d}\pi + 10\int h^2 \frac{(1+th)^2}{(1+th)^2 + \delta}\mathrm{d}\pi$$
$$- 4\int h^2 \frac{(1+th)^4}{[(1+th)^2 + \delta]^2}\mathrm{d}\pi - 2 - \frac{4t^2}{1+t^2}.$$

By Jensen's inequality,

$$\int h^2 \log \frac{(1+th)^2 + \delta}{(1+t^2)h^2}\mathrm{d}\pi \leqslant \log \int \frac{(1+th)^2 + \delta}{1+t^2}\mathrm{d}\pi = \log\left(1 + \frac{\delta}{1+t^2}\right),$$

since $\pi(h) = 0$ and $\|h\| = 1$. On the other hand, by the Schwarz inequality,

$$A(t,\delta)^2 \leqslant \int h^2 \frac{(1+th)^4}{[(1+th)^2 + \delta]^2}\mathrm{d}\pi, \quad A(t,\delta) := \int h^2 \frac{(1+th)^2}{(1+th)^2 + \delta}\mathrm{d}\pi \in [0,1].$$

Hence

$$h''_\delta(t) \leqslant 2\log\left(1 + \frac{\delta}{1+t^2}\right) - [4A(t,\delta)^2 - 10A(t,\delta)] - 2 \leqslant 2\log(1+\delta) + 4.$$

Noting that $h_\delta(0) = \log(1+\delta)$ and $h'_\delta(0) = 0$, by Taylor expansion, we get

$$h_\delta(t) \leqslant \log(1+\delta) + [2 + \log(1+\delta)]t^2.$$

The assertion now follows by letting $\delta \to 0$.   □

4.7   Upper bounds

The upper bound given by Theorem 4.12 is usually very rough. Here we introduce two results which are often rather effective. The results show that order one (respectively, two) of exponential integrability is required for $\lambda_1 > 0$ (respectively, $\sigma > 0$).

Theorem 4.15 (Chen and F.Y. Wang, 1998). Suppose that the function r used in (4.2) is J-a.e. positive. If there exists $\varphi \geqslant 0$ such that

$$\text{ess sup}_J |\varphi(x) - \varphi(y)|^2 r(x,y) \leqslant 1, \tag{4.18}$$

then

$$\lambda_1 \leqslant \inf\left\{\varepsilon^2/4 : \varepsilon \geqslant 0, \pi(e^{\varepsilon\varphi}) = \infty\right\}. \tag{4.19}$$

Consequently, $\lambda_1 = 0$ if there exists $\varphi \geqslant 0$ satisfying (4.18) such that $\pi(e^{\varepsilon\varphi}) = \infty$ for all $\varepsilon > 0$.

Proof. We need to show that if $\pi(e^{\varepsilon\varphi}) = \infty$, then $\lambda_1 \leqslant \varepsilon^2/4$. For $n \geqslant 1$, define $f_n = \exp[\varepsilon(\varphi \wedge n)/2]$. Then we have

$$\lambda_1 \leqslant D(f_n)/\left[\pi(f_n^2) - \pi(f_n)^2\right]. \tag{4.20}$$

Since $\{\varphi \geqslant m\} \downarrow \emptyset$ as $m \to \infty$, for every $m \geqslant 1$, we can choose $r_m > 0$ such that $\pi(\varphi \geqslant r_m) \leqslant 1/m$. Then

$$\pi\left(I_{[\varphi \geqslant r_m]} f_n^2\right)^{1/2} \geqslant \sqrt{m}\,\pi\left(I_{[\varphi \geqslant r_m]} f_n\right) \geqslant \sqrt{m}\,\pi(f_n) - \sqrt{m}\,e^{\varepsilon r_m/2}.$$

Hence

$$\pi(f_n)^2 \leqslant \left[\sqrt{\pi(f_n^2)}/\sqrt{m} + e^{\varepsilon r_m/2}\right]^2. \tag{4.21}$$

On the other hand, by the mean value theorem,

$$|e^A - e^B| \leqslant |A - B|e^{A\vee B} = |A - B|(e^A \vee e^B)$$

for all $A, B \geq 0$. Hence

$$
\begin{aligned}
D(f_n) &= \frac{1}{2} \int J(\mathrm{d}x, \mathrm{d}y)[f_n(x) - f_n(y)]^2 \\
&\leq \frac{\varepsilon^2}{8} \int J^{(1)}(\mathrm{d}x, \mathrm{d}y)[\varphi(x) - \varphi(y)]^2 r(x, y) \big[f_n(x) \vee f_n(y)\big]^2 \quad (4.22) \\
&\leq \frac{\varepsilon^2}{4} \pi(f_n^2).
\end{aligned}
$$

Noticing that $\pi(f_n^2) \uparrow \infty$ as $n \to \infty$, combining (4.22) with (4.20) and (4.21) and then letting $n \uparrow \infty$, we obtain $\lambda_1 \leq \varepsilon^2/[4(1 - m^{-1})]$. The proof is completed by setting $m \uparrow \infty$. $\square$

Theorem 4.16 (F.Y. Wang, 2000a). Suppose that (4.18) holds. If $\sigma > 0$, then

$$
\pi\big(e^{\varepsilon\sigma\varphi^2}\big) \leq \exp\left[\frac{\varepsilon\sigma\pi(\varphi^2)}{1 - 2\varepsilon}\right] < \infty, \qquad \varepsilon \in [0, 1/2).
$$

Proof. (a) Given $n \geq 1$, let $\varphi_n = \varphi \wedge n$, $f_n = \exp\big[r\varphi_n^2/2\big]$, and $h_n(r) = \pi\big(e^{r\varphi_n^2}\big)$. Then, by (4.2), (4.18), and applying the mean value theorem to the function $\exp[rx^2/2]$, we get

$$
\begin{aligned}
D(f_n) &= \frac{1}{2} \int J(\mathrm{d}x, \mathrm{d}y)[f_n(x) - f_n(y)]^2 \\
&\leq \frac{r^2}{2} \int J^{(1)}(\mathrm{d}x, \mathrm{d}y)[\varphi(x) - \varphi(y)]^2 r(x, y) \\
&\quad \times \max\big\{\varphi_n(x)f_n(x), \varphi_n(y)f_n(y)\big\}^2 \\
&\leq r^2 \int J^{(1)}(\mathrm{d}x, \mathrm{d}y)\varphi_n(x)^2 f_n(x)^2 \\
&\leq r^2 h_n'.
\end{aligned}
$$

(b) Next, applying the logarithmic Sobolev inequality to the function f_n and using (a), it follows that

$$
rh_n'(r) \leq h_n(r) \log h_n(r) + 2r^2 h_n'(r)/\sigma, \qquad r \geq 0.
$$

That is,

$$
h_n'(r) \leq \frac{1}{r(1 - 2r/\sigma)} h_n(r) \log h_n(r), \qquad r \in [0, \sigma/2).
$$

Since $h_n(r) \geq 1$ for all $r \geq \varepsilon > 0$, applying Corollary A.5 to the interval $[\varepsilon, 2/\sigma)$, it follows that

$$
h_n(r) \leq h_n(\varepsilon)^{r(1 - 2\varepsilon/\sigma)/\varepsilon(1 - 2r/\sigma)} \to \exp\left[\frac{r\pi(\varphi_n^2)}{1 - 2r/\sigma}\right] \qquad \text{as } \varepsilon \to 0.
$$

The required assertion then follows by letting $n \to \infty$. $\square$

4.8 Nash inequality

Theorem 4.17 (Chen, 1999b). Define the isoperimetric constant I_ν as follows:

$$I_\nu = \inf_{0 < \pi(A) \leqslant 1/2} \frac{J^{(1/2)}(A \times A^c)}{\pi(A)^{(\nu-1)/\nu}} = \inf_{0 < \pi(A) < 1} \frac{J^{(1/2)}(A \times A^c)}{\left[\pi(A) \wedge \pi(A^c)\right]^{(\nu-1)/\nu}}, \qquad \nu > 1.$$

Then

$$\text{Var}(f)^{1+2/\nu} \leqslant 2I_\nu^{-2}D(f)\|f\|_1^{4/\nu}, \qquad f \in L^2(\pi). \tag{4.23}$$

Proof. The proof below is quite close to that of L. Saloff-Coste (1997). Fix a bounded $g \in \mathscr{D}(D)$. Let c be a median of g. Set $f = \text{sgn}(g - c)|g - c|^2$. Then f has median 0. On the one hand, by using the functional form of I_ν,

$$I_\nu = \inf \left\{ \frac{\frac{1}{2}\int J^{(1/2)}(dx, dy)|f(y) - f(x)|}{\inf_{c:\ c \text{ is a median of } f}\|f - c\|_{\nu/(\nu-1)}} :\right.$$
$$\left. f \in L^1(\pi) \text{ is nonconstant} \right\}, \tag{4.24}$$

which will be proved later, and setting $q = \nu/(\nu - 1)$, we obtain

$$\|g - c\|_{2q}^2 = \|f\|_q \leqslant \frac{1}{2}I_\nu^{-1}\int J^{(1/2)}(dx, dy)|f(y) - f(x)|. \tag{4.25}$$

On the other hand, since

$$|a - b|\,(|a| + |b|) = \begin{cases} |a^2 - b^2| & \text{if } ab > 0, \\ (|a| + |b|)^2 & \text{if } ab < 0, \end{cases}$$

we have $|f(y) - f(x)| \leqslant |g(y) - g(x)|\,(|g(y) - c| + |g(x) - c|)$. By using this inequality and following the last part of the proof of Theorem 4.11, we get

$$\int J^{(1/2)}(dx, dy)|f(y) - f(x)|$$
$$\leqslant \sqrt{2D(g)}\left[\int J^{(1)}(dx, dy)\left[|g(y) - c| + |g(x) - c|\right]^2\right]^{1/2} \tag{4.26}$$
$$\leqslant 2\sqrt{2D(g)}\,\|g - c\|_2.$$

Combining (4.25) with (4.26), we get

$$\|g - c\|_{2q}^2 \leqslant I_\nu^{-1}\sqrt{2D(g)}\,\|g - c\|_2.$$

On the other hand, writing $g^2 = g^{2/(\nu+1)} \cdot g^{2\nu/(\nu+1)}$ and applying the Hölder inequality with $p' = (\nu + 1)/2$ and $q' = (\nu + 1)/(\nu - 1)$, we obtain

$$\|g\|_2 \leqslant \|g\|_1^{1/(\nu+1)}\|g\|_{2q}^{\nu/(\nu+1)}.$$

From these facts, it follows that

$$\|g - c\|_2 \leqslant \left[I_\nu^{-1}\sqrt{2D(g)}\,\|g - c\|_2\right]^{\nu/2(\nu+1)}\|g - c\|_1^{1/(\nu+1)}.$$

Thus

$$\|g - c\|_2^{2(1+2/\nu)} \leqslant 2I_\nu^{-2}D(g)\,\|g - c\|_1^{4/\nu},$$

and hence

$$\mathrm{Var}(g)^{1+2/\nu} \leqslant 2I_\nu^{-2}D(g)\,\|g\|_1^{4/\nu}.$$

We now return to prove (4.24). Denote by J_ν the right-hand side of (4.24). Set $q = \nu/(\nu - 1)$ and ignore the superscript "(1/2)" in $J^{(1/2)}$ everywhere for simplicity. Take $f = I_A$ with $0 < \pi(A) \leqslant 1/2$. Then f has median 0. Moreover,

$$\int J(\mathrm{d}x, \mathrm{d}y)|f(y) - f(x)| = 2J(A \times A^c), \qquad \|f\|_q = \pi(A)^{1/q}.$$

This proves that $I_\nu \geqslant J_\nu$.

Conversely, fix f with median c. Set $f_\pm = (f - c)^\pm$. Then $f_+ + f_- = |f - c|$ and $|f(y) - f(x)| = |f_+(y) - f_+(x)| + |f_-(y) - f_-(x)|$. Put $F_t^\pm = \{f_\pm \geqslant t\}$. Then

$$\frac{1}{2}\int J(\mathrm{d}x, \mathrm{d}y)|f(y) - f(x)| = \frac{1}{2}\int J(\mathrm{d}x, \mathrm{d}y)\big[|f_+(y) - f_+(x)| + |f_-(y) - f_-(x)|\big]$$

$$= \int_0^{\|f\|_u}\big[J\big(F_t^+ \times (F_t^+)^c\big) + J\big(F_t^- \times (F_t^-)^c\big)\big]\mathrm{d}t$$

(by the coarea formula)

$$\geqslant I_\nu \int_0^{\|f\|_u}\big[\pi(F_t^+)^{1/q} + \pi(F_t^-)^{1/q}\big]\mathrm{d}t.$$

Next, we need the following simple result.

Claim. Let $p \geqslant 1$. Then $\|f\|_p \leqslant F$ iff $\|fg\|_1 \leqslant FG$ holds for all g satisfying $\|g\|_q \leqslant G$, where $1/p + 1/q = 1$.

It follows that

$$\pi(F_t^\pm)^{1/q} = \big\|I_{F_t^\pm}\big\|_q = \sup_{\|g\|_r \leqslant 1}\big\langle I_{F_t^\pm}, g\big\rangle, \qquad 1/r + 1/q = 1.$$

Thus, for every g with $\|g\|_r \leqslant 1$, we have

$$\frac{1}{2}\int J(\mathrm{d}x, \mathrm{d}y)|f(y) - f(x)| \geqslant I_\nu \int_0^\infty\big[\langle I_{F_t^+}, g\rangle + \langle I_{F_t^-}, g\rangle\big]\mathrm{d}t$$

$$= I_\nu\big[\langle f_+, g\rangle + \langle f_-, g\rangle\big]$$

$$= I_\nu\langle|f - c|, g\rangle.$$

Taking the supremum with respect to g, we get

$$\frac{1}{2}\int J(\mathrm{d}x, \mathrm{d}y)|f(y) - f(x)| \geqslant I_\nu\|f - c\|_q. \qquad \square$$

4.9 Birth–death processes

Finally, we apply the above results to birth–death processes to illustrate the power of Cheeger's approach. Consider a regular birth–death process on $\mathbb{Z}_+$ with birth rates (b_i) and death rates (a_i). Then $J_{ij} = \pi_i b_i$ if $j = i + 1$, $J_{ij} = \pi_i a_i$ if $j = i - 1$, and $J_{ij} = 0$ otherwise. We have the following result.

Theorem 4.18. For a birth–death process, take $r_{ij} = (a_i + b_i) \vee (a_j + b_j)$ $(i \neq j)$. Then the following assertions hold:

(1) For the Nash inequality, $I_\nu > 0$ for some $\nu \geqslant 1$ iff there exists a constant $c > 0$ such that

$$
\frac{\pi_i a_i}{\sqrt{r_{i,i-1}}} \geqslant c \left[\sum_{j \geqslant i} \pi_j \right]^{(\nu-1)/\nu}, \qquad i \geqslant 1.
$$

If so, we indeed have $I_\nu \geqslant c$.

(2) For the logarithmic Sobolev inequality, $\xi^{(\infty)} > 0$ iff

$$
\inf_{i \geqslant 1} \frac{\pi_i a_i}{\sqrt{r_{i,i-1}}} \Big/ \left(\sum_{j \geqslant i} \pi_j \right) \sqrt{- \log \sum_{j \geqslant i} \pi_j} > 0;
$$

$\kappa^{(\alpha)} > 0$ iff

$$
\inf_{i \geqslant 1} \frac{\pi_i a_i}{r_{i,i-1}^{\alpha}} \Big/ \left(- \sum_{j \geqslant i} \pi_j \right) \log \sum_{j \geqslant i} \pi_j > 0.
$$

(3) For the Poincaré inequality, $k^{(\alpha)} > 0$ iff there exists a constant $c > 0$ such that

$$
\frac{\pi_i a_i}{r_{i,i-1}^{\alpha}} \geqslant c \sum_{j \geqslant i} \pi_j, \qquad i \geqslant 1.
$$

Then we indeed have $k^{(\alpha)} \geqslant c$.

In particular, for the Poincaré inequality, the sufficient condition given by the last assertion can be rewritten as

$$
\sup_{n \geqslant 1} \mu[n, \infty) \frac{\sqrt{r_{n,n-1}}}{\mu_{n-1} b_{n-1}} < \infty. \tag{4.27}
$$

The criterion given in Table 1.4 is

$$
\sup_{n \geqslant 1} \mu[n, \infty) \sum_{j \leqslant n-1} \frac{1}{\mu_j b_j} < \infty. \tag{4.28}
$$

From these, the difference of the methods should be clear. Further comparison is given in Section 7.3.

Example 4.19. Consider the birth–death process with $a_{2i-1} = (2i-1)^2$, $a_{2i} = (2i)^4$, and $b_i = a_i$ for all $i \geqslant 1$. Then (4.28) holds but (4.27) fails.

Proof. Since $b_i = a_i$, we have $\mu_n b_n \sim 1$ and so

$$\sum_{j \leqslant n-1} \frac{1}{\mu_j b_j} \sim n \qquad \text{as } n \to \infty.$$

But $\mu[n, \infty) \sim 1/n$; hence (4.28) holds. Next, since $r_{n,n-1} \sim n^4$, it is clear that (4.27) fails. $\square$

   The details of this chapter are included in Chapter 9 of the second edition of Chen (1992a).

Chapter 5

Ten Explicit Criteria in Dimension One

Traditional ergodicity constitutes a crucial part of the theory of stochastic processes and plays a key role in practical applications. The theory of ergodicity has been much refined recently, due to the study of some inequalities, which are especially powerful in the infinite-dimensional situation. The explicit criteria for various types of ergodicity for birth–death processes and one-dimensional diffusions are collected in Table 1.4 and Table 5.1 below, respectively. In this chapter, we explain in detail an interesting story about how to obtain one of the criteria for birth–death processes.

This chapter is organized as follows. First, we recall the study of exponential convergence from different points of view in different subjects: probability theory, spectral theory, and harmonic analysis (Sections 5.1 and 5.2). Then we introduce an explicit criterion for convergence, variational formulas, and explicit estimates for the convergence rates. Some comparisons with known results and an application to geometry are included (Section 5.3). Next, we present ten (eleven) criteria for the two classes of processes (birth–death processes and one-dimensional diffusions), respectively (Section 5.4). The technical proofs are collected in the last two sections. Section 5.5 is devoted to exponential ergodicity for the discrete case. Section 5.6 is devoted to strong ergodicity, using both analytic and coupling proofs.

Let us begin with the chapter by recalling the three traditional types of ergodicity.

5.1 Three traditional types of ergodicity

Let $Q = (q_{ij})$ be a regular Q-matrix on a countable set $E = \{i, j, k, \ldots\}$. That is, $q_{ij} \geqslant 0$ for all $i \neq j$, $q_i := -q_{ii} = \sum_{j \neq i} q_{ij} < \infty$ for all $i \in E$, and Q determines uniquely a transition probability matrix $P_t = (p_{ij}(t))$ (which is also called a *Q-process* or a *Markov chain*). Denote by $\pi = (\pi_i)$ a stationary distribution of P_t: $\pi P_t = \pi$ for all $t \geqslant 0$. From now on, assume that the Q-matrix is irreducible and hence the stationary distribution π is unique. Then the three types of ergodicity are defined as follows:

$$\text{Ordinary ergodicity}: \qquad \lim_{t \to \infty} |p_{ij}(t) - \pi_j| = 0, \tag{5.1}$$

$$\text{Exponential ergodicity}: \qquad \lim_{t \to \infty} e^{\hat{\alpha}t} |p_{ij}(t) - \pi_j| = 0, \tag{5.2}$$

$$\text{Strong ergodicity}: \qquad \lim_{t \to \infty} \sup_i |p_{ij}(t) - \pi_j| = 0 \tag{5.3}$$

$$\Longleftrightarrow \lim_{t \to \infty} e^{\hat{\beta}t} \sup_i |p_{ij}(t) - \pi_j| = 0, \tag{5.4}$$

where $\hat{\alpha}$ and $\hat{\beta}$ are (the largest) positive constants and i, j vary over all of E. The equivalence in (5.4) is well known, but one may refer to the second part of Section 5.6. These definitions are meaningful for general Markov processes once pointwise convergence is replaced by convergence in total variation norm. The three types of ergodicity were studied a great deal during the period 1953–1981. Especially, it was proved that

strong ergodicity $\Longrightarrow$ exponential ergodicity $\Longrightarrow$ ordinary ergodicity.

Refer to W.J. Anderson (1991), Chen (1992a, Chapter 4), S.P. Meyn and R.L. Tweedie (1993b) for details and related references. We will return to this topic in Chapter 8. The study is quite complete in the sense that we have the following criteria, which are described by the Q-matrix plus a test sequence (y_i) only, except for the exponential ergodicity, for which one requires an additional parameter λ.

Theorem 5.1 (Criteria). Let $H \neq \emptyset$ be an arbitrary but fixed finite subset of E. Then the following conclusions hold:

(1) The process P_t is ergodic iff the system of inequalities

$$\begin{cases} \sum_j q_{ij} y_j \leqslant -1, & i \notin H, \\ \sum_{i \in H} \sum_{j \neq i} q_{ij} y_j < \infty, \end{cases} \tag{5.5}$$

has a nonnegative finite solution (y_i).

(2) The process P_t is exponentially ergodic iff for some $\lambda > 0$ with $\lambda < q_i$ for all i, the system of inequalities

$$\begin{cases} \sum_j q_{ij} y_j \leqslant -\lambda y_i - 1, & i \notin H, \\ \sum_{i \in H} \sum_{j \neq i} q_{ij} y_j < \infty, \end{cases} \tag{5.6}$$

has a nonnegative finite solution (y_i).

(3) The process P_t is strongly ergodic iff the system (5.5) of inequalities has a bounded nonnegative solution (y_i).

Replacing (y_i) in (5.6) with $(\tilde{y}_i = \lambda y_i + 1)$, one can rewrite (5.6) as

$$\begin{cases} y_i \geqslant 1, & i \in E, \\ \sum_j q_{ij} y_j \leqslant -\lambda y_i, & i \notin H, \\ \sum_{i \in H} \sum_{j \neq i} q_{ij} y_j < \infty. \end{cases} \tag{5.6'}$$

The probabilistic meaning of the criteria are respectively as follows:

$$\max_{i \in H} \mathbb{E}_i \sigma_H < \infty, \quad \max_{i \in H} \mathbb{E}_i e^{\lambda \sigma_H} < \infty \quad \text{and} \quad \sup_{i \in E} \mathbb{E}_i \sigma_H < \infty,$$

where $\sigma_H = \inf\{t \geqslant \text{the first jumping time} : X_t \in H\}$ and λ is the same as in (5.6). The criteria are not completely explicit, since they depend on the test sequences (y_i), and in general it is often nontrivial to solve a system of infinitely many inequalities. Hence, one expects to discover some explicit criteria for some specific processes. Clearly, for this, the first candidate should be the birth–death processes. Recall that for a *birth–death process* with state space $E = \mathbb{Z}_+ = \{0, 1, 2, \ldots\}$, its Q-matrix has the form: $q_{i,i+1} = b_i > 0$ for all $i \geqslant 0$, $q_{i,i-1} = a_i > 0$ for all $i \geqslant 1$, and $q_{ij} = 0$ for all other $i \neq j$. Along this line, it was proved by R.L. Tweedie (1981) [see also W.J. Anderson (1991) or Chen (1992a)] that

$$S := \sum_{n \geqslant 1} \mu_n \sum_{j \leqslant n-1} \frac{1}{\mu_j b_j} < \infty \implies \text{Exponential ergodicity}, \tag{5.7}$$

where $\mu_0 = 1$ and $\mu_n = b_0 \cdots b_{n-1}/a_1 \cdots a_n$ for all $n \geqslant 1$. Refer to Z.K. Wang (1980), X.Q. Yang (1986), or Z.T. Hou et al. (2000) for the probabilistic meaning of S. The condition is explicit, since it depends only on the rates a_i and b_i. However, the condition is not necessary. A simple example is as follows. Let $a_i = b_i = i^\gamma$ $(i \geqslant 1)$ and $b_0 = 1$. Then the process is exponentially ergodic iff $\gamma \geqslant 2$ but $S < \infty$ iff $\gamma > 2$. See Chen (1996) or Examples 8.2 and 8.3. Surprisingly, the condition is correct for strong ergodicity.

Theorem 5.2 (H.J. Zhang, X. Lin, and Z.T. Hou, 2000). Strong ergodicity holds iff $S < \infty$.

Refer to Z.T. Hou et al (2000). With a different proof, the result has been extended by Y.H. Zhang (2001) to the *single birth processes* with state space $\mathbb{Z}_+$ (the details are presented in Section 5.6 below). Here, the term "single birth" means that $q_{i,i+1} > 0$ for all $i \geqslant 0$ but $q_{ij} \geqslant 0$ can be arbitrary for $j < i$. Introducing this class of Q-processes is due to the following observation. If the first inequality in (5.5) is replaced by equality, then we get a recursion formula for (y_i) with one parameter only. Hence, there should exist an explicit criterion for ergodicity (respectively, uniqueness, recurrence, and strong ergodicity). For (5.6), there is also a recursion formula, but now two parameters are involved, and so it is unclear whether there exists an explicit criterion for exponential ergodicity.

Note that the criteria are not enough to estimate the convergence rate $\hat{\alpha}$ or $\hat{\beta}$ [cf. Chen (2000a)]. That is the main reason why we have to return to the well-developed theory of Markov chains. For birth–death processes, the estimation of $\hat{\alpha}$ was studied by E.A. van Doorn in a book (1981) and in a series of papers (1985; 1987; 1991; 2002). He proved, for instance, the lower bound

$$\hat{\alpha} \geqslant \inf_{i \geqslant 0}\left\{a_{i+1} + b_i - \sqrt{a_i b_i} - \sqrt{a_{i+1}b_{i+1}}\right\},$$

which is exact when a_i and b_i are constant. The following formula for the lower bounds was implied in his papers and rediscovered from a different point of view (in a study of the spectral gap) by Chen (1996):

$$\hat{\alpha} = \sup_{v>0} \inf_{i \geqslant 0}\{a_{i+1} + b_i - a_i/v_{i-1} - b_{i+1}v_i\}.$$

Furthermore, the precise value of $\hat{\alpha}$ was determined by E.A. van Doorn for four practical models. The main tool used in van Doorn's study is the Karlin–Mcgregor representation theorem, a specific spectral representation involving heavy techniques. No explicit criterion for $\hat{\alpha} > 0$ ever appeared until Chen (2000c).

5.2 The first (nontrivial) eigenvalue (spectral gap)

Birth–death processes have a nice property, symmetrizability: $\mu_i p_{ij}(t) = \mu_j p_{ji}(t)$ for all i, j and $t \geqslant 0$. Then the matrix Q can be regarded as a self-adjoint operator on the real L^2-space $L^2(\mu)$ with norm $\|\cdot\|$. In other words, one can use the well-developed L^2-theory. For instance, one can study the L^2-exponential convergence given below, assuming that $Z = \sum_i \mu_i < \infty$ and then setting $\pi_i = \mu_i/Z$. Then convergence means that

$$\|P_t f - \pi(f)\| \leqslant \|f - \pi(f)\|e^{-\lambda_1 t} \tag{5.8}$$

for all $t \geqslant 0$, where $\pi(f) = \int f\,d\pi$ and λ_1 is the first nontrivial eigenvalue (or spectral gap) of $(-Q)$ [cf. Chen (1992a, Chapter 9) or Theorem 5.10 below].

The estimation of λ_1 for birth–death processes was studied by W.G. Sullivan (1984), T.M. Liggett (1989), and C. Landim, S. Sethuraman, and S.R.S. Varadhan (1996) [see also C. Kipnis and C. Lamdin (1999)]. It was used as a comparison tool to deal with the convergence rate for some interacting particle systems, which are infinite-dimensional Markov processes. Here we recall three results.

Theorem 5.3 (W.G. Sullivan, 1984). Let c_1 and c_2 be two constants satisfying

$$c_1 \geqslant \sup_{i \geqslant 1} \frac{\sum_{j \geqslant i} \mu_j}{\mu_i}, \qquad c_2 \geqslant \sup_{i \geqslant 1} \frac{1}{a_i}.$$

Then $\lambda_1 \geqslant 1/4c_1^2 c_2$.

Theorem 5.4 (T.M. Liggett, 1989). Let c_1 and c_2 be two constants satisfying

$$c_1 \geqslant \sup_{i \geqslant 1} \frac{\sum_{j \geqslant i} \mu_j}{\mu_i a_i}, \qquad c_2 \geqslant \sup_{i \geqslant 1} \frac{\sum_{j \geqslant i} \mu_j a_j}{\mu_i a_i}.$$

Then $\lambda_1 \geqslant 1/4c_1 c_2$.

Theorem 5.5 (T.M. Liggett, 1989). Let $\inf_{i \geqslant 1} a_i > 0$, $\inf_{i \geqslant 0} b_i > 0$ and $\sup_{i \geqslant 0} b_i < \infty$. Then $\lambda_1 > 0$ iff (μ_i) has an exponential tail: $\sum_{j \geqslant i} \mu_j \leqslant C \mu_i$ for all i and some constant $C < \infty$.

The reason we are mainly interested in the lower bounds is that on the one hand, they are more useful in practice, and on the other hand, the upper bounds are usually easier to obtain from the following classical variational formula:

$$\lambda_1 = \inf \left\{ D(f) : \mu(f) = 0, \mu(f^2) = 1 \right\},$$

where

$$D(f) = \frac{1}{2} \sum_{i,j} \mu_i q_{ij} (f_j - f_i)^2, \qquad \mathscr{D}(D) = \{ f \in L^2(\mu) : D(f) < \infty \},$$

and $\mu(f) = \int f \mathrm{d}\mu$.

Let us now leave Markov chains for a while and turn to diffusions.

One-dimensional diffusions

As a parallel of birth–death process, we now consider an elliptic operator $L = a(x)\mathrm{d}^2/\mathrm{d}x^2 + b(x)\mathrm{d}/\mathrm{d}x$ on the half-line $[0, \infty)$ with $a(x) > 0$ everywhere and with reflecting boundary at the origin. Again, we are interested in estimation of the principal eigenvalues, of which the study is a typical, well-known Sturm–Liouville eigenvalue problem in the spectral theory. Refer to Y. Egorov and V. Kondratiev (1996) for the present status of the study and references. Here, we mention two results, which are the most general ones we have ever seen.

From now on, we often omit the integral variable when it is integrated with respect to the Lebesgue measure.

Theorem 5.6. Let $b(x) \equiv 0$ (which corresponds to the birth–death process with $a_i = b_i$ for all $i \geqslant 1$) and set $\delta = \sup_{x > 0} x \int_x^\infty a^{-1}$. Then we have the following results.

(1) I.S. Kac and M.G. Krein (1958): $\delta^{-1} \geqslant \lambda_0 \geqslant (4\delta)^{-1}$, where λ_0 is the first eigenvalue corresponding to the Dirichlet boundary $f(0) = 0$.
(2) S. Kotani and S. Watanabe (1982): $\delta^{-1} \geqslant \lambda_1 \geqslant (4\delta)^{-1}$.

It is a simple matter to rewrite the classical variational formula as (5.9) below. Similarly, we have (5.10) for λ_0.

Poincaré inequalities

$$\lambda_1: \qquad \|f - \pi(f)\|^2 \leqslant \lambda_1^{-1} D(f), \qquad\qquad (5.9)$$

$$\lambda_0: \qquad \|f\|^2 \leqslant \lambda_0^{-1} D(f), \quad f(0) = 0. \qquad\qquad (5.10)$$

It is interesting that inequality (5.10) is a special but typical case of the weighted Hardy inequality discussed in the next subsection.

Weighted Hardy inequality

The classical Hardy inequality goes back to G.H. Hardy (1920):

$$\int_0^\infty \left(\frac{f}{x}\right)^p \leqslant \left(\frac{p}{p-1}\right)^p \int_0^\infty f'^p, \quad f(0) = 0, \ f' \geqslant 0, \ p > 1,$$

where the optimal constant was determined by Landau (1926). When $p = 2$, this corresponds to the operator $L = x \mathrm{d}^2/\mathrm{d}x^2$. Then the result says that $\lambda_0 = 1/4$. Clearly, this operator is too special. The extension by I. Kac and G. Krein (1958) to the operator $L = a(x)\mathrm{d}^2/\mathrm{d}x^2$, mentioned in Theorem 5.6 (1), provides nice lower and upper bounds up to a factor 4, rather than an explicit formula for λ_0. After a long period of efforts by analysts, the inequality was finally extended to the following form, called the *weighted Hardy inequality*, by B. Muckenhoupt (1972):

$$\int_0^\infty f^p \mathrm{d}\nu \leqslant A \int_0^\infty f'^p \mathrm{d}\lambda, \quad f \in C^1, \ f(0) = 0, \ f' \geqslant 0, \qquad (5.11)$$

where ν and λ are nonnegative Borel measures. This is the most general form in dimension one.

Another direction to generalize the Hardy inequality is to go to higher dimensions. A general result was stated in Theorem 4.8.

The Hardy-type inequalities play a very important role in harmonic analysis and potential theory, and have been treated in many publications. Refer to the books B. Opic and A. Kufner (1990), V.G. Maz'ya (1985), A. Kufner and L.E. Persson (2003), and the articles E.M. Dynkin (1991), E.B. Davies (1999), and A. Wedestig (2003) for more details. We will return to this inequality soon.

We have finished an overview of exponential convergence (equivalently, the Poincaré inequalities) in different areas. The difficulties of the topic are illustrated in Section 1.1.

5.3   The first eigenvalues and exponentially ergodic rate

We are now in a position to state our results. To do so, define

$$\mathscr{W} = \{w : w_0 = 0, w_i \uparrow\uparrow\}, \qquad Z = \sum_i \mu_i, \qquad \delta = \sup_{i>0} \sum_{j \leqslant i-1} \frac{1}{\mu_j b_j} \sum_{j \geqslant i} \mu_j,$$

where "$\uparrow\uparrow$" means strictly increasing. Recall the notation $\bar{w}_i = w_i - \pi(w)$. By suitable modification, we can define $\widetilde{\mathscr{W}}$ (cf. the last subsection of Section 1.2) and explicit sequences $\{\delta_n\}$ and $\{\tilde{\delta}_n\}$ (the diffusion analogues are given in Theorem 6.1 (3)). Refer to Chen (2001a) for details.

The next result provides a complete answer to the question proposed in Section 5.1. In particular, we have gone a long way to arrive at an explicit criterion (parts (3) and (4) below). A direct proof of the criterion will be presented in Section 5.5. The assertion (4) below is now a consequence of Theorem 8.18 (4).

Theorem 5.7. For birth–death processes, the following assertions hold:

(1) *Dual variational formulas.*

$$\lambda_1 = \sup_{w \in \mathscr{W}} \inf_{i \geqslant 0} \mu_i b_i (w_{i+1} - w_i) \Big/ \sum_{j \geqslant i+1} \mu_j \bar{w}_j \qquad \text{[Chen (1996)]} \qquad (5.12)$$

$$= \inf_{w \in \widetilde{\mathscr{W}}} \sup_{i \geqslant 0} \mu_i b_i (w_{i+1} - w_i) \Big/ \sum_{j \geqslant i+1} \mu_j \bar{w}_j \qquad \text{[Chen (2001a)]} \qquad (5.13)$$

(2) *Approximation procedure and explicit bounds.*

$$Z\delta^{-1} \geqslant \tilde{\delta}_n^{-1} \geqslant \lambda_1 \geqslant \delta_n^{-1} \geqslant (4\delta)^{-1} \quad \text{for all } n \qquad \text{[Chen (2000b; 2001a)]}.$$

(3) *Explicit criterion.* $\lambda_1 > 0$ iff $\delta < \infty$ [L. Miclo (1999b), Chen (2000b)].
(4) *Relation:* $\hat{\alpha} = \lambda_1$ [Chen (1991b)].

The formula (5.12) is nothing new, but is simply Theorem 3.2 (Section 3.7). The proofs of parts (2) and (3) are similar to those for the continuous case (stated as Theorem 5.8 below), and will be presented in Section 6.7. In part (1) of Theorem 5.7, only two notations are used: the sets $\mathscr{W}$ and $\widetilde{\mathscr{W}}$ of test functions (sequences). Clearly, for each test function, (5.12) gives us a lower bound of λ_1. This explains the meaning of "variational." Because of (5.12), it is now easy to obtain some lower estimates of λ_1, and in particular, one obtains all the lower bounds given by Theorems 5.3–5.5. Next, by exchanging the orders of "sup" and "inf," we get (5.13) from (5.12), ignoring a slight modification of $\mathscr{W}$. In other words, (5.12) and (5.13) are dual to each other. For the explicit estimates "$\delta^{-1} \geqslant \lambda_0 \geqslant (4\delta)^{-1}$," and in particular for the criterion, one needs to find a representative test function w among all $w \in \mathscr{W}$. This is certainly not obvious, because the test function w used in the formula is indeed a mimic eigenfunction (eigenvector) of λ_1, and in general, the eigenvalues and the corresponding eigenfunctions can be very sensitive, as we have seen from the examples given in Section 1.1. Fortunately, there exists such a representative function with a simple form. We will illustrate the function in the context of diffusions in the second to the last paragraph of this section.

In parallel, for diffusions on $[0, \infty]$, define

$$C(x) = \int_0^x b/a, \qquad \delta = \sup_{x>0} \int_0^x e^{-C} \int_x^\infty e^C/a,$$
$$\mathscr{F} = \{f \in C[0,\infty) \cap C^1(0,\infty) : f(0) = 0 \text{ and } f'|_{(0,\infty)} > 0\}.$$

Again, denote by $\widetilde{\mathscr{F}}$ a suitable modification of $\mathscr{F}$ [cf. (6.8) below].

Theorem 5.8 (Chen (1999a; 2000b; 2001a)). For diffusion on $[0,\infty)$, the following assertions hold;

(1) *Dual variational formulas.*

$$\lambda_0 \geqslant \sup_{f \in \mathscr{F}} \inf_{x>0} e^{C(x)} f'(x) \Big/ \int_x^\infty f e^C/a, \qquad (5.14)$$

$$\lambda_0 \leqslant \inf_{f \in \widetilde{\mathscr{F}}} \sup_{x>0} e^{C(x)} f'(x) \Big/ \int_x^\infty f e^C/a. \qquad (5.15)$$

Furthermore, the equality holds in (5.14) and (5.15) if both a and b are continuous on $[0,\infty)$.

(2) *Approximation procedure and explicit bounds.* A decreasing sequence $\{\delta_n\}$ and an increasing sequence $\{\tilde{\delta}_n\}$ are constructed explicitly such that

$$\delta^{-1} \geqslant \tilde{\delta}_n^{-1} \geqslant \lambda_0 \geqslant \delta_n^{-1} \geqslant (4\delta)^{-1} \qquad \text{for all } n.$$

(3) *Explicit criterion.* λ_0 (respectively, λ_1) > 0 iff $\delta < \infty$.

We mention that the above two theorems are also based on Chen and F.Y. Wang (1997a).

To see the power of the dual variational formulas, let us return to the weighted Hardy inequality.

Theorem 5.9 (B. Muckenhoupt, 1972). The optimal constant A in the inequality

$$\int_0^\infty f^2 d\nu \leqslant A \int_0^\infty f'^2 d\lambda, \qquad f \in C^1, \ f(0) = 0, \qquad (5.16)$$

satisfies $B \leqslant A \leqslant 4B$, where

$$B = \sup_{x>0} \nu[x,\infty) \int_0^x (d\lambda_{\text{abs}}/d\text{Leb})^{-1}$$

and $\mathrm{d}\lambda_{\mathrm{abs}}/\mathrm{dLeb}$ is the derivative of the absolutely continuous part of λ with respect to Lebesgue measure.

By setting $\nu = \pi$ and $\lambda = e^C \mathrm{d}x$, it follows that the criterion in Theorem 5.8 is a consequence of Muckenhoupt's theorem. Along this line, the criteria in Theorems 5.7 and 5.8 for a typical class of the processes were also obtained by S.G. Bobkov and F. Götze (1999a; 1999b), in which, the contribution of an earlier paper by J.H. Luo (1992) was noted. Refer also to L.H.Y. Chen (1985a), and his joint paper with J.H. Lou (1987) for a different approach to some Poincaré-type inequalities.

We now point out that the explicit estimates "$\delta^{-1} \geqslant \lambda_0 \geqslant (4\delta)^{-1}$" in Theorems 5.8 and 5.9 follow from our variational formulas immediately. Here we consider the lower bound "$(4\delta)^{-1}$" only; the proof for the upper bound "δ^{-1}" is also easy, in terms of (5.15) (cf. Section 6.2).

Recall that $\delta = \sup_{x>0} \int_0^x e^{-C} \int_x^\infty e^C/a$. Set $\varphi(x) = \int_0^x e^{-C}$. Using the integration by parts formula, it follows that

$$\int_x^\infty \frac{\sqrt{\varphi}\, e^C}{a} = -\int_x^\infty \sqrt{\varphi}\, \mathrm{d}\left(\int_\bullet^\infty \frac{e^C}{a}\right)$$
$$\leqslant \frac{\delta}{\sqrt{\varphi(x)}} + \frac{\delta}{2}\int_x^\infty \frac{\varphi'}{\varphi^{3/2}} \leqslant \frac{2\delta}{\sqrt{\varphi(x)}}.$$

Hence

$$\frac{e^{-C(x)}}{(\sqrt{\varphi})'(x)}\int_x^\infty \frac{\sqrt{\varphi}\, e^C}{a} \leqslant \frac{e^{-C(x)}\sqrt{\varphi(x)}}{(1/2)e^{-C(x)}} \cdot \frac{2\delta}{\sqrt{\varphi(x)}} = 4\delta.$$

This gives us the required bound by (5.14), since $\sqrt{\varphi} \in \mathscr{F}$.

Theorem 5.8 can be immediately applied to the whole line or higher-dimensional situations. For instance, for the Laplacian on compact Riemannian manifolds, it was proved by Chen and F.Y. Wang (1997b) that

$$\lambda_1 \geqslant \sup_{f \in \mathscr{F}}\ \inf_{r \in (0,D)}\ \frac{4f(r)}{\int_0^r C(s)^{-1}\mathrm{d}s \int_s^D C(u)f(u)\mathrm{d}u} =: \xi_1,$$

for a specific function $C(r)$ (Theorem 1.1). Thanks are given to the coupling technique, which reduces the higher-dimensional case to dimension one. We now have $\delta^{-1} \geqslant \tilde{\delta}_n^{-1} \downarrow \geqslant \xi_1 \geqslant \delta_n^{-1} \uparrow \geqslant (4\delta)^{-1}$, similar to Theorem 5.8. Refer to Chen (2000b; 2001a), or Theorem 6.1 (3) and Theorem 6.2 (3), for details. As we mentioned before, the use of test functions is necessary for producing sharp estimates. In fact, the variational formula enables us to improve a number of best known estimates obtained previously by geometers, but none of them can be deduced from the estimates "$\delta^{-1} \geqslant \xi_1 \geqslant (4\delta)^{-1}$." Furthermore, the approximation procedure enables us to determine the optimal linear approximation of ξ_1 in K: $\xi_1 \geqslant \pi^2/D^2 + K/2$, where D is the diameter of the manifold and K is the lower bound of Ricci curvature, as stated in Corollary 1.3 [cf. Chen, E. Scacciatelli and L. Yao (2002)]. We have thus shown the value of our dual variational formulas.

5.4 Explicit criteria

Three basic inequalities

In the last three sections, we have mainly studied the Poincaré inequality, i.e., (5.17) below. Naturally, one may study other inequalities, for instance, the logarithmic Sobolev inequality or the Nash inequality listed below:

$$\textit{Poincaré inequality}: \quad \|f - \pi(f)\|^2 \leqslant \lambda_1^{-1} D(f); \tag{5.17}$$

$$\textit{Logarithmic Sobolev inequality}: \quad \int f^2 \log(|f|/\|f\|) \mathrm{d}\pi \leqslant \sigma^{-1} D(f); \tag{5.18}$$

$$\textit{Nash inequality}: \quad \|f - \pi(f)\|^{2+4/\nu} \leqslant \eta^{-1} D(f) \|f\|_1^{4/\nu} \quad \text{(for some } \nu > 0). \tag{5.19}$$

Here, to save notation, σ (respectively, η) denotes the largest constant such that (5.18) (respectively, (5.19)) holds.

The next inequality is a generalization of the Nash one.

$$\textit{Liggett–Stroock inequality}: \quad \|f - \pi(f)\|^2 \leqslant CD(f)^{1/p} V(f)^{1/q}, \tag{5.20}$$

where V is homogeneous of degree two:

$$V(cf + d) = c^2 V(f), \qquad c, d \in \mathbb{R}. \tag{5.21}$$

The importance of these inequalities is due to the fact that each inequality describes a type of ergodicity.

Theorem 5.10 ([Chen (1991b; 1992a), T.M. Liggett (1989; 1991)]). Let V satisfy (5.21) and let $(P_t)_{t \geqslant 0}$ be the semigroup determined by the Dirichlet form $(D, \mathscr{D}(D))$.

(1) Let $V(P_t f) \leqslant V(f)$ for all $t \geqslant 0$ and $f \in L^2(\pi)$ (which is automatic when $V(f) = \|f\|_r^2$). Then the Liggett–Stroock inequality implies that

$$\mathrm{Var}(P_t f) = \|P_t f - \pi(f)\|^2 \leqslant CV(f)/t^{q-1}, \qquad t > 0.$$

(2) Conversely, in the reversible case, the last inequality in (1) implies the Liggett–Stroock inequality.

(3) Poincaré inequality $\Longleftrightarrow \mathrm{Var}(P_t f) \leqslant \mathrm{Var}(f) \exp[-2\lambda_1 t]$.

Proof. Here we prove parts (1) and (2) only. The proof of part (3) is similar [cf. Chen (1992a, Theorem 9.1)].

(a) Assume that (5.20) holds. Let $f \in \mathscr{D}(D)$ and $\pi(f) = 0$. Then $f_t := P(t)f \in \mathscr{D}(D)$. Set $F_t = \pi(f_t^2)$. Since

$$F_t' = -2D(f_t) \leqslant -2C^{-p} V(f)^{-p/q} \|f_t\|^{2p} = -2C^{-p} F_t^p V(f)^{-p/q},$$

part (1) follows from Corollary A.4. Note that the proof of this step does not need reversibility.

(b) Conversely, since the process is reversible, the spectral representation theorem (cf. Section 7.4) gives us

$$\frac{1}{t}(f - P(t)f, \, f) \uparrow D(f) \qquad \text{as } t \downarrow 0.$$

Hence

$$\|f\|^2 - tD(f) \leqslant (P(t)f, \, f) \leqslant \|P(t)f\| \, \|f\| \leqslant \|f\| \sqrt{CV(f)t^{1-q}}, \qquad \pi(f) = 0.$$

Put $A = D(f)$, $B = \|f\| \sqrt{CV(f)}$, and $C_1 = \|f\|^2$. It follows that $C_1 - At \leqslant Bt^{(1-q)/2}$. The function $h(t) := At + Bt^{(1-q)/2} - C_1$ ($\geqslant 0$ for all $t > 0$) achieves its minimum

$$h(t_0) = \left[\left(\frac{q-1}{2}\right)^{2/(q+1)} + \left(\frac{2}{q-1}\right)^{(q-1)/(q+1)}\right] A^{(q-1)/(q+1)} B^{2/(q+1)} - C_1$$

at the point

$$t_0 = \left[\frac{2A}{B(q-1)}\right]^{-2/(q+1)} > 0.$$

Now, since $h(t_0) \geqslant 0$, it follows that $\|f\|^2 \leqslant C_2 \, D(f)^{1/p} V(f)^{1/q}$ for some constant $C_2 > 0$, and so we have proved part (2). □

Criteria

Recently, the criteria for the logarithmic Sobolev and Nash inequalities as well as for the discrete spectrum (which means that there is no continuous spectrum and moreover, all eigenvalues have finite multiplicity) were obtained by Y.H. Mao (2002a; 2002b; 2004), based on the weighted Hardy inequality. On the other hand, the main parts of Theorems 5.7 and 5.8 were extended to a general class of Banach spaces by Chen (2002a; 2003a; 2003b), which unify a large class of inequalities and in particular provide a unified criterion. This is the aim of the next chapter. We can now summarize the results in Tables 1.4 and 5.1. The tables are arranged in such an order that the property in each line is stronger than the previous one, the only exception being that even though the strong ergodicity is often stronger than the logarithmic Sobolev inequality, they are not comparable in general (Chen, 2002b); refer to Chapter 8 for more details.

For birth–death processes, ten criteria are presented in Table 1.4. For two of the criteria, the proofs are given in the next two sections.

For diffusion processes on $[0, \infty)$ with reflecting boundary and operator

$$L = a(x)\frac{\mathrm{d}^2}{\mathrm{d}x^2} + b(x)\frac{\mathrm{d}}{\mathrm{d}x},$$

define

$$C(x) = \int_0^x b/a, \qquad \mu[x,y] = \int_x^y e^C/a.$$

Then we have criteria listed in Table 5.1. Here, "$(*)$ & $\cdots$" means that one

Table 5.1   Eleven criteria for one-dimensional diffusions

Property	Criterion
Uniqueness	$\displaystyle\int_0^\infty \mu[0,x]e^{-C(x)} = \infty \quad (*)$
Recurrence	$\displaystyle\int_0^\infty e^{-C(x)} = \infty$
Ergodicity	$(*)$ & $\mu[0,\infty) < \infty$
Exponential ergodicity L^2-exp. convergence	$(*)$ & $\displaystyle\sup_{x>0} \mu[x,\infty)\int_0^x e^{-C} < \infty$
Discrete spectrum	$(*)$ & $\displaystyle\lim_{n\to\infty}\sup_{x>n} \mu[x,\infty)\int_n^x e^{-C} = 0$
Log. Sobolev inequality Exp. convergence in entropy	$(*)$ & $\displaystyle\sup_{x>0} \mu[x,\infty)\log[\mu[x,\infty)^{-1}]\int_0^x e^{-C} < \infty$
Strong ergodicity L^1-exp. convergence	$(*)$ & $\displaystyle\int_0^\infty \mu[x,\infty)e^{-C(x)} < \infty$
Nash inequality	$(*)$ & $\displaystyle\sup_{x>0} \mu[x,\infty)^{(\nu-2)/\nu}\int_0^x e^{-C} < \infty \ (\varepsilon)$

requires the uniqueness condition in the first line plus the condition "$\cdots$". The "(ε)" in the last line means that there is still a small gap from being necessary. In other words, when $\nu \in (0,2]$, there is still no criterion for the Nash inequality. The reason we have one more criterion here than in Table 1.4 is due to the equivalence of the logarithmic Sobolev inequality and exponential convergence in entropy. However, this is no longer true in the discrete case. In general, the logarithmic Sobolev inequality is stronger than exponential convergence in entropy. A criterion for the exponential convergence in entropy for birth–death processes remains open (*open problem*) [cf. S.Y. Zhang and Y.H. Mao (2000), Y.H. Mao and S.Y. Zhang (2000)]. The other two equivalences or coincidences in the tables come from Figure 1.1.

5.5 Exponential ergodicity for single birth processes

In this section, we study exponential ergodicity for single birth processes, which are in general irreversible. In particular, we prove the criterion for the ergodicity of birth–death processes presented in Table 1.4. The strong ergodicity for this class of processes will be studied in the next section.

The Q-matrix of a *single birth process* $Q = (q_{ij} : i, j \in \mathbb{Z}_+)$ is as follows: $q_{i,i+1} > 0$, $q_{i,i+j} = 0$ for all $i \in \mathbb{Z}_+ := \{0, 1, 2, \ldots\}$ and $j \geqslant 2$. Throughout the chapter, we consider only a totally stable and conservative Q-matrix: $q_i = -q_{ii} = \sum_{j \neq i} q_{ij} < \infty$ for all $i \in \mathbb{Z}_+$. Define $q_n^{(k)} = \sum_{j=0}^{k} q_{nj}$ for $0 \leqslant k < n \, (k, n \in \mathbb{Z}_+)$ and

$$
m_0 = \frac{1}{q_{01}}, \quad m_n = \frac{1}{q_{n,n+1}}\left(1 + \sum_{k=0}^{n-1} q_n^{(k)} m_k\right), \qquad n \geqslant 1,
$$

$$
F_n^{(n)} = 1, \quad F_n^{(i)} = \frac{1}{q_{n,n+1}} \sum_{k=i}^{n-1} q_n^{(k)} F_k^{(i)}, \qquad 0 \leqslant i < n,
$$

$$
d_0 = 0, \quad d_n = \frac{1}{q_{n,n+1}}\left(1 + \sum_{k=0}^{n-1} q_n^{(k)} d_k\right), \qquad n \geqslant 1.
$$

More simply,

$$
d_n = \sum_{k=1}^{n} \frac{F_n^{(k)}}{q_{k,k+1}}, \quad m_n = \sum_{k=0}^{n} \frac{F_n^{(k)}}{q_{k,k+1}} = \frac{F_n^{(0)}}{q_{01}} + d_n, \qquad n \geqslant 0.
$$

Here, as usual, $\sum_{\emptyset} = 0$. For birth–death processes, these quantities take a simpler form:

$$
m_n = \frac{1}{\mu_n b_n}\mu[0, n], \quad F_n^{(0)} = \frac{b_0}{\mu_n b_n}, \quad d_n = \frac{1}{\mu_n b_n}\mu[1, n], \qquad n \geqslant 1,
$$

where $\mu[i, k] = \sum_{i \leqslant j \leqslant k} \mu_j$. The main advantage of single birth processes is that the exit boundary consists of at most one single extremal point, and so explicit criteria can be expected. Here are the criteria for the three classical problems.

- Uniqueness (regularity) $\iff \sum_{n=0}^{\infty} m_n = \infty$. Next, assume that the Q-matrix is irreducible; then
- recurrence $\iff \sum_{n=0}^{\infty} F_n^{(0)} = \infty$. In the regular case,
- ergodicity $\iff d := \sup_{k \in \mathbb{Z}_+}\left(\sum_{n=0}^{k} d_n\right)/\left(\sum_{n=0}^{k} F_n^{(0)}\right) < \infty$

[cf. Chen (1992a, Theorems 3.16 and 4.54)].

Unfortunately, a criterion for the exponential ergodicity of general single birth processes remains unknown (*open problem*). Here is a sufficient condition, due to Y.H. Mao and Y.H. Zhang (2004) with an addition (Proposition 5.13), which is a generalization of the criterion for birth–death processes.

Theorem 5.11. Let the single birth Q-matrix be regular and irreducible. If

$$
\inf_i q_i > 0 \quad \text{and} \quad M := \sup_{i>0} \sum_{j=0}^{i-1} F_j^{(0)} \sum_{j=i}^{\infty} \frac{1}{q_{j,j+1} F_j^{(0)}} < \infty, \tag{5.22}
$$

then the process is exponentially ergodic. Condition (5.22) is necessary for the exponential ergodicity of birth–death processes for which

$$M = \delta := \sup_{i>0} \sum_{j=0}^{i-1} \frac{1}{\mu_j b_j} \sum_{j=i}^{\infty} \mu_j < \infty.$$

Proof. In view of Theorem 5.1, the condition $\inf_i q_i > 0$ is indeed necessary.

We prove first the sufficiency of (5.22).

(a) Let $H = \{0\}$. We need to construct a solution (g_i) to the equation (5.6') for a fixed λ: $0 < \lambda < \inf_i q_i$. First, define an operator

$$II_i(f) = \frac{1}{f_i} \sum_{j=0}^{i-1} F_j^{(0)} \sum_{k=j+1}^{\infty} \frac{f_k}{q_{k,k+1} F_k^{(0)}}, \qquad i \geqslant 1.$$

This is an analogue of the operator $I(f)$ used several times before and will be discussed in more detail in the next chapter. It indicates a key point in this proof, which comes from a study of the first eigenvalue. Next, define

$$\varphi_0 = 0, \qquad \varphi_i = \frac{1}{q_{01}} \sum_{j=0}^{i-1} F_j^{(0)}, \quad i \geqslant 1.$$

Then φ is increasing in i, and $\varphi_1 = q_{01}^{-1}$. Let $f = cq_{10}\sqrt{q_{01}\varphi}$ for some $c > 1$. Then f is increasing and $f_1 = cq_{10}$. Finally, define $g = fII(f)$. Then g is increasing and

$$g_1 = \sum_{k=1}^{\infty} \frac{f_k}{q_{k,k+1} F_k^{(0)}} \geqslant \frac{f_1}{q_{12} F_1^{(0)}} = c > 1.$$

We now need a technical result, to be proved later, taken from Chen (2000c).

Lemma 5.12. Let (m_i) and (n_i) be nonnegative sequences, $n_i \not\equiv 0$, satisfying

$$c := \sup_{i>0} \sum_{j=0}^{i-1} n_j \sum_{j=i}^{\infty} m_j < \infty.$$

Define $\varphi_0 = 0$, $\varphi_k = \sum_{j=0}^{k-1} n_j$. Then for every $\gamma \in (0,1)$, we have

$$\sum_{j=i}^{\infty} \varphi_j^{\gamma} m_j \leqslant c(1-\gamma)^{-1} \varphi_i^{\gamma-1}.$$

By Lemma 5.12, it follows that

$$g_i = cq_{10}\sqrt{q_{01}} \sum_{j=0}^{i-1} F_j^{(0)} \sum_{k=j+1}^{\infty} \frac{\sqrt{\varphi_k}}{q_{k,k+1}F_k^{(0)}}$$

$$\leqslant \frac{2Mcq_{10}}{\sqrt{q_{01}}} \sum_{j=0}^{i-1} F_j^{(0)} \varphi_{j+1}^{-1/2}$$

$$\leqslant \frac{2Mcq_{10}}{\sqrt{q_{01}\varphi_1}} \sum_{j=0}^{i-1} F_j^{(0)} < \infty, \qquad i \geqslant 1.$$

Let $g_0 = 1$. Then $1 \leqslant g_i < \infty$ for all $i \geqslant 0$. We now determine λ in terms of equation (5.6′). When $i = 1$, we get $\lambda \leqslant (c-1)c^{-1}\varPi_1(f)^{-1}$. When $i \geqslant 2$, we should have

$$\lambda g_i \leqslant \sum_{k=0}^{i-1} q_i^{(k)} F_k^{(0)} \sum_{j=k+1}^{\infty} \frac{f_j}{q_{j,j+1}F_j^{(0)}} - q_{i,i+1}F_i^{(0)} \sum_{k=i+1}^{\infty} \frac{f_k}{q_{k,k+1}F_k^{(0)}}.$$

For this, it suffices that

$$\lambda g_i \leqslant \sum_{k=0}^{i-1} q_i^{(k)} F_k^{(0)} \sum_{j=i}^{\infty} \frac{f_j}{q_{j,j+1}F_j^{(0)}} - q_{i,i+1}F_i^{(0)} \sum_{k=i+1}^{\infty} \frac{f_k}{q_{k,k+1}F_k^{(0)}}$$

$$= q_{i,i+1}F_i^{(0)} \sum_{k=i}^{\infty} \frac{f_k}{q_{k,k+1}F_k^{(0)}} - q_{i,i+1}F_i^{(0)} \sum_{k=i+1}^{\infty} \frac{f_k}{q_{k,k+1}F_k^{(0)}}$$

$$= f_i.$$

In other words, for (5.6′), we need only $\lambda \leqslant f_i/g_i = \varPi_i(f)^{-1}$ for all $i \geqslant 2$ and $\lambda \leqslant (c-1)c^{-1}\varPi_1(f)^{-1}$. Then we can choose any λ satisfying

$$0 < \lambda < \left(\frac{c-1}{c}\varPi_1(f)^{-1}\right) \bigwedge \left(\inf_{i\geqslant 2} \varPi_i(f)^{-1}\right) \bigwedge \left(\inf_i q_i\right), \qquad (5.23)$$

provided the right-hand side of (5.23) is positive, or equivalently, $\sup_{i\geqslant 2}\varPi_i(f) < \infty$. To prove the last property, define another operator

$$I_i(f) = \frac{F_i^{(0)}}{f_{i+1} - f_i} \sum_{k=i+1}^{\infty} \frac{f_k}{q_{k,k+1}F_k^{(0)}}, \qquad i \geqslant 1,$$

which is exactly the analogue of the one we have used many times before. By the proportion property, we get

$$\sup_{i\geqslant 1} \varPi_i(f) \leqslant \sup_{i\geqslant 1} I_i(f).$$

By Lemma 5.12 and the condition $M < \infty$, it follows that

$$I_i(f) = \frac{F_i^{(0)}}{\sqrt{\varphi_{i+1}} - \sqrt{\varphi_i}} \sum_{k=i+1}^{\infty} \frac{\sqrt{\varphi_k}}{q_{k,k+1} F_k^{(0)}} \leqslant \frac{2MF_i^{(0)}}{q_{01}(\sqrt{\varphi_{i+1}} - \sqrt{\varphi_i})\sqrt{\varphi_{i+1}}} \leqslant 4M$$

for all $i \geqslant 1$. Therefore, $\sup_{i \geqslant 1} II_i(f) \leqslant 4M < \infty$ as required. We have thus constructed a solution (g_i) to equation (5.6$'$) with $1 \leqslant g_i < \infty$ for all i. This implies the exponential ergodicity of the process.

For the remainder of this section, we consider birth–death processes only. We need only prove the necessity of (5.22).

(b) Denote σ_H by σ_0. Suppose that the process is exponentially ergodic. As mentioned below Theorem 5.1, there exists a λ with $0 < \lambda < q_i$ for all i such that $\mathbb{E}_0 e^{\lambda \sigma_0} < \infty$. Define

$$e_{i0}(\lambda) = \int_0^{\infty} e^{\lambda t} \mathbb{P}_i[\sigma_0 > t]\mathrm{d}t, \qquad i \in \mathbb{Z}_+.$$

Then $\mathbb{E}_i e^{\lambda \sigma_0} = \lambda e_{i0}(\lambda) + 1$. By Chen (1992a, Proof (a) of Theorems 4.45 and 4.44), one gets $e_{i0}(\lambda) < \infty$ for all $i \geqslant 1$. Furthermore, $\mathbb{E}_i e^{\lambda \sigma_0} < \infty$ for all $i \geqslant 1$. Note that if the starting point is not 0, then σ_0 is equal to the first hitting time:

$$\tau_0 = \inf\{t > 0 : X(t) = 0\}.$$

Hence $\mathbb{E}_i e^{\lambda \tau_0} < \infty$ for all $i \geqslant 1$. Define $m_i^{(n)} = \mathbb{E}_i \tau_0^n$. The Taylor's expansion

$$\infty > \mathbb{E}_i e^{\lambda \tau_0} = \sum_{n=0}^{\infty} \frac{\lambda^n}{n!} m_i^{(n)} \tag{5.24}$$

leads us to estimate the moments $m_i^{(n)}$. By a result due to Z.K. Wang [cf. Wang (1980, Chapter 3), or Z.T. Hou and Q.F. Guo (1978, Chapter 9), or Z.K. Wang and X.Q. Yang (1992)], we have

$$m_i^{(1)} = \sum_{j=0}^{i-1} \frac{1}{\mu_j b_j} \sum_{k=j+1}^{\infty} \mu_k,$$

$$m_i^{(n)} = n \sum_{j=0}^{i-1} \frac{1}{\mu_j b_j} \sum_{k=j+1}^{\infty} \mu_k m_k^{(n-1)}, \qquad n \geqslant 2. \tag{5.25}$$

Obviously, $m_k^{(n)} \geqslant m_i^{(n)}$ if $k \geqslant i$. By (5.25), it follows that

$$m_i^{(n)} \geqslant n \sum_{j=0}^{i-1} \frac{1}{\mu_j b_j} \sum_{k=i}^{\infty} \mu_k m_k^{(n-1)} \geqslant n \left(\sum_{j=0}^{i-1} \frac{1}{\mu_j b_j} \sum_{k=i}^{\infty} \mu_k \right) m_i^{(n-1)}, \qquad n \geqslant 2,$$

and

$$m_i^{(1)} \geqslant \sum_{j=0}^{i-1} \frac{1}{\mu_j b_j} \sum_{k=i}^{\infty} \mu_k.$$

Hence, by induction, one gets

$$m_i^{(n)} \geqslant n! \left(\sum_{j=0}^{i-1} \frac{1}{\mu_j b_j} \sum_{k=i}^{\infty} \mu_k \right)^n, \qquad n \geqslant 1.$$

Combining this with (5.24), we obtain

$$\sum_{n=1}^{\infty} \left(\lambda \sum_{j=0}^{i-1} \frac{1}{\mu_j b_j} \sum_{k=i}^{\infty} \mu_k \right)^n < \infty,$$

which implies that

$$\lambda \sum_{j=0}^{i-1} \frac{1}{\mu_j b_j} \sum_{k=i}^{\infty} \mu_k < 1.$$

Taking the supremum over i, we obtain $\delta \leqslant \lambda^{-1} < \infty$. Hence, the necessity of the second condition in (5.22) is proven.

(c) To complete the proof of the theorem, it suffices to show that

$$\inf_i q_i = 0 \implies \delta = \infty. \tag{5.26}$$

To do so, we need the following result.

Proposition 5.13. For a general reversible Markov chain, if $\inf_i q_i = 0$, then $\lambda_1 = 0$.

Having the result at hand, the proof of (5.26) is trivial, because for birth–death processes, by Theorem 5.7 (2), we have $(4\delta)^{-1} \leqslant \lambda_1 \leqslant Z\delta^{-1}$. We have thus completed the proof of Theorem 5.11. $\square$

Proof of Lemma 5.12. Let $M_n = \sum_{j \geqslant n} m_j$. Fix $N > i$. Then by the summation by parts formula and the assumption $M_n \leqslant c\varphi_n^{-1}$, we get

$$\sum_{j=i}^{N} \varphi_j^\gamma m_j \leqslant \varphi_i^\gamma M_i + \sum_{j=i}^{N} [\varphi_{j+1}^\gamma - \varphi_j^\gamma] M_{j+1} \leqslant c \left\{ \varphi_i^{\gamma-1} + \sum_{j=i}^{N} [\varphi_{j+1}^\gamma - \varphi_j^\gamma] / \varphi_{j+1} \right\}.$$

By using the elementary inequality $\gamma(1-\gamma)^{-1}(x^{\gamma-1} - 1) + x^\gamma \geqslant 1$ $(x > 0)$, it is easy to check that

$$\varphi_{j+1}^{\gamma-1} - \varphi_j^\gamma / \varphi_{j+1} \leqslant \gamma(1-\gamma)^{-1} [\varphi_j^{\gamma-1} - \varphi_{j+1}^{\gamma-1}].$$

Combining this with the last estimate gives us the required assertion. $\square$

Proof of Proposition 5.13. Without loss of generality, let the state space E be $\{0, 1, \ldots\}$. Consider the test function $f = c_1 I_{\{k\}} + c_2$, where c_1 and c_2 are constants such that $\pi(f) = 0$ and $\pi(f^2) = 1$:

$$c_2 = -c_1 \pi_k, \qquad c_1 = 1/\sqrt{\pi_k(1 - \pi_k)}.$$

Then

$$D(f) = \sum_{(i,j):\, i<j} \pi_i q_{ij} (f_j - f_i)^2$$

$$= \sum_{j:\, k<j} \pi_k q_{kj} (f_j - f_k)^2 + \sum_{i:\, i<k} \pi_i q_{ik} (f_k - f_i)^2$$

$$= \sum_{j:\, k<j} \frac{q_{kj}}{1 - \pi_k} + \sum_{i:\, i<k} \frac{q_{ki}}{1 - \pi_k}$$

$$= \frac{q_k}{1 - \pi_k}.$$

Applying the classical variational formula (5.9) to this f, we obtain

$$\lambda_1 \leqslant \frac{q_k}{1 - \pi_k},$$

since for large enough k_0, we have $\inf_{k \geqslant k_0} \pi_k \leqslant 1/2$, and therefore

$$\inf_k q_k = \left(\inf_{k \geqslant k_0} q_k \right) \bigwedge \left(\min_{k < k_0} q_k \right) \geqslant \left(\frac{\lambda_1}{2} \right) \bigwedge \left(\min_{k < k_0} q_k \right) > 0. \quad \square$$

5.6   Strong ergodicity

This section is devoted to strong ergodicity for general Markov processes. We will adopt both the analytic method and the coupling method. Let us begin with the analytic method for single birth processes.

For birth–death processes, the next result was first obtained by Hou et al. (2000) with a different proof. The general case is due to Y.H. Zhang (2001). We adopt the notation introduced at the beginning of the last section.

Analytic method

Theorem 5.14. Let $Q = (q_{ij})$ be a regular, irreducible single birth Q-matrix. Then the Q-process is strongly ergodic iff

$$\sup_{k \in \mathbb{Z}_+} \sum_{j=0}^{k} \left(F_j^{(0)} d - d_j \right) < \infty. \tag{5.27}$$

For birth–death processes, the criterion becomes

$$S := \sum_{n \geqslant 0} \frac{1}{\mu_n b_n} \mu[n+1, \infty) = \sum_{n \geqslant 1} \mu_n \sum_{j \leqslant n-1} \frac{1}{\mu_j b_j} < \infty$$

as stated in Table 1.4.

Proof. (a) We prove that the equation

$$y_i = \sum_{j \neq i} \frac{q_{ij}}{q_i} y_j + \frac{1}{q_i}, \quad i \geqslant 1; \qquad y_0 = 0, \tag{5.28}$$

has a bounded nonnegative solution iff (5.27) holds. If so,

$$d := \sup_{k \in \mathbb{Z}_+} \sum_{n=0}^{k} d_n \Big/ \sum_{n=0}^{k} F_n^{(0)} = \lim_{k \to \infty} \sum_{n=0}^{k} d_n \Big/ \sum_{n=0}^{k} F_n^{(0)},$$

and the unique solution to (5.28) is as follows:

$$y_0 = 0, \quad y_1 = d, \quad y_{n+1} = y_n + F_n^{(0)} y_1 - d_n, \qquad n \geqslant 1. \tag{5.29}$$

First, assume that (5.27) holds and define (y_i) by (5.29). Then, it should be easy to verify that (y_i) is a bounded nonnegative solution of (5.28).

Next, let (y_i) be a bounded nonnegative solution of (5.28) and define $v_n = y_{n+1} - y_n$ for $n \geqslant 0$. From (5.28), it is not difficult to derive

$$v_n = \frac{1}{q_{n,n+1}} \left(\sum_{k=0}^{n-1} q_n^{(k)} v_k - 1 \right), \qquad n \geqslant 1.$$

By induction, we can easily prove that $v_n = F_n^{(0)} v_0 - d_n$ for all $n \geqslant 0$. Note that $v_0 = y_1$. From these facts, it follows that

$$y_{k+1} = \sum_{n=0}^{k} v_n = \sum_{n=0}^{k} \left(F_n^{(0)} v_0 - d_n \right), \qquad k \in \mathbb{Z}_+. \tag{5.30}$$

Now, on the one hand, by (5.30) and $y_{k+1} \geqslant 0$, it follows that

$$v_0 \geqslant \sum_{n=0}^{k} d_n \Big/ \sum_{n=0}^{k} F_n^{(0)}, \qquad k \in \mathbb{Z}_+.$$

Hence $v_0 \geqslant d = \sup_{k \in \mathbb{Z}_+} \sum_{n=0}^{k} d_n \big/ \sum_{n=0}^{k} F_n^{(0)}$. On the other hand, by (5.30) again,

$$\frac{y_{k+1}}{\sum_{n=0}^{k} F_n^{(0)}} = v_0 - \frac{\sum_{n=0}^{k} d_n}{\sum_{n=0}^{k} F_n^{(0)}}, \qquad k \in \mathbb{Z}_+. \tag{5.31}$$

Note that (y_i) is bounded and $\sum_{n=0}^{k} F_n^{(0)} \to +\infty$ as $k \to \infty$ (by recurrence). Letting $k \to \infty$ in (5.31), we see that the right-hand side of (5.31) tends to a limit $v_0 - d'$, where

$$d' = \lim_{k \to \infty} \sum_{n=0}^{k} d_n \Big/ \sum_{n=0}^{k} F_n^{(0)}.$$

Furthermore $v_0 \leqslant d' \leqslant d$. Hence, we have $y_1 = v_0 = d = d'$. Combining this with (5.30), it follows that the solution (y_i) to (5.28) must have the representation (5.29) and hence is unique. Finally, by the boundedness of (y_i) and (5.30), condition (5.27) follows.

(b) By Theorem 5.1 (3), we know that the Q-process is strongly ergodic iff the following equation has a bounded nonnegative solution:

$$\sum_j q_{ij} y_j \leqslant -1, \quad i \notin H; \qquad \sum_{i \in H} \sum_{j \neq i} q_{ij} y_j < \infty,$$

where H is a nonempty finite subset of $\mathbb{Z}_+$. Let $H = \{0\}$. For single birth processes, the last equation is reduced to

$$\sum_j q_{ij} y_j \leqslant -1, \qquad i \neq 0, \tag{5.32}$$

since $\sum_{j \neq 0} q_{0j} y_j = q_{01} y_1 < \infty$.

Assume that the single birth process is strongly ergodic. Then there exists a bounded nonnegative solution (u_i) of (5.32), i.e.,

$$u_i \geqslant \sum_{j \neq i} \frac{q_{ij}}{q_i} u_j + \frac{1}{q_i}, \quad i \geqslant 1; \qquad u_0 \geqslant 0.$$

Denote by (u_i^*) the minimal nonnegative solution of (5.28). By the comparison theorem below, we have $u_i \geqslant u_i^*$ for all $i \geqslant 0$. Thus, (u_i^*) is bounded, and (5.28) has a bounded nonnegative solution. By (a), (5.27) holds.

Conversely, let (5.27) hold. Define (y_i) by (5.29). By (a), (y_i) is a bounded nonnegative solution of (5.28). Clearly, (y_i) is also a bounded nonnegative solution of (5.32). This implies strong ergodicity by the criterion quoted above. $\square$

To conclude this subsection, we introduce some elementary facts, taken from Z.T. Hou and Q.F. Guo (1978), or Chen (1992a), also needed in Chapter 9, about the theory of the minimal nonnegative solutions for systems of equations with nonnegative coefficients. All the results below can be easily proved using induction.

Theorem 5.15 (Existence and uniqueness theorem). Let $c_{ij} \geqslant 0$, $b_i \geqslant 0$. Then there exists a unique minimal solution $(x_i^* : i \in E)$ to the equations

$$x_i = \sum_{k \in E} c_{ik} x_k + b_i, \qquad i \in E.$$

More precisely, define

$$x_i^{(0)} = 0, \quad x_i^{(n+1)} = \sum_{k \in E} c_{ik} x_k^{(n)} + b_i, \qquad i \in E, \ n \geqslant 0.$$

Then $x_i^{(n)} \uparrow x_i^*$ for all $i \in E$ as $n \to \infty$.

Theorem 5.16 (Comparison theorem). Let $(x_i : i \in E)$ satisfy

$$x_i \geqslant \sum_{k \in E} c_{ik} x_k + b_i, \qquad i \in E.$$

Then $x_i \geqslant x_i^*$ for all $i \in E$.

Theorem 5.17 (Linear combination theorem). Let G be a countable set and $c_\alpha \geqslant 0$ for all $\alpha \in G$. Denote by $\left(x_i^{*\,(\alpha)} : i \in E\right)$ the minimal solution to

$$x_i = \sum_{k \in E} c_{ik} x_k + b_i^{(\alpha)}, \qquad i \in E.$$

Then $\left(\sum_{\alpha \in G} c_\alpha x_i^{*\,(\alpha)} : i \subset E\right)$ is the minimal solution to the equations

$$x_i = \sum_{k \in E} c_{ik} x_k + \sum_{\alpha \in G} c_\alpha b_i^{(\alpha)}, \qquad i \in E.$$

Coupling method

For a general Markov process with transition probability $P(t, x, dy)$ on $(E, \mathscr{E})$ and having a stationary distribution π, strong ergodicity means that

$$\sup_{x \in E} \|P(t, x, \cdot) - \pi\|_{\mathrm{Var}} \to 0 \qquad \text{as } t \to \infty.$$

Such convergence must be exponential. Indeed, note that for a signed measure ν, we have

$$\|\nu\|_{\mathrm{Var}} = \sup_{f: |f| \leqslant 1} |\nu(f)|.$$

It is easy to check that

$$\sup_x \|P(t + s, x, \cdot) - \pi\|_{\mathrm{Var}} \leqslant \frac{1}{2} \sup_x \|P(t, x, \cdot) - \pi\|_{\mathrm{Var}} \cdot \sup_x \|P(s, x, \cdot) - \pi\|_{\mathrm{Var}}.$$

Hence $\frac{1}{2} \sup_x \|P(t, x, \cdot) - \pi\|_{\mathrm{Var}}$ must have exponential decay. The various applications to the convergence rate in the total variation of the coupling methods is based on the following simple observation. Let (X_t, Y_t) be a coupling of the Markov processes starting from x and y, respectively, and define the coupling time as follows:

$$T = \inf\{t \geqslant 0 : X_t = Y_t\}.$$

Then we have

$$\|P(t, x, \cdot) - \pi\|_{\mathrm{Var}} \leqslant 2 \int_E \pi(dy) \widetilde{\mathbb{E}}^{x,y} I_{[X_t \neq Y_t]} = 2 \int_E \pi(dy) \widetilde{\mathbb{P}}^{x,y}[T > t].$$

Here we assume that $X_t = Y_t$ for all $t \geqslant T$. In particular, by the Chebyshev inequality, we have

$$\sup_x \|P(t, x, \cdot) - \pi\|_{\mathrm{Var}} \leqslant \sup_{x \neq y} \widetilde{\mathbb{E}}^{x,y}(T^n)/t^n \quad \text{and}$$

$$\sup_x \|P(t, x, \cdot) - \pi\|_{\mathrm{Var}} \leqslant \sup_{x \neq y} \widetilde{\mathbb{E}}^{x,y}(e^{\lambda T})e^{-\lambda t}, \qquad \lambda > 0.$$

We have thus obtained some strongly ergodic rates in terms of the coupling time. Along this line, there are a number of publications. See for instance Chen (1992a), T. Lindvall (1992) and the references within, or more recent papers by Y.Z. Wang (1999), Y.H. Mao (2002d; 2002e).

We now study the estimation of the moments of the coupling time [refer to Chen and S.F. Li (1989, Theorem 5.7) for a refined result].

Theorem 5.18. Let (X_t, Y_t) be a Markovian coupling with operator $\widetilde{L}$, $F \in \mathscr{D}_w(\widetilde{L})$, $F \geqslant 0$ and $F(x, x) = 0$ for all x. Suppose that

$$\widetilde{L}F(x, y) \leqslant -1, \qquad x \neq y. \tag{5.33}$$

Then for every nonnegative $\varphi \in C^1[0, \infty)$, we have

$$\widetilde{\mathbb{E}}^{x,y} \int_0^{t \wedge T} \varphi(s)\mathrm{d}s \leqslant \varphi(0)F(x, y) + \widetilde{\mathbb{E}}^{x,y} \int_0^{t \wedge T} \varphi'(s)F(X_s, Y_s)\mathrm{d}s. \tag{5.34}$$

Proof. Let $f(t; x, y) = \varphi(t)F(x, y)$. By using the martingale formulation,

$$f(t; X_t, Y_t) - \int_0^t \left[\partial_s f(s; X_s, Y_s) + \widetilde{L}f(s; X_s, Y_s)\right]\mathrm{d}s$$

is a $\widetilde{\mathbb{P}}^{x,y}$-martingale with respect to the natural flow of σ-algebras. Hence

$$\varphi(0)F(x, y) = \varphi(t)\widetilde{\mathbb{E}}^{x,y}F(X_{t \wedge T}, Y_{t \wedge T})$$
$$- \widetilde{\mathbb{E}}^{x,y} \int_0^{t \wedge T} \left[\varphi'(s)F(X_s, Y_s) + \varphi(s)\widetilde{L}F(X_s, Y_s)\right]\mathrm{d}s$$
$$\geqslant -\widetilde{\mathbb{E}}^{x,y} \int_0^{t \wedge T} \varphi'(s)F(X_s, Y_s)\mathrm{d}s + \widetilde{\mathbb{E}}^{x,y} \int_0^{t \wedge T} \varphi(s)\mathrm{d}s,$$

by (5.33). This gives us the required assertion. $\quad\square$

Applying (5.34) to $\varphi = 1$, we get

$$\widetilde{\mathbb{E}}^{x,y}T \leqslant F(x, y).$$

Next, applying (5.34) to $\varphi(t) = t^m (m \geqslant 1)$, we get

$$\widetilde{\mathbb{E}}^{x,y}T^{m+1} \leqslant (m + 1)\|F\|_\infty \widetilde{\mathbb{E}}^{x,y}T^m.$$

Hence, we have

$$\widetilde{\mathbb{E}}^{x,y}T^m \leqslant m! \, \|F\|_\infty^{m-1} F(x,y) \leqslant m! \, \|F\|_\infty^m, \qquad m \geqslant 1. \qquad (5.35)$$

Finally, applying (5.34) to $\varphi(t) = e^{\lambda t}$ $(\lambda > 0)$, we obtain

$$\lambda^{-1} \widetilde{\mathbb{E}}^{x,y} \left[e^{\lambda(t \wedge T)} - 1 \right] \leqslant F(x,y) + \|F\|_\infty \widetilde{\mathbb{E}}^{x,y} \left[e^{\lambda(t \wedge T)} - 1 \right].$$

Thus,

$$\widetilde{\mathbb{E}}^{x,y} e^{\lambda T} \leqslant 1 + \frac{\|F\|_\infty}{1/\lambda - \|F\|_\infty} = \frac{1}{1 - \lambda \|F\|_\infty}, \qquad 0 < \lambda < \|F\|_\infty < \infty. \qquad (5.36)$$

Certainly, one can also deduce (5.36) from (5.35) by Taylor expansion.

For a compact Riemannian manifold, simply take $F = H \circ \rho$, where ρ is the Riemannian distance and

$$H(r) = \int_0^r C(s)^{-1} \mathrm{d}s \int_s^D C(u) \mathrm{d}u, \qquad C(r) = \cosh^{d-1}\left[\frac{r}{2}\sqrt{\frac{-K}{d-1}}\right], \ r \in (0, D),$$

and adopt the coupling by reflection. Then (5.36) holds. This was done by Y.Z. Wang (1999) and Y.H. Mao (2002d; 2002e). In the one-dimensional case, since there is a linear order, one can simply use the classical coupling, which is order–preserving. Note that T is controlled by the hitting time $\inf\{t \geqslant 0 : Y_t = 0\}$. One can study its moments in the same way as above, replacing the coupling operator by the marginal one. For birth–death processes, one takes

$$F_i = \sum_{j=0}^{i-1} \frac{1}{b_j \mu_j} \sum_{k=j}^{\infty} \mu_k.$$

Then (5.36) holds once $\|F\|_\infty = \lim_{i \to \infty} F_i < \infty$. Equivalently, $S < \infty$, which is indeed necessary by Theorem 5.14. This was done by Y.H. Mao (2002d).

Chapter 6

Poincaré-Type Inequalities in Dimension One

This chapter offers a brief and elementary overview of recent progress on a large class of Poincaré-type inequalities in dimension one. The higher-dimensional case will be discussed in the next chapter. The explicit criteria for the inequalities, the variational formulas, and explicit bounds of the corresponding constants are presented, first for ordinary Poincaré inequalities (Section 6.2) and then for Poincaré-type inequalities (Sections 6.3 and 6.4). As typical applications, the Nash inequalities (Section 6.5) and logarithmic Sobolev inequalities (Section 6.6) are examined. To illustrate the main ideas, some short proofs are included. In the last section (Section 6.7), partial proofs are presented for the main dual variational formulas given in Theorem 6.1, which is the starting point of this chapter.

6.1 Introduction

First, we explain the problems we are going to study in this chapter.

The one-dimensional case in this chapter means either the second-order elliptic operators (one-dimensional diffusions) or the tridiagonal matrices (birth-death Markov processes). Let us begin with diffusions.

Let $L = a(x)\mathrm{d}^2/\mathrm{d}x^2 + b(x)\mathrm{d}/\mathrm{d}x$ be an elliptic operator on an interval $(0, D)$ $(D \leqslant \infty)$ with Dirichlet boundary at 0 and Neumann boundary at D when $D < \infty$, where a and b are Borel measurable functions and a is positive everywhere. Set $C(x) = \int_0^x b/a$. Here and in what follows, the Lebesgue measure $\mathrm{d}x$ is often omitted. Throughout this chapter, assume that

$$Z := \int_0^D e^C/a < \infty. \qquad (6.1)$$

Hence, $\mathrm{d}\mu := a^{-1}e^C \mathrm{d}x$ is a finite measure, which is crucial in the chapter. We are interested in the first Poincaré inequality

$$\|f\|^2 := \int_0^D f^2 \mathrm{d}\mu \leqslant A \int_0^D f'^2 e^C := AD(f), \quad f \in \mathbb{C}_d[0, D],\ f(0) = 0, \quad (6.2)$$

where $\mathbb{C}_d$ is the set of all continuous functions, differentiable almost everywhere and having compact support. When $D = \infty$, one should replace $[0, D]$ by $[0, D)$, but we will not mention this again in what follows. Next, we are also interested in the second Poincaré inequality

$$\|f - \pi(f)\|^2 := \int_0^D (f - \pi(f))^2 \mathrm{d}\mu \leqslant \overline{A} D(f), \qquad f \in \mathbb{C}_d[0, D], \quad (6.3)$$

where $\pi(f) = \mu(f)/Z = \int f \mathrm{d}\mu/Z$. To save on notation, we use the same A (respectively, $\overline{A}$,) to denote the optimal constant in (6.2) (respectively, (6.3)).

The aim of the study of these inequalities is to find a criterion under which (6.2) (respectively, (6.3)) holds, i.e., the optimal constant $A < \infty$ (respectively, $\overline{A} < \infty$), and furthermore to estimate A (respectively, $\overline{A}$). The reason we are restricting our attention to dimension one is that we are looking for some explicit criteria and explicit estimates. Actually, we have dual variational formulas for the upper and lower bounds of these constants. Generally speaking, such an explicit story does not exist in the higher-dimensional situation (see Section 7.1).

Next, replacing the L^2-norm on the left-hand sides of (6.2) and (6.3) with a general norm $\|\cdot\|_{\mathbb{B}}$ in a suitable normed linear space (the details are given in Section 6.3), we obtain the following Poincaré-type inequalities

$$\|f^2\|_{\mathbb{B}} \leqslant A_{\mathbb{B}} D(f), \qquad f \in \mathbb{C}_d[0, D],\ f(0) = 0, \quad (6.4)$$

$$\|(f - \pi(f))^2\|_{\mathbb{B}} \leqslant \overline{A}_{\mathbb{B}} D(f), \qquad f \in \mathbb{C}_d[0, D], \quad (6.5)$$

for which it is natural to study the same problems as above. The main purpose of this chapter is to solve these problems. By using this general setup, we are able to deal with the Nash inequalities (J. Nash, 1958)

$$\|f - \pi(f)\|^{2+4/\nu} \leqslant A_N D(f) \|f\|_1^{4/\nu} \quad (6.6)$$

in the case of $\nu > 2$, and the logarithmic Sobolev inequality (L. Gross, 1976)

$$\mathrm{Ent}\left(f^2\right) := \int_0^D f^2 \log \frac{f^2}{\pi(f^2)} \mathrm{d}\mu \leqslant A_{LS} D(f). \quad (6.7)$$

The remainder of the chapter is organized as follows. In the next section, we review the criteria for (6.2) and (6.3), the dual variational formulas, and explicit estimates of A and $\overline{A}$. Then, we extend a large part of these results to normed linear spaces, first for the Dirichlet case and then for the Neumann one. For a very general setup of normed linear spaces, the resulting conclusions are still rather satisfactory. Next, we specify the results to Orlicz spaces and finally apply them to the Nash inequalities and the logarithmic Sobolev inequality.

6.2 Ordinary Poincaré inequalities

In this section, we introduce the criteria for (6.2) and (6.3), the dual variational formulas, and explicit estimates of A and $\overline{A}$, which strengthen Theorem
1.5 and the results listed in Section 5.3.

   To state the main results, we need some notation. Write $x \wedge y = \min\{x, y\}$
and similarly, $x \vee y = \max\{x, y\}$. Define

$$
\begin{aligned}
\mathscr{F} &= \{f \in C[0, D] \cap C^1(0, D) : f(0) = 0,\ f'|_{(0,D)} > 0\}, \\
\widetilde{\mathscr{F}} &= \{f \in C[0, D] : f(0) = 0,\ \text{there exists } x_0 \in (0, D] \text{ such that} \\
&\qquad f = f(\cdot \wedge x_0),\ f \in C^1(0, x_0),\ \text{and } f'|_{(0,x_0)} > 0\}, \\
\mathscr{F}' &= \{f \in C[0, D] : f(0) = 0, f|_{(0,D)} > 0\}, \\
\widetilde{\mathscr{F}}' &= \{f \in C[0, D] : f(0) = 0,\ \text{there exists } x_0 \in (0, D] \text{ such that} \\
&\qquad f = f(\cdot \wedge x_0) \text{ and } f|_{(0,x_0)} > 0\}.
\end{aligned} \tag{6.8}
$$

Here the sets $\mathscr{F}$ and $\mathscr{F}'$ are essential; they are used, respectively, to define
below the operators of single and double integrals, and are used for the upper
bounds. The sets $\widetilde{\mathscr{F}}$ and $\widetilde{\mathscr{F}}'$ are less essential, being simply modifications of
$\mathscr{F}$ and $\mathscr{F}'$, respectively, to avoid the problem of integrability, and are used
for the lower bounds. Define

$$
I(f)(x) = \frac{e^{-C(x)}}{f'(x)} \int_x^D \left[fe^C/a\right](u)du, \qquad f \in \mathscr{F}, \tag{6.9}
$$

$$
II(f)(x) = \frac{1}{f(x)} \int_0^x dy\, e^{-C(y)} \int_y^D \left[fe^C/a\right](u)du, \qquad f \in \mathscr{F}'. \tag{6.10}
$$

   The next result is taken from Chen (2001b, Theorems 1.1 and 1.2). The
word "dual" below means that the upper and lower bounds are interchangeable if one exchanges the orders of "sup" and "inf" with a slight modification
of the set $\mathscr{F}$ (respectively, $\mathscr{F}'$) of test functions.

Theorem 6.1. Let (6.1) hold. Define $\varphi(x) = \int_0^x e^{-C}$ and $B = \sup_{x \in (0,D)} \varphi(x)$
$\times \int_x^D \frac{e^C}{a}$. Then we have the following assertions:

(1) *Explicit criterion.* $A < \infty$ iff $B < \infty$.
(2) *Dual variational formulas.*

$$
A \leqslant \inf_{f \in \mathscr{F}'} \sup_{x \in (0,D)} II(f)(x) = \inf_{f \in \mathscr{F}} \sup_{x \in (0,D)} I(f)(x), \tag{6.11}
$$

$$
A \geqslant \sup_{f \in \widetilde{\mathscr{F}}'} \inf_{x \in (0,D)} II(f)(x) = \sup_{f \in \widetilde{\mathscr{F}}} \inf_{x \in (0,D)} I(f)(x). \tag{6.12}
$$

The two inequalities all become equalities whenever both a and b are continuous on $[0, D]$.

(3) *Approximation procedure and explicit bounds.*

(a) Define $f_1 = \sqrt{\varphi}$, $f_n = f_{n-1}II(f_{n-1})$, and $D_n = \sup_{x \in (0,D)} II(f_n)(x)$. Then D_n is decreasing in n and $A \leqslant D_n \leqslant 4B$ for all $n \geqslant 1$.

(b) Fix $x_0 \in (0, D)$. Define

$$f_1^{(x_0)} = \varphi(\cdot \wedge x_0), \qquad f_n^{(x_0)} = f_{n-1}^{(x_0)}(\cdot \wedge x_0)II\big(f_{n-1}^{(x_0)}(\cdot \wedge x_0)\big),$$

and $C_n = \sup_{x_0 \in (0,D)} \inf_{x \in (0,D)} II\big(f_n^{(x_0)}(\cdot \wedge x_0)\big)(x)$. Then C_n is increasing in n and $A \geqslant C_n \geqslant B$ for all $n \geqslant 1$.

We mention that the explicit estimates "$B \leqslant A \leqslant 4B$" were obtained previously in the study of the weighted Hardy inequality by B. Muckenhoupt (1972). A short proof of "$A \leqslant 4B$" was presented in Section 5.3. The proof of "$A \geqslant B$" is also easy. Indeed, fix $x \in (0, D)$ and take

$$f(y) = f_x(y) = \int_0^{x \wedge y} e^{-C(s)} ds, \qquad y \in (0, D).$$

Then $f'(y) = e^{-C(y)}$ if $y < x$ and $f'(y) = 0$ if $y \in (x, D)$. Furthermore,

$$\|f\|^2 = \int_0^x f(y)^2 \pi(dy) + f(x)^2 \pi[x, D],$$

$$D(f) = \int_0^x e^{-2C(y)} e^{C(y)} dy / Z = f(x)/Z,$$

where $\pi[p, q] = \int_p^q d\pi$ and $Z = \mu[0, D]$. Hence

$$A \geqslant \|f\|^2 / D(f) = Zf(x)^{-1} \int_0^x f(y)^2 \pi(dy) + Zf(x)\pi[x, D] \geqslant f(x)\mu[x, D].$$

Taking the supremum with respect to x, it follows that $A \geqslant B$.

The proofs of parts (2) and (3) of Theorem 6.1 are more technical; see Section 6.7 for details.

We now turn to study $\overline{A}$, for which it is necessary to assume that

$$\int_0^D e^{-C(s)} ds \int_0^s a(u)^{-1} e^{C(u)} du = \infty, \tag{6.13}$$

since we are working in the ergodic situation.

Theorem 6.2. Let (6.1) and (6.13) hold and set $\bar{f} = f - \pi(f)$. Then we have the following assertions:

(1) *Explicit criterion.* $\overline{A} < \infty$ iff $B < \infty$, where B is given by Theorem 6.1.

(2) *Dual variational formulas.*

$$\sup_{f \in \tilde{\mathscr{F}}} \inf_{x \in (0,D)} I(\bar{f})(x) \leqslant \overline{A} \leqslant \inf_{f \in \mathscr{F}} \sup_{x \in (0,D)} I(\bar{f})(x). \tag{6.14}$$

The two inequalities all become equalities whenever both a and b are continuous on $[0, D]$.

(3) *Approximation procedure and explicit bounds.*

(a) Define $f_1 = \sqrt{\varphi}$, $f_n = \bar{f}_{n-1} II(\bar{f}_{n-1})$, and $\overline{D}_n = \sup_{x \in (0,D)} II(\bar{f}_n)(x)$. Then $\overline{A} \leqslant \overline{D}_n \leqslant 4B$ for all $n \geqslant 1$.

(b) Fix $x_0 \in (0, D)$. Define
$$f_1^{(x_0)} = \varphi(\cdot \wedge x_0), \quad f_n^{(x_0)} = \bar{f}_{n-1}^{(x_0)}(\cdot \wedge x_0) II\big(\bar{f}_{n-1}^{(x_0)}(\cdot \wedge x_0)\big),$$
and $\overline{C}_n = \sup_{x_0 \in (0,D)} \inf_{x \in (0,D)} II\big(\bar{f}_n^{(x_0)}(\cdot \wedge x_0)\big)(x)$. Then $\overline{A} \geqslant \overline{C}_n$ for all $n \geqslant 2$. By convention, $1/0 = \infty$.

Part (1) of the theorem is taken from Chen (2000c, Theorem 3.7). The upper bound in (6.14) is due to Chen and F.Y. Wang (1997b). The other parts are taken from Chen (2001b, Theorems 1.3 and 1.4).

Finally, we consider inequality (6.3) on a general interval (p, q) $(-\infty \leqslant p < q \leqslant \infty)$. When p (respectively, q) is finite, at which the Neumann boundary condition is endowed, we adopt a splitting technique. The intuitive idea goes as follows. Since the eigenfunction corresponding to $\overline{A}$, if it exists, must change sign, it should vanish somewhere in the present continuous situation, say at θ. Thus, it is natural to divide the interval (p, q) into two parts: (p, θ) and (θ, q). Then, one compares $\overline{A}$ with the optimal constants in the inequality (6.2), denoted by $A_{1\theta}$ and $A_{2\theta}$, respectively, on (θ, q) and (p, θ) having a common Dirichlet boundary at θ. Actually, we do not care about the existence of the vanishing point θ. Such a θ is unknown, even if it exists. In practice, we regard θ as a reference point and then apply an optimization procedure with respect to θ. We now redefine $C(x) = \int_\theta^x b/a$. Again, since we are in the ergodic situation, we assume the following (nonexplosive) conditions:

$$Z_{1\theta} := \int_\theta^q e^C/a < \infty, \qquad Z_{2\theta} := \int_p^\theta e^C/a < \infty,$$

$$\int_p^\theta e^{-C(s)} ds \int_s^\theta e^C/a = \infty \text{ if } p = -\infty, \quad \text{and} \qquad (6.15)$$

$$\int_\theta^q e^{-C(s)} ds \int_\theta^s e^C/a = \infty \text{ if } q = \infty,$$

for some (equivalently, all) $\theta \in (p, q)$. Corresponding to the intervals (θ, q) and (p, θ), respectively, we have constants $B_{1\theta}$ and $B_{2\theta}$, given by Theorem 6.1.

Theorem 6.3. Let (6.15) hold. Then we have the following:

(1) $\sup_{\theta \in (p,q)} (A_{1\theta} \wedge A_{2\theta}) \leqslant \overline{A} \leqslant \inf_{\theta \in (p,q)} (A_{1\theta} \vee A_{2\theta})$.

(2) Let θ be the median of μ; then $(A_{1\theta} \vee A_{2\theta})/2 \leqslant \overline{A} \leqslant A_{1\theta} \vee A_{2\theta}$.

In particular, $\overline{A} < \infty$ iff $B_{1\theta} \vee B_{2\theta} < \infty$.

Comparing the variational formulas (6.11), (6.12), and (6.14) with the classical variational formulas

$$\lambda_0 = \inf \left\{ D(f) : f \in C^1(0, D) \cap C[0, D], f(0) = 0, \pi(f^2) = 1 \right\},$$
$$\lambda_1 = \inf \left\{ D(f) : f \in C^1(0, D) \cap C[0, D], \pi(f) = 0, \pi(f^2) = 1 \right\},$$

one sees that there are no common points. This explains why the new formulas (6.12) and (6.14) have not appeared before. The key here is the discovery of the formulas rather than their proofs, which are usually not hard, due to the advantage of dimension one. As an illustration, here we present a part of the proofs.

Proof of the upper bound in (6.14)

Originally, the assertion was proved in Chen and F.Y. Wang (1997b) using coupling methods. Here we adopt the analytic proof given in Chen (1999a). The discrete analogue was presented in Section 3.2.

Let $g \in C[0, D] \cap C^1(0, D)$, $\pi(g) = 0$, and $\pi(g^2) = 1$. Then, for every $f \in \mathscr{F}$ with $\pi(f) \geqslant 0$, we have

$$1 = \frac{1}{2} \int_0^D \pi(\mathrm{d}x) \pi(\mathrm{d}y) [g(y) - g(x)]^2$$

$$= \int_{\{x \leqslant y\}} \pi(\mathrm{d}x) \pi(\mathrm{d}y) \left(\int_x^y \frac{g'(u) \sqrt{f'(u)}}{\sqrt{f'(u)}} \mathrm{d}u \right)^2$$

$$\leqslant \int_{\{x \leqslant y\}} \pi(\mathrm{d}x) \pi(\mathrm{d}y) \int_x^y \frac{g'(u)^2}{f'(u)} \mathrm{d}u \int_x^y f'(\xi) \mathrm{d}\xi$$

(by the Cauchy–Schwarz inequality)

$$= \int_{\{x \leqslant y\}} \pi(\mathrm{d}x) \pi(\mathrm{d}y) \int_x^y g'(u)^2 e^{C(u)} \frac{e^{-C(u)}}{f'(u)} \mathrm{d}u [f(y) - f(x)]$$

$$= \int_0^D a(u) g'(u)^2 \pi(\mathrm{d}u) \frac{Z e^{-C(u)}}{f'(u)} \int_0^u \pi(\mathrm{d}x) \int_u^D \pi(\mathrm{d}y) [f(y) - f(x)]$$

$$\leqslant D(g) \sup_{u \in (0, D)} \frac{Z e^{-C(u)}}{f'(u)} \int_0^u \pi(\mathrm{d}x) \int_u^D \pi(\mathrm{d}y) [f(y) - f(x)]$$

$$\leqslant D(g) \sup_{x \in (0, D)} I(f)(x) \quad (\text{since } \pi(f) \geqslant 0).$$

Thus, $D(g)^{-1} \leqslant \sup_{x \in (0, D)} I(\bar{f})(x)$, and so

$$\overline{A} = \sup_{g:\, \pi(g) = 0,\, \pi(g^2) = 1} D(g)^{-1} \leqslant \sup_{x \in (0, D)} I(\bar{f})(x).$$

This gives us the required assertion:

$$\overline{A} \leqslant \inf_{f \in \mathscr{F}} \sup_{x \in (0, D)} I(\bar{f})(x).$$

The proof that equality holds for continuous a and b needs more work, since it requires some more precise properties of the corresponding eigenfunctions, as shown in Sections 3.7 and 3.8 for the discrete case. □

6.3 Extension: normed linear spaces

Beginning in this section, we introduce recent results obtained in Chen (2002a; 2003a). We will not quote these papers time by time subsequently.

In this section, we study the Poincaré-type inequality (6.4). Clearly, the normed linear spaces used here cannot be completely arbitrary, since we are dealing with a topic of hard mathematics. From now on, let $(\mathbb{B}, \|\cdot\|_{\mathbb{B}}, \mu)$ be a normed linear space of functions $f : [0, D] \to \mathbb{R}$ satisfying the following conditions:

(1) $1 \in \mathbb{B}$;

(2) $\mathbb{B}$ is ideal: if $h \in \mathbb{B}$ and $|f| \leqslant |h|$, then $f \in \mathbb{B}$;

(3) $\|f\|_{\mathbb{B}} = \sup\limits_{g \in \mathscr{G}} \int_0^D |f| g \mathrm{d}\mu,$ (6.16)

(4) $\mathscr{G} \ni g_0$ with $\inf g_0 > 0$,

where $\mathscr{G}$ is a fixed set, to be specified case by case later, of nonnegative functions on $[0, D]$. The first two conditions mean that $\mathbb{B}$ is rich enough, and the last one means that $\mathscr{G}$ is not trivial: it contains at least one strictly positive function. The third condition is essential in this chapter, which means that the norm $\|\cdot\|_{\mathbb{B}}$ has a "dual" representation. A typical example of such a normed linear space is $\mathbb{B} = L^r(\mu)$; then $\mathscr{G} =$ the unit ball in $L_+^{r'}(\mu)$, $1/r + 1/r' = 1$.

The optimal constant $A_{\mathbb{B}}$ in (6.4) can be expressed as a variational formula as follows:

$$A_{\mathbb{B}} = \sup\left\{ \frac{\|f^2\|_{\mathbb{B}}}{D(f)} : f \in \mathbb{C}_{\mathrm{d}}[0, D], f(0) = 0, \, 0 < D(f) < \infty \right\}. \quad (6.17)$$

Clearly, this formula is powerful mainly for the lower bounds of $A_{\mathbb{B}}$. However, the upper bounds are more useful in practice but much harder to handle. Fortunately, we have quite complete results.

Define $\varphi(x) = \int_0^x e^{-C}$ as before and let

$$B_{\mathbb{B}} = \sup_{x \in (0,D)} \varphi(x) \|I_{(x,D)}\|_{\mathbb{B}}, \qquad C_{\mathbb{B}} = \sup_{x \in (0,D)} \frac{\|\varphi(x \wedge \cdot)^2\|_{\mathbb{B}}}{\varphi(x)},$$

$$D_{\mathbb{B}} = \sup_{x \in (0,D)} \frac{\|\sqrt{\varphi}\,\varphi(x \wedge \cdot)^2\|_{\mathbb{B}}}{\sqrt{\varphi(x)}}. \qquad (6.18)$$

Theorem 6.4. Let (6.1) and (6.16) hold. Then we have the following asser-

tions:

(1) *Explicit criterion.* $A_\mathbb{B} < \infty$ iff $B_\mathbb{B} < \infty$.

(2) *Variational formulas for the upper bounds.*

$$A_\mathbb{B} \leqslant \inf_{f \in \mathscr{F}'} \sup_{x \in (0,D)} f(x)^{-1} \| f\varphi(x \wedge \cdot) \|_\mathbb{B}$$

$$\leqslant \inf_{f \in \mathscr{F}} \sup_{x \in (0,D)} \frac{e^{-C(x)}}{f'(x)} \| f I_{(x,D)} \|_\mathbb{B}. \tag{6.19}$$

(3) *Approximation procedure and explicit bounds.* Let $B_\mathbb{B} < \infty$. Define $f_0 = \sqrt{\varphi}$, $f_n(x) = \| f_{n-1}\varphi(x \wedge \cdot) \|_\mathbb{B}$, and $D_\mathbb{B}(n) = \sup_{x \in (0,D)} f_n/f_{n-1}$ for $n \geqslant 1$. Then $D_\mathbb{B}(n)$ is decreasing in n and

$$B_\mathbb{B} \leqslant C_\mathbb{B} \leqslant A_\mathbb{B} \leqslant D_\mathbb{B}(n) \leqslant D_\mathbb{B} \leqslant 4B_\mathbb{B} \tag{6.20}$$

for all $n \geqslant 1$.

We are now going to sketch the proof of the second variational formula in (6.19), from which the explicit upper bound $A_\mathbb{B} \leqslant 4B_\mathbb{B}$ follows immediately, as we did at the end of the last section. The explicit estimates "$B_\mathbb{B} \leqslant A_\mathbb{B} \leqslant 4B_\mathbb{B}$" were previously obtained in S.G. Bobkov and F. Götze (1999b) in terms of the weighted Hardy inequality [cf. B. Muckenhoupt (1972), i.e., Theorem 5.9]. The lower bounds follows easily from (6.17).

Sketch of the proof of the second variational formula in (6.19)

The starting point is the variational formula for A (cf. (6.11)):

$$A \leqslant \inf_{f \in \mathscr{F}} \sup_{x \in (0,D)} \frac{e^{-C(x)}}{f'(x)} \int_x^D \frac{fe^C}{a} = \inf_{f \in \mathscr{F}} \sup_{x \in (0,D)} \frac{e^{-C(x)}}{f'(x)} \int_x^D f d\mu.$$

Fix $g > 0$ and introduce a transform

$$b \to b/g, \qquad a \to a/g > 0, \tag{6.21}$$

under which $C(x)$ is transformed into

$$C_g(x) = \int_0^x \frac{b/g}{a/g} = C(x).$$

This means that the function C is invariant with respect to the transform, and so is the Dirichlet form $D(f)$. The left-hand side of (6.2) is changed into

$$\int_0^D f^2 g e^C / a = \int_0^D f^2 g d\mu.$$

At the same time, the constant A is changed into

$$A_g \leqslant \inf_{f \in \mathscr{F}} \sup_{x \in (0,D)} \frac{e^{-C(x)}}{f'(x)} \int_x^D fg d\mu.$$

Taking the supremum with respect to $g \in \mathscr{G}$, the left-hand side becomes

$$\sup_{g \in \mathscr{G}} \int_0^D f^2 g d\mu = \|f^2\|_{\mathbb{B}},$$

and the constant becomes

$$A_{\mathbb{B}} = \sup_g A_g \leqslant \sup_g \inf_f \sup_x \frac{e^{-C(x)}}{f'(x)} \int_x^D fg d\mu$$

$$\leqslant \inf_f \sup_g \sup_x \frac{e^{-C(x)}}{f'(x)} \int_x^D fg d\mu = \inf_f \sup_x \frac{e^{-C(x)}}{f'(x)} \sup_g \int_0^D fI_{(x,D)} g d\mu.$$

$$= \inf_f \sup_x \frac{e^{-C(x)}}{f'(x)} \|fI_{(x,D)}\|_{\mathbb{B}}.$$

We are done! Of course, more details are required for completing the proof. For instance, one may use $g + 1/n$ instead of g to avoid the condition "$g > 0$" and then pass to the limit. $\quad\square$

The lucky point in the proof is that "sup inf $\leqslant$ inf sup," which goes in the correct direction. However, we do not know at the moment how to generalize the dual variational formula for lower bounds given in (6.12) to general normed linear spaces, since the same procedure goes in the opposite direction.

6.4 Neumann case: Orlicz spaces

In the Neumann case, the boundary condition becomes $f'(0) = 0$, rather than $f(0) = 0$. Then $\lambda_0 = 0$ is trivial. Hence, we study λ_1 (called the *spectral gap* of L), that is, the inequality (6.3). We now consider its generalization (6.5). Naturally, one may play the same game as in the last section extending (6.14) to normed linear spaces. However, it does not work this time. Note that on the left-hand side of (6.5), the term $\pi(f)$ is not invariant under the transform (6.21). Moreover, since $\pi(\bar{f}) = 0$, it is easy to check that for each fixed $f \in \mathscr{F}$, $I(\bar{f})(x)$ is positive for all $x \in (0, D)$. But this property is no longer true when $d\mu$ is replaced by $g d\mu$. Our goal is to adopt the splitting technique explained in Section 6.2.

Let $\theta \in (p, q)$ be a reference point and let $A_{\mathbb{B}}^{k\theta}$, $B_{\mathbb{B}}^{k\theta}$, $C_{\mathbb{B}}^{k\theta}$, $D_{\mathbb{B}}^{k\theta}$ ($k = 1, 2$) be the constants defined in (6.17) and (6.18) corresponding to the intervals (θ, q) and (p, θ), respectively. By Theorem 6.4, we have

$$B_{\mathbb{B}}^{k\theta} \leqslant C_{\mathbb{B}}^{k\theta} \leqslant A_{\mathbb{B}}^{k\theta} \leqslant D_{\mathbb{B}}^{k\theta} \leqslant 4B_{\mathbb{B}}^{k\theta}, \qquad k = 1, 2.$$

Theorem 6.5. Let (6.15) and (6.16) hold. Then we have the following assertions:

(1) *Explicit criterion.* $\overline{A}_{\mathbb{B}} < \infty$ iff $B_{\mathbb{B}}^{1\theta} \vee B_{\mathbb{B}}^{2\theta} < \infty$.
(2) *Explicit estimates.*

$$\max \left\{ \frac{1}{2} \big(A_{\mathbb{B}}^{1\theta} \wedge A_{\mathbb{B}}^{2\theta} \big), \; K_{\theta} \big(A_{\mathbb{B}}^{1\theta} \vee A_{\mathbb{B}}^{2\theta} \big) \right\} \leqslant \overline{A}_{\mathbb{B}} \leqslant A_{\mathbb{B}}^{1\theta} \vee A_{\mathbb{B}}^{2\theta},$$

where K_{θ} is a constant.

We may now consider briefly the discrete case, i.e., the birth–death processes. Let $b_i \, (i \geqslant 0)$ be the birth rates and $a_i \, (i \geqslant 1)$ the death rates of a birth–death process. Define

$$\mu_0 = 1, \quad \mu_n = \frac{b_0 \cdots b_{n-1}}{a_1 \cdots a_n}, \quad Z = \sum_{n=0}^{\infty} \mu_n, \quad \pi_n = \frac{\mu_n}{Z}, \; n \geqslant 1.$$

Consider a normed linear space $(\mathbb{B}, \|\cdot\|_{\mathbb{B}}, \mu)$ of functions $E := \{0, 1, 2, \ldots\} \to \mathbb{R}$ satisfying (6.16). Define

$$\varphi_i = \sum_{j=1}^{i} \frac{1}{\mu_j a_j}, \; i \geqslant 1; \qquad B_{\mathbb{B}} = \sup_{i \geqslant 1} \varphi_i \big\| I_{\{i, i+1, \ldots\}} \big\|_{\mathbb{B}}.$$

Clearly, the inequalities (6.4) and (6.5) are meaningful with a slight modification.

Theorem 6.6. Consider birth–death processes with state space E. Assume that $Z < \infty$.

(1) *Explicit criterion for (6.4).* $A_{\mathbb{B}} < \infty$ iff $B_{\mathbb{B}} < \infty$.
(2) *Explicit bounds for $A_{\mathbb{B}}$.* $B_{\mathbb{B}} \leqslant A_{\mathbb{B}} \leqslant 4 B_{\mathbb{B}}$.
(3) *Explicit criterion for (6.5).* Let the birth–death process be nonexplosive:

$$\sum_{i=0}^{\infty} \frac{1}{\mu_i b_i} \sum_{j=0}^{i} \mu_j = \infty. \tag{6.22}$$

Then $\overline{A}_{\mathbb{B}} < \infty$ iff $B_{\mathbb{B}} < \infty$.
(4) *Explicit estimates for $\overline{A}_{\mathbb{B}}$:* Let $E_1 = \{1, 2, \ldots\}$ and let c_1 and c_2 be two constants such that $|\pi(f)| \leqslant c_1 \|f\|_{\mathbb{B}}$ and $|\pi(f I_{E_1})| \leqslant c_2 \|f I_{E_1}\|_{\mathbb{B}}$ for all $f \in \mathbb{B}$. Then

$$\overline{A}_{\mathbb{B}} \geqslant \max \left\{ \|1\|_{\mathbb{B}}^{-1}, \; \left(1 - \sqrt{c_2 (1 - \pi_0) \|1\|_{\mathbb{B}}} \right)^2 A_{\mathbb{B}} \right\},$$
$$\overline{A}_{\mathbb{B}} \leqslant \left(1 + \sqrt{c_1 \|1\|_{\mathbb{B}}} \right)^2 A_{\mathbb{B}}. \tag{6.23}$$

Similarly, one can handle the birth–death processes on $\mathbb{Z}$.

An interesting point here is that the first lower bound in (6.23) is meaningful only in the discrete situation.

Orlicz spaces. The results obtained so far can be specialized to Orlicz spaces. This is, as far as we know, the only way to deduce the criteria for the Nash and logarithmic Sobolev inequalities, given in the next two sections, respectively. The idea also goes back to S.G. Bobkov and F. Götze (1999b). A function Φ: $\mathbb{R} \to \mathbb{R}$ is called an N-*function* if it is nonnegative, continuous, convex, even (i.e., $\Phi(-x) = \Phi(x)$), and satisfies the following conditions:

$$\Phi(x) = 0 \ \text{iff} \ x = 0, \qquad \lim_{x \to 0} \Phi(x)/x = 0, \qquad \lim_{x \to \infty} \Phi(x)/x = \infty.$$

In what follows, we assume the following *growth condition* (or Δ_2-*condition*) for Φ:

$$\sup_{x \gg 1} \Phi(2x)/\Phi(x) < \infty \ \left(\Longleftrightarrow \sup_{x \gg 1} x\Phi'_-(x)/\Phi(x) < \infty \right),$$

where Φ'_- is the left derivative of Φ. Corresponding to each N-function, we have a complementary N-function:

$$\Phi_c(y) := \sup\{x|y| - \Phi(x) : x \geqslant 0\}, \qquad y \in \mathbb{R}.$$

Alternatively, let φ_c be the inverse function of Φ'_-; then $\Phi_c(y) = \int_0^{|y|} \varphi_c$ [cf. M.M. Rao and Z.D. Ren (1991)].

Given an N-function and a finite measure μ on $E := (p, q) \subset \mathbb{R}$, define an *Orlicz space* as follows:

$$L^\Phi(\mu) = \left\{ f \ (E \to \mathbb{R}) : \int_E \Phi(f)\mathrm{d}\mu < \infty \right\}, \qquad \|f\|_\Phi = \sup_{g \in \mathscr{G}} \int_E |f| g \mathrm{d}\mu, \qquad (6.24)$$

where $\mathscr{G} = \left\{ g \geqslant 0 : \int_E \Phi_c(g)\mathrm{d}\mu \leqslant 1 \right\}$, which is the set of nonnegative functions in the unit ball of $L^{\Phi_c}(\mu)$. Under the Δ_2-condition, $(L^\Phi(\mu), \|\cdot\|_\Phi, \mu)$ is a Banach space. For this, the Δ_2-condition is indeed necessary. Clearly, $L^\Phi(\mu) \ni 1$ and is ideal. Obviously, $(L^\Phi(\mu), \|\cdot\|_\Phi, \mu)$ satisfies condition (6.16), and so we have the following result.

Corollary 6.7. For any N-function Φ satisfying the growth condition, if (6.1) (respectively, (6.15)) holds, then Theorem 6.4 (respectively, 6.5) holds for the Orlicz space $(L^\Phi(\mu), \|\cdot\|_\Phi, \mu)$.

6.5 Nash inequality and Sobolev-type inequality

It is known that when $\nu > 2$, the Nash inequality (6.6)

$$\|f - \pi(f)\|^{2+4/\nu} \leqslant A_N D(f) \|f\|_1^{4/\nu}$$

is equivalent to the Sobolev-type inequality

$$\|f - \pi(f)\|_{\nu/(\nu-2)}^2 \leqslant A_S D(f),$$

where $\|\cdot\|_r$ is the $L^r(\mu)$-norm. Refer to D. Bakry et al. (1995), E.A. Carlen et al. (1987), and N. Varopoulos (1985). This leads to the use of the Orlicz space $L^{\Phi}(\mu)$ with $\Phi(x) = |x|^r/r$, $r = \nu/(\nu-2)$:

$$\left\|(f - \pi(f))^2\right\|_{\Phi} \leqslant \overline{A}_{\nu} D(f). \tag{6.25}$$

The results in this section were obtained in Y.H. Mao (2002b), based on the weighted Hardy inequalities (Theorem 5.9).

Define $C(x) = \int_{\theta}^{x} b/a$, $\mu(m, n) = \int_{m}^{n} e^{C}/a$ and

$$\varphi^{1\theta}(x) = \int_{\theta}^{x} e^{-C}, \qquad B_{\nu}^{1\theta} = \sup_{x > \theta} \varphi^{1\theta}(x)\,\mu(x, q)^{(\nu-2)/\nu},$$

$$\varphi^{2\theta}(x) = \int_{x}^{\theta} e^{-C}, \qquad B_{\nu}^{2\theta} = \sup_{x < \theta} \varphi^{2\theta}(x)\,\mu(p, x)^{(\nu-2)/\nu}.$$

Here $B_{\nu}^{k\theta}$ ($k = 1, 2$) is specified from $B_{\mathbb{B}}$ given in (6.18) with $\mathbb{B} = L^{\Phi}((\theta, q), \mu)$ or $\mathbb{B} = L^{\Phi}((p, \theta), \mu)$, since $\|\cdot\|_{\Phi} = (r')^{1/r'}\|\cdot\|_r$, $1/r + 1/r' = 1$.

Theorem 6.8. Let (6.15) hold and assume $\nu > 2$.

(1) *Explicit criterion.* The Nash inequality (equivalently, (6.25)) holds on (p, q) iff $B_{\nu}^{1\theta} \vee B_{\nu}^{2\theta} < \infty$.

(2) *Explicit bounds.*

$$\overline{A}_{\nu} \geqslant \max\left\{\frac{1}{2}(B_{\nu}^{1\theta} \wedge B_{\nu}^{2\theta}), \left[1 - \left(\frac{Z_{1\theta} \vee Z_{2\theta}}{Z_{1\theta} + Z_{2\theta}}\right)^{1/2+1/\nu}\right]^2 (B_{\nu}^{1\theta} \vee B_{\nu}^{2\theta})\right\},$$

$$\overline{A}_{\nu} \leqslant 4(B_{\nu}^{1\theta} \vee B_{\nu}^{2\theta}). \tag{6.26}$$

In particular, if θ is the median of μ, then

$$\left[1 - (1/2)^{1/2+1/\nu}\right]^2 (B_{\nu}^{1\theta} \vee B_{\nu}^{2\theta}) \leqslant \overline{A}_{\nu} \leqslant 4(B_{\nu}^{1\theta} \vee B_{\nu}^{2\theta}).$$

We now consider birth–death processes with state space $\{0, 1, 2, \ldots\}$. Define

$$\varphi_i = \sum_{j=1}^{i} \frac{1}{\mu_j a_j}, \quad i \geqslant 1; \qquad B_{\nu} = \sup_{i \geqslant 1} \varphi_i \left(\sum_{j=i}^{\infty} \mu_j\right)^{(\nu-2)/\nu}.$$

Theorem 6.9. For birth–death processes, let (6.22) hold and assume that $Z < \infty$. Then we have

$$\max\left\{\left(\frac{2}{\nu Z^{\nu/2-1}}\right)^{2/\nu}, \left[1 - \left(\frac{Z-1}{Z}\right)^{1/2+1/\nu}\right]^2\right\} B_{\nu} \leqslant \overline{A}_{\nu} \leqslant 16 B_{\nu}. \tag{6.27}$$

Hence, when $\nu > 2$, the Nash inequality holds iff $B_{\nu} < \infty$.

6.6 Logarithmic Sobolev inequality

The starting point of our study is the following observation.

Lemma 6.10. Let $\Phi(x) = |x|\log(1+|x|)$, $\mathscr{L}(f) = \sup_{c\in\mathbb{R}} \text{Ent}\left((f+c)^2\right)$, and $\text{Ent}(f) = \int_{\mathbb{R}} f\log\frac{f}{\pi(f)}d\mu$, $f \geqslant 0$. Then we have

$$\frac{2}{5}\left\|(f - \pi(f))^2\right\|_\Phi \leqslant \mathscr{L}(f) \leqslant \frac{51}{20}\left\|(f - \pi(f))^2\right\|_\Phi. \tag{6.28}$$

The proof of Lemma 6.10 is given at the end of this section.

The observation (Lemma 6.10) leads to the use of the Orlicz space $\mathbb{B} = L^\Phi(\mu)$ with $\Phi(x) = |x|\log(1+|x|)$. The results in this section were obtained by S.G. Bobkov and F. Götze (1999b), and Y.H. Mao (2002a), based again on the weighted Hardy inequalities (Theorem 5.9). Refer also to L. Miclo (1999b) for a related study, and F. Barthe and C. Roberto (2003) for a refinement.

Define

$$C(x) = \int_\theta^x \frac{b}{a}, \qquad \mu(m,n) = \int_m^n e^C/a;$$

$$\varphi^{1\theta}(x) = \int_\theta^x e^{-C}, \qquad \varphi^{2\theta}(x) = \int_x^\theta e^{-C};$$

$$M(x) = x\left[\frac{2}{1+\sqrt{1+4x}} + \log\left(1 + \frac{1+\sqrt{1+4x}}{2x}\right)\right]; \tag{6.29}$$

$$B_\Phi^{1\theta} = \sup_{x\in(\theta,q)} \varphi^{1\theta}(x)M(\mu(\theta,x)), \qquad B_\Phi^{2\theta} = \sup_{x\in(p,\theta)} \varphi^{2\theta}(x)M(\mu(x,\theta)).$$

Again, here $B_\Phi^{k\theta}$ ($k = 1,2$) is specified from $B_\mathbb{B}$ given in (6.18).

Theorem 6.11. Let (6.15) hold.

(1) *Explicit criterion.* The logarithmic Sobolev inequality on $(p,q) \subset \mathbb{R}$ holds iff

$$\sup_{x\in(\theta,q)} \mu(x,q)\log\frac{1}{\mu(x,q)} \int_\theta^x e^{-C} < \infty \quad \text{and}$$

$$\sup_{x\in(p,\theta)} \mu(p,x)\log\frac{1}{\mu(p,x)} \int_x^\theta e^{-C} < \infty \tag{6.30}$$

hold for some (equivalently, all) $\theta \in (p,q)$.

(2) *Explicit bounds.* Let $\bar{\theta}$ be the root of $B_\Phi^{1\theta} = B_\Phi^{2\theta}$, $\theta \in [p,q]$. Then we have

$$\frac{1}{5}B_\Phi^{1\bar\theta} \leqslant A_{LS} \leqslant \frac{51}{5}B_\Phi^{1\bar\theta}. \tag{6.31}$$

By a translation if necessary, assume that $\theta = 0$ is the median of μ. Then we have

$$\frac{\left(\sqrt{2}-1\right)^2}{5}\left(B_\Phi^{1\theta} \vee B_\Phi^{2\theta}\right) \leqslant A_{LS} \leqslant \frac{51}{5}\left(B_\Phi^{1\theta} \vee B_\Phi^{2\theta}\right). \tag{6.32}$$

We now consider birth–death processes with state space $\{0, 1, 2, \ldots\}$. Define

$$\varphi_i = \sum_{j=1}^{i} \frac{1}{\mu_j a_j}, \; i \geqslant 1; \qquad B_\Phi = \sup_{i \geqslant 1} \varphi_i M(\mu[i, \infty)),$$

where $\mu[i, \infty) = \sum_{j \geqslant i} \mu_j$ and $M(x)$ is defined in (6.29).

Theorem 6.12. For birth–death processes, let (6.22) hold and assume that $Z < \infty$. Then we have

$$A_{LS} \geqslant \frac{2}{5} \max \left\{ \frac{\sqrt{4Z + 1} - 1}{2}, \; \left(1 - \frac{Z_1 \Psi^{-1}(Z_1^{-1})}{Z \Psi^{-1}(Z^{-1})} \right)^2 \right\} B_\Phi,$$

$$A_{LS} \leqslant \frac{51 \times 4}{5} B_\Phi,$$

where $Z_1 = Z - 1$ and Ψ^{-1} is the inverse function of Ψ: $\Psi(x) = x^2 \log(1 + x^2)$. In particular, $A_{LS} < \infty$ iff

$$\sup_{i \geqslant 1} \varphi_i \, \mu[i, \infty) \log \frac{1}{\mu[i, \infty)} < \infty.$$

Proof of Lemma 6.10. We follow the proof given in S.G. Bobkov and F. Götze (1999b), and Chen (2001b).

Without loss of generality, we may replace μ with π in definitions of $\mathscr{L}(f)$, $\mathrm{Ent}(f)$, and $\| \cdot \|_\Phi$. In other words, we may assume that $\mu = \pi$.

For convenience, we adopt a more practical but equivalent norm as follows:

$$\|f\|_{(\Phi)} = \inf \left\{ \alpha > 0 : \int_E \Phi(f/\alpha) \mathrm{d}\mu \leqslant 1 \right\}. \tag{6.33}$$

The comparison of these two norms is as follows:

$$\|f\|_{(\Phi)} \leqslant \|f\|_\Phi \leqslant 2\|f\|_{(\Phi)} \tag{6.34}$$

[cf. M.M. Rao and Z.D. Ren (1991, Section 3.3, Proposition 4)]. Because $\|f^2\|_{(\Phi)} = \|f\|_{(\Psi)}^2$, it suffices to prove that

$$\frac{4}{5} \|f - \pi(f)\|_{(\Psi)}^2 \leqslant \mathscr{L}(f) \leqslant \frac{51}{20} \|f - \pi(f)\|_{(\Psi)}^2. \tag{6.35}$$

Let $\|f\|_{(\Psi)} = 1$ and $\pi(f) = 0$. By Lemma 4.14, we have

$$\mathscr{L}(f) \leqslant \mathrm{Ent}(f^2) + 2\pi(f^2). \tag{6.36}$$

Express the right-hand side as

$$\int f^2 (\delta + \log f^2) \mathrm{d}\pi + \pi(f^2) [2 - \delta - \log \pi(f^2)]$$

for some $\delta \in [0, 2]$. Note that $x(2 - \delta - \log x) \leqslant e^{1-\delta}$ for all $x > 0$. Let $c(\delta)$ be the bound such that $\delta + \log x \leqslant c(\delta) \log(1 + x)$ for all $x > 0$. By solving this inequality, it follows that the smallest $c(\delta)$ should satisfy the equation

$$c \log c - (c - 1) \log(c - 1) = \delta, \qquad c > 1.$$

Then we have

$$\mathscr{L}(f) \leqslant c(\delta) \int f^2 \log\left(1 + f^2\right) d\pi + e^{1-\delta} \leqslant c(\delta) + e^{1-\delta}.$$

Minimizing the right-hand side in δ under the above constraint, we obtain $\delta \approx 1.02118$, $c(\delta) \approx 1.56271$. Then $c(\delta) + e^{1-\delta} < 2.542 < 2.55 = 51/20$, and the required upper bound follows.

For the lower bound, let $\pi(f) = 0$ and $\mathscr{L}(f) = 2$. Because

$$\pi\left(f^2\right) - \pi(f)^2 = \frac{1}{2} \lim_{|a| \to \infty} \mathrm{Ent}\left((f + a)^2\right) \leqslant \frac{1}{2} \mathscr{L}(f),$$

we get $\pi\left(f^2\right) \leqslant 1$. Hence $\pi\left(f^2\right) \log \pi\left(f^2\right) \leqslant 0$, and moreover,

$$\pi\left(f^2 \log(f^2)\right) = \mathrm{Ent}(f^2) + \pi\left(f^2\right) \log \pi\left(f^2\right) \leqslant \mathrm{Ent}\left(f^2\right) \leqslant \mathscr{L}(f) = 2. \qquad (6.37)$$

The idea is to find the smallest constant $\delta \approx 0.4408$ such that

$$x \log\left(1 + x/(2 + \delta)\right) \leqslant \delta + x \log x$$

for all $x > 0$. Then

$$\int (f^2/(2 + \delta)) \log\left(1 + f^2/(2 + \delta)\right) d\pi \leqslant \left(\delta + \int f^2 \log f^2 d\pi\right)/(2 + \delta) \leqslant 1,$$

by (6.37). Thus, $\|f/\sqrt{2 + \delta}\|_{(\Psi)} \leqslant 1$, and so $\|f\|_{(\Psi)}^2 \leqslant 2 + \delta < 5\mathscr{L}(f)/4$. $\quad\square$

6.7 Partial proofs of Theorem 6.1

In this section, we prove Theorem 6.1 except for the conclusion that equality in (6.11) and (6.12) holds for continuous coefficients a and b. The proof of the last assertion requires some finer properties of the eigenfunction in the weak sense, as seen from Sections 3.7 and 3.8 in the discrete case.

Proof of (6.11)

(a) First we prove that $A \leqslant \inf_{f \in \mathscr{F}'} \sup_{x \in (0,D)} II(f)(x)$. Given h with $h|_{(0,D)} > 0$, for every g: $g(0) = 0$, $\|g\| = 1$, by the Cauchy–Schwarz inequality, we have

$$1 = \int_0^D g(x)^2 \pi(\mathrm{d}x) = \int_0^D \pi(\mathrm{d}x) \left[\int_0^x g'(u)\mathrm{d}u \right]^2$$

$$\leqslant \int_0^D \pi(\mathrm{d}x) \int_0^x [g'^2 e^C h^{-1}](u)\mathrm{d}u \int_0^x [he^{-C}](\xi)\mathrm{d}\xi$$

$$= \int_0^D a(u)g'(u)^2 \pi(\mathrm{d}u) \frac{Z}{h(u)} \int_u^D \pi(\mathrm{d}x) \int_0^x he^{-C} \qquad (6.38)$$

$$\leqslant D(g) \sup_{x \in (0,D)} \frac{1}{h(x)} \int_x^D \frac{e^{C(y)}}{a(y)} \mathrm{d}y \int_0^y he^{-C}$$

$$=: D(g) \sup_{x \in (0,D)} H(x).$$

Now, let $f \in \mathscr{F}'$ satisfy $\sup_{x \in (0,D)} II(f)(x) < \infty$. Take $h(x) = \int_x^D f a^{-1} e^C$. By the mean value theorem, we get

$$\sup_{x \in (0,D)} H(x) \leqslant \sup_{x \in (0,D)} \left[-\frac{e^C}{ah'}(x) \right] \int_0^x he^{-C} = \sup_{x \in (0,D)} II(f)(x). \qquad (6.39)$$

Because g is arbitrary, by (6.38) and (6.39), we obtain the required assertion. When $D = \infty$, there is a problem about the integrability of h. To avoid this, replacing D with finite M in the definitions of h and H, the above proof is still valid. Then the conclusion follows by letting $M \uparrow \infty$.

(b) Next, we prove that

$$\inf_{f \in \mathscr{F}'} \sup_{x \in (0,D)} II(f)(x) = \inf_{f \in \mathscr{F}} \sup_{x \in (0,D)} I(f)(x).$$

Given $f \in \mathscr{F}$, without loss of generality, assume that $\sup_{x \in (0,D)} I(f)(x) < \infty$. By the mean value theorem, $\sup_{x \in (0,D)} II(f)(x) \leqslant \sup_{x \in (0,D)} I(f)(x)$. But $\mathscr{F}' \supset \mathscr{F}$, so

$$\inf_{f \in \mathscr{F}'} \sup_{x \in (0,D)} II(f)(x) \leqslant \inf_{f \in \mathscr{F}} \sup_{x \in (0,D)} I(f)(x).$$

Conversely, for a given $f \in \mathscr{F}'$ with $\sup_{x \in (0,D)} II(f)(x) < \infty$, let $g = fII(f)$. Then $g \in \mathscr{F}$. By using the mean value theorem again, we obtain

$$I(g)(x) = \left[\int_x^D f a^{-1} e^C \right]^{-1} \int_x^D g a^{-1} e^C \leqslant \sup_{x \in (0,D)} (g/f)(x) = \sup_{x \in (0,D)} II(f)(x).$$

When $D = \infty$, there is again a problem about the integrability of g, which can be solved by using the method mentioned in the last paragraph. Hence

$$\sup_{x \in (0,D)} II(f)(x) \geqslant \sup_{x \in (0,D)} I(g)(x) \geqslant \inf_{f \in \mathscr{F}} \sup_{x \in (0,D)} I(f)(x).$$

Taking the supremum with respect to $f \in \mathscr{F}'$, it follows that

$$\inf_{f \in \mathscr{F}'} \sup_{x \in (0,D)} II(f)(x) \geqslant \inf_{f \subset \mathscr{F}} \sup_{x \in (0,D)} I(f)(x).$$

An alternative proof of this inequality can be obtained by using the identity

$$(e^C g')' = -f e^C / a. \tag{6.40}$$

We have thus proved the required assertion.

Combining (a) with (b), we get (6.11). □

Proof of (6.12)

(a) For $\sup_{f \in \widetilde{\mathscr{F}}'} \inf_{x \in (0, D)} II(f)(x) = \sup_{f \in \widetilde{\mathscr{F}}} \inf_{x \in (0, D)} I(f)(x)$, the proof is a dual of the above one, exchanging supremum and infimum, inverting the order of the inequalities, and redefining $g = [f II(f)](\cdot \wedge x_0)$.

(b) Let $f \in \widetilde{\mathscr{F}}'$ satisfy $f = f(\cdot \wedge x_0)$ and $c := \sup_{x \in (0, D)} II(f)(x)^{-1} < \infty$ and let $g_0 = [f II(f)](\cdot \wedge x_0)$. Then g_0 is bounded and (6.40) holds on $(0, x_0)$. By the integration by parts formula, we get

$$\int_0^D g_0'^2 e^C = [g_0 g_0' e^C](x_0-) - \int_0^{x_0} g_0 (e^C g_0')'$$

$$= \int_0^{x_0} g_0 f e^C / a + g_0(x_0) \int_{x_0}^D f e^C / a$$

$$= \int_0^D g_0 f e^C / a \leqslant \int_0^D (g_0^2 e^C / a) \sup_{x \in (0, D)} f / g_0$$

$$= c \int_0^D g_0^2 e^C / a.$$

Hence $A \geqslant c^{-1}$, and furthermore, $A \geqslant \sup_{f \in \widetilde{\mathscr{F}}'} \inf_{x \in (0, D)} II(f)(x)$.

Combining (a) with (b), we get (6.12). □

Proof of part (3) of Theorem 6.1

First, we consider the case (a). Condition $B < \infty$ implies that

$$\int_0^D \sqrt{\varphi} e^C / a \leqslant \sqrt{B} \int_0^D \left(\int_x^D e^C / a \right)^{-1/2} e^C / a = 2\sqrt{B\widetilde{Z}} < \infty.$$

Hence $f_1 = \sqrt{\varphi} \in L^1(\pi)$ [these two conditions are needed for the initial function f_1. In practice, one can certainly choose some more convenient functions]. Assume that $f_{n-1} \in L^1(\pi)$. Then

$$f_n(x) = f_{n-1}(x) II(f_{n-1})(x) = \int_0^x dy e^{-C(y)} \int_y^D f_{n-1} e^C / a$$

$$\leqslant Z \|f_{n-1}\|_1 \int_0^x e^{-C} < \infty.$$

By induction, it follows that $f_n \in L^1(\pi)$ for all n. Furthermore, since $f_n \in \mathscr{F}'$, by (6.11),

$$A \leqslant \inf_{f \in \mathscr{F}'} \sup_{x \in (0,D)} II(f)(x) \leqslant \sup_{x \in (0,D)} II(f_n)(x) = D_n.$$

Then by the mean value theorem and the proof of the upper bound given in Section 5.3, we get $D_1 \leqslant 4B$. On the other hand, by the definition of f_n and (6.40), we have

$$\left(-e^C f_n' \right)' = a^{-1} e^C f_{n-1} \geqslant a^{-1} f_n e^C D_{n-1}^{-1}. \tag{6.41}$$

That is, $f_n e^C / a \leqslant D_{n-1} \left(-e^C f_n' \right)'$. Hence

$$f_{n+1}(x) \leqslant D_{n-1} \int_0^x e^{-C(y)} dy \int_y^D \left(-e^C f_n' \right)'(u) du = D_{n-1} f_n(x). \tag{6.42}$$

From this, one deduces that $D_n \leqslant D_{n-1}$.

We now consider the case (b). By the identity

$$[f II(f)](x) = \int_0^D f\varphi(\cdot \wedge x) e^C / a = \int_0^x f\varphi e^C / a + \varphi(x) \int_x^D f e^C / a,$$

we get

$$f_2^{(x_0)}(x \wedge x_0) \geqslant \varphi(x \wedge x_0)\varphi(x_0) \int_{x_0}^D e^C / a,$$

and so

$$\sup_{x \in (0,D)} \left[f_1^{(x_0)} / f_2^{(x_0)} \right](x \wedge x_0) = \sup_{x \in (0,x_0)} \left[f_1^{(x_0)} / f_2^{(x_0)} \right](x) \leqslant \left[\varphi(x_0) \int_{x_0}^D e^C / a \right]^{-1}.$$

This implies that $C_1 \geqslant B$. Here, the reason one needs the local procedure "stopping at x_0" is the possibility of $\varphi \notin L^1(\pi)$, which then implies that $D(\varphi) = \infty$.

Finally, we prove the monotonicity of the C_n. Applying the mean value theorem twice, we obtain

$$\sup_{x \in (0,D)} \left[f_n^{(x_0)} / f_{n+1}^{(x_0)} \right](x \wedge x_0)$$

$$= \sup_{x \in (0,x_0)} \left[f_n^{(x_0)} / f_{n+1}^{(x_0)} \right](x) \leqslant \sup_{x \in (0,x_0)} \left[f_n^{(x_0)'} / f_{n+1}^{(x_0)'} \right](x)$$

$$\leqslant \sup_{x \in (0,x_0)} \int_x^D f_{n-1}^{(x_0)}(\cdot \wedge x_0) e^C a^{-1} \bigg/ \int_x^D f_n^{(x_0)}(\cdot \wedge x_0) e^C a^{-1}$$

$$\leqslant \sup_{x \in (0,D)} \left[f_{n-1}^{(x_0)} / f_n^{(x_0)} \right](x \wedge x_0).$$

This implies that $C_n \geqslant C_{n-1}$. The inequality $A \geqslant C_n \geqslant B$ comes from (6.12). $\square$

Chapter 7

Functional Inequalities

This chapter deals with some stronger and weaker inequalities than the Poincaré one, called *functional inequalities* for simplicity. Equivalently, we are studying some stronger and weaker types of convergence than the exponential one. Correspondingly, this chapter is divided into two parts.

In the first part, we discuss some types of stronger convergence. We show how to go to normed linear (Orlicz) spaces, starting from Hilbert space (L^2-space), in the higher–dimensional situation. This part is an extension of the main results obtained in the last chapter. There are three sections. In Section 7.1, we state the results. Their proofs are sketched in Section 7.2. In Section 7.3, we compare the capacitary method used here with Cheeger's method.

In the second part, we discuss some weaker (slower) types of convergence. There are four sections. In Section 7.4, we examine the general convergence speed, and then two functional inequalities are introduced in Section 7.5. In Section 7.6, we discuss algebraic convergence. The general (irreversible) case is discussed in the last section, Section 7.7.

7.1 Statement of results

Let E be a locally compact separable metric space with Borel σ-algebra $\mathscr{E}$, μ an everywhere dense Radon measure on E, and $(D, \mathscr{D}(D))$ a regular Dirichlet form on $L^2(\mu) = L^2(E; \mu)$. Regularity means that $\mathscr{D}(D) \cap C_0(E)$ is dense with respect to the norm $\sqrt{D(f) + \|f\|^2}$, where $C_0(E)$ is the set of continuous functions with compact support. The starting point of our study is the following result, which is a copy of Theorem 4.8.

Theorem 7.1. For a regular transient Dirichlet form $(D, \mathscr{D}(D))$, the optimal constant A in the *Poincaré inequality*

$$\|f\|^2 = \int_E f^2 d\mu \leqslant AD(f), \qquad f \in \mathscr{D}(D) \cap C_0(E), \tag{7.1}$$

satisfies $B \leqslant A \leqslant 4B$, where

$$B = \sup_{\text{compact } K} \frac{\mu(K)}{\text{Cap}(K)}. \tag{7.2}$$

The transiency here has the usual probabilistic meaning. Recall that

$$\text{Cap}(K) = \inf \left\{ D(f) : f \in \mathscr{D}(D) \cap C_0(E), f|_K \geqslant 1 \right\}.$$

Certainly, in (7.1), one may replace "$\mathscr{D}(D) \cap C_0(E)$" by "$\mathscr{D}(D)$" or by the extended Dirichlet space "$\mathscr{D}_e(D)$," which is the set of $\mathscr{E}$-measurable functions f:

> $|f| < \infty$, μ-a.e., there exists a sequence $\{f_n\} \subset \mathscr{D}(D)$ such that $D(f_n - f_m) \to 0$ as $n, m \to \infty$, and $\lim_{n \to \infty} f_n = f$, μ-a.e.

Refer to the standard books by M. Fukushima, Y. Oshima, and M. Takeda (1994), and by Z.M. Ma and M. Röckner (1992) for some preliminary facts about the theory of Dirichlet forms.

As mentioned in Chapter 5, the proof of inequality (7.1) on the half-line $(E = \mathbb{R}_+)$ was begun by G.H. Hardy in 1920 and completed by B. Muckenhoupt in 1972 [see also B. Opic and A. Kufner (1990)] with explicitly isoperimetric constant B.

The first goal of this chapter is extending (7.1) to the *Poincaré-type inequality*

$$\left\| f^2 \right\|_{\mathbb{B}} \leqslant A_{\mathbb{B}} D(f), \qquad f \in \mathscr{D}(D) \cap C_0(E), \tag{7.3}$$

for a class of *normed linear spaces* $(\mathbb{B}, \| \cdot \|_{\mathbb{B}}, \mu)$ of real functions on E. To do so, we need the following assumptions on $(\mathbb{B}, \| \cdot \|_{\mathbb{B}}, \mu)$:

(H1) Transient case: $I_K \in \mathbb{B}$ for all compact K. Ergodic case: $1 \in \mathbb{B}$.
(H2) If $h \in \mathbb{B}$ and $|f| \leqslant h$, then $f \in \mathbb{B}$.
(H3) $\|f\|_{\mathbb{B}} = \sup_{g \in \mathscr{G}} \int_E |f| g \, d\mu$,

where $\mathscr{G}$, to be specified case by case, is a class of nonnegative $\mathscr{E}$-measurable functions. Unless otherwise stated, these assumptions will be used throughout this and the next sections. Note that part (4) of condition (6.16) is ignored here.

We can now state our first result as follows.

Theorem 7.2. For a regular transient Dirichlet form $(D, \mathscr{D}(D))$, the optimal constant $A_{\mathbb{B}}$ in (7.3) satisfies

$$B_{\mathbb{B}} \leqslant A_{\mathbb{B}} \leqslant 4 B_{\mathbb{B}}, \tag{7.4}$$

where the *isoperimetric constant* $B_{\mathbb{B}}$ is given by

$$B_{\mathbb{B}} = \sup_{\text{compact } K} \frac{\|I_K\|_{\mathbb{B}}}{\text{Cap}(K)}. \tag{7.5}$$

Next, we go to the ergodic case. Assume that $\mu(E) < \infty$ and set $\pi = \mu/\mu(E)$. Throughout this chapter, we use a simplified notation: $\bar{f} = f - \pi(f)$, where $\pi(f) = \int f \mathrm{d}\pi$. We adopt a splitting technique. Let $E_1 \subset E$ be open with $\pi(E_1) \in (0,1)$ and write $E_2 = E_1^c \setminus \partial E_1$.

The restriction of $\mathbb{B}$ to E_i gives us $(\mathbb{B}^i, \|\cdot\|_{\mathbb{B}^i}, \mu^i)$:

$$\mathbb{B}^i = \{f|_{E_i} : f \in \mathbb{B}\}, \quad \mu^i = \mu|_{E_i}, \quad \mathscr{G}^i = \{g|_{E_i} : g \in \mathscr{G}\},$$

$$\|f\|_{\mathbb{B}^i} = \sup_{g \in \mathscr{G}^i} \int_{E_i} |f|g\mathrm{d}\mu^i = \sup_{g \in \mathscr{G}} \int_{E_i} |f|g\mathrm{d}\mu, \qquad i = 1, 2.$$

Correspondingly, we have a restricted Dirichlet form $(D, \mathscr{D}_i)$ on $L^2(E_i, \mu^i)$, where $\mathscr{D}_i = \{f \in \mathscr{D}(D) : \text{the quasi-version of } f \text{ equals } 0 \text{ on } E_i^c, \text{ q.e.}\}$. The corresponding constants given by Theorem 7.2 are denoted by $A_{\mathbb{B}^i}$ and $B_{\mathbb{B}^i}$ $(i = 1, 2)$, respectively.

Denote by c_1 a constant such that

$$|\pi(f)| \leqslant c_1\|f\|_{\mathbb{B}}, \qquad f \in \mathbb{B}. \tag{7.6}$$

For each $G \subset E$, denote by $c_2(G)$ a constant such that

$$|\pi(fI_G)| \leqslant c_2(G)\|fI_G\|_{\mathbb{B}}, \qquad f \in \mathbb{B}. \tag{7.7}$$

Theorem 7.3. Let $(D, \mathscr{D}(D))$ be a regular irreducible Dirichlet form. Assume that $\mu(E) < \infty$ and $\sup_{\text{open } E_1: \pi(E_1) \in (0,1/2]} c_2(E_1)\pi(E_1)\|1\|_{\mathbb{B}} < 1$. Then the optimal constant $\overline{A}_{\mathbb{B}}$ in the *Poincaré-type inequality*

$$\left\|\bar{f}^2\right\|_{\mathbb{B}} \leqslant \overline{A}_{\mathbb{B}} D(f), \qquad f \in \mathscr{D}(D) \cap C_0(E), \tag{7.8}$$

satisfies

$$\overline{A}_{\mathbb{B}} \geqslant \underline{\kappa} \sup_{\text{open } E_1: \pi(E_1) \in (0,1/2]} A_{\mathbb{B}^1} \geqslant \underline{\kappa} \sup_{\text{open } E_1: \pi(E_1) \in (0,1/2]} B_{\mathbb{B}^1}, \tag{7.9}$$

$$\overline{A}_{\mathbb{B}} \leqslant \bar{\kappa} \sup_{\text{open } E_1: \pi(E_1) \in (0,1/2]} A_{\mathbb{B}^1} \leqslant 4\bar{\kappa} \sup_{\text{open } E_1: \pi(E_1) \in (0,1/2]} B_{\mathbb{B}^1}, \tag{7.10}$$

where $\underline{\kappa} = \left(1 - \sup_{\text{open } E_1: \pi(E_1) \in (0,1/2]} \sqrt{c_2(E_1)\pi(E_1)\|1\|_{\mathbb{B}}}\right)^2$ and $\bar{\kappa} = \left(1 + \sqrt{c_1\|1\|_{\mathbb{B}}}\right)^2$.

For the *logarithmic Sobolev inequality*,

$$\int_E f^2 \log\left[f^2/\pi(f^2)\right]\mathrm{d}\mu \leqslant A_{\text{LS}} D(f), \qquad f \in \mathscr{D}(D) \cap C_0(E), \tag{7.11}$$

the next theorem is a generalization of the one-dimensional result given in Barthe F. and Roberto, C. (2003).

Theorem 7.4. Let $(D, \mathscr{D}(D))$ be a regular irreducible Dirichlet form. Assume that $\mu(E) < \infty$. Then the optimal A_{LS} in (7.11) satisfies

$$B_{\mathrm{LS}}(e^2)/4 \leqslant B_{\mathrm{LS}}(1/2) \leqslant A_{\mathrm{LS}} \leqslant 4\,B_{\mathrm{LS}}(e^2), \tag{7.12}$$

where

$$B_{\mathrm{LS}}(\gamma) = \sup_{\substack{\text{open } O\,:\, \pi(O) \in (0,\,1/2] \\ \text{compact } K \subset O}} \frac{\mu(K)}{\mathrm{Cap}(K)} \log\left(1 + \frac{\gamma}{\pi(K)}\right). \tag{7.13}$$

In particular, for one-dimensional diffusions, the assertion holds with $B_{\mathrm{LS}}(\gamma) = B_+(\gamma) \vee B_-(\gamma)$, where

$$B_+(\gamma) = \sup_{x>m} \mu[x, \infty) \log\left(1 + \frac{\gamma}{\pi[x,\infty)}\right) \int_m^x e^{-C} \tag{7.14}$$

$$B_-(\gamma) = \sup_{x<m} \mu(-\infty, x] \log\left(1 + \frac{\gamma}{\pi(-\infty, x]}\right) \int_x^m e^{-C} \tag{7.15}$$

and m is the median of π.

The results stated in this section are mainly taken from Chen (2002d).

Computation of the isoperimetric constant in dimension one

It is known that in general, the optimal constant $A_{\mathbb{B}}$ is not explicitly computable even in dimension one. However, the next results show that the isoperimetric constant $B_{\mathbb{B}}$ in dimension one is computable and coincides with those given in Chapter 6. Here we consider the ergodic case only. Then $\overline{A}_{\mathbb{B}}$ is controlled by $B_{\mathbb{B}}$ in view of Theorem 7.3 and Theorem 6.6 (3).

Corollary 7.5. Consider an ergodic birth–death process with birth rates b_i ($i \geqslant 0$) and death rates a_i ($i \geqslant 1$). Define (μ_n) as before. Then the isoperimetric constant $B_{\mathbb{B}}$ with Dirichlet boundary at 0 can be expressed as follows:

$$B_{\mathbb{B}} = \sup_{n \geqslant 1} \|I_{[n,\infty)}\|_{\mathbb{B}} \sum_{i=0}^{n-1} \frac{1}{\mu_i b_i}.$$

Proof. (a) We show that in the definition of $\mathrm{Cap}(K)$, one can replace "$f|_K \geqslant 1$" by "$f|_K = 1$."

Because $1 \in \mathscr{D}(D)$, we have $f \wedge 1 \in \mathscr{D}(D) \cap C_0(E)$ if so is f. Then the assertion follows from $D(f) \geqslant D(f \wedge 1)$.

(b) Next, let K_i ($i = 1, 2, \ldots, k$) be disjoint intervals with natural order. Set $K = [\min K_1, \max K_k]$, where $\min K = \min\{i : i \in K\}$ and $\max K = \max\{i : i \in K\}$. We show that

$$\frac{\|I_K\|_{\mathbb{B}}}{\mathrm{Cap}(K)} \geqslant \frac{\|I_{K_1+\cdots+K_k}\|_{\mathbb{B}}}{\mathrm{Cap}(K_1 + \cdots + K_k)}.$$

In other words, the ratio for a disconnected compact set is less than or equal to that of the corresponding connected one. For f with $f|_{K_1+\cdots+K_k} = 1$, the restriction of f on the intervals $[\max K_i, \min K_{i+1}]$ may not be a constant. Thus, if we define $\tilde{f} = f$ on K^c and $\tilde{f}|_K = 1$, then $D(\tilde{f}) \geqslant D(f)$, due to the character of birth–death processes. This means that $\mathrm{Cap}(K) \leqslant \mathrm{Cap}(K_1 + \cdots + K_k)$. In fact, equality holds, because for f with $f|_K = 1$, we must have $f|_{K_1+\cdots+K_k} = 1$, and so the inverse inequality is trivial. Since $K \supset K_1 + \cdots + K_k$ and (H3), we have $\|I_K\|_{\mathbb{B}} \geqslant \|I_{K_1+\cdots+K_k}\|_{\mathbb{B}}$. This proves the required assertion.

(c) Because of (b), to compute the isoperimetric constant, it suffices to consider the compact sets having the form $K = \{n, n+1, \ldots, m\}$ for $m \geqslant n \geqslant 1$. We now fix such a compact set K and compute $\mathrm{Cap}(K)$.

Given f with $f|_K = 1$ and $\mathrm{supp}(f) = \{1, \ldots, N\}$, $N \geqslant m$, we have

$$D(f) = \sum_{i=0}^{n-1} \mu_i b_i (f_{i+1} - f_i)^2 + \sum_{i=m}^{N} \mu_i b_i (f_{i+1} - f_i)^2, \qquad (7.16)$$

where $f_0 = 0$ and $f_{N+1} = 0$. Then

$$\frac{\partial D}{\partial f_j} = -2\mu_j b_j (f_{j+1} - f_j) + 2\mu_{j-1} b_{j-1} (f_j - f_{j-1})$$

$$= -2\mu_j b_j v_j + 2\mu_{j-1} b_{j-1} v_{j-1}, \qquad 1 \leqslant j \leqslant n-1 \ \text{ or } \ m+1 \leqslant j \leqslant N,$$

where $v_i = f_{i+1} - f_i$. The condition $\partial D / \partial f_j = 0$ gives us

$$v_j = \frac{\mu_{j-1} b_{j-1}}{\mu_j b_j} v_{j-1}, \qquad 1 \leqslant j \leqslant n-1 \ \text{ or } \ m+1 \leqslant j \leqslant N.$$

Hence

$$v_j = \frac{\mu_0 b_0 v_0}{\mu_j b_j}, \quad 0 \leqslant j \leqslant n-1, \ \text{ and } \ v_j = \frac{\mu_m b_m v_m}{\mu_j b_j}, \quad m \leqslant j \leqslant N. \quad (7.17)$$

Therefore

$$f_j = \sum_{i=0}^{j-1} v_j = \mu_0 b_0 v_0 \sum_{i=0}^{j-1} \frac{1}{\mu_i b_i}, \qquad 0 \leqslant j \leqslant n,$$

$$f_j = \sum_{i=m}^{j-1} v_j + 1 = \mu_m b_m v_m \sum_{i=m}^{j-1} \frac{1}{\mu_i b_i} + 1, \qquad m \leqslant j \leqslant N.$$

On the other hand, since $f_n = 1$ and $v_N = f_{N+1} - f_N = -f_N$, we get

$$1 = \mu_0 b_0 v_0 \sum_{i=0}^{n-1} \frac{1}{\mu_i b_i}, \qquad \frac{\mu_m b_m v_m}{\mu_N b_N} = -\mu_m b_m v_m \sum_{i=m}^{N-1} \frac{1}{\mu_i b_i} - 1.$$

Then

$$\mu_0 b_0 v_0 = \left(\sum_{i=0}^{n-1} \frac{1}{\mu_i b_i} \right)^{-1}, \qquad \mu_m b_m v_m = - \left(\sum_{i=m}^{N} \frac{1}{\mu_i b_i} \right)^{-1}. \tag{7.18}$$

Inserting (7.17) and (7.18) into (7.16), we obtain

$$D(f) = \sum_{i=0}^{n-1} \mu_i b_i v_i^2 + \sum_{i=m}^{N} \mu_i b_i v_i^2$$

$$= (\mu_0 b_0 v_0)^2 \sum_{i=0}^{n-1} \frac{1}{\mu_i b_i} + (\mu_m b_m v_m)^2 \sum_{i=m}^{N} \frac{1}{\mu_i b_i}$$

$$= \left(\sum_{i=0}^{n-1} \frac{1}{\mu_i b_i} \right)^{-1} + \left(\sum_{i=m}^{N} \frac{1}{\mu_i b_i} \right)^{-1}.$$

Since the process is recurrent, $\sum_{i=m}^{\infty} 1/\mu_i b_i = \infty$, we have

$$\mathrm{Cap}(K) = \inf\{D(f), f_0 = 0, f \text{ has finite support}, f|_K \geqslant 1\} = \left(\sum_{i=0}^{n-1} \frac{1}{\mu_i b_i} \right)^{-1},$$

which is independent of m. Therefore

$$B_{\mathbb{B}} = \sup_K \frac{\|I_K\|_{\mathbb{B}}}{\mathrm{Cap}(K)} = \sup_{1 \leqslant n \leqslant m} \frac{\|I_{[n,m]}\|_{\mathbb{B}}}{\mathrm{Cap}([n,m])} = \sup_{n \geqslant 1} \|I_{[n,\infty)}\|_{\mathbb{B}} \sum_{i=0}^{n-1} \frac{1}{\mu_i b_i}$$

as required. $\square$

We remark that once we know the solution f that minimizes $D(f)$, the proof (c) above can be done in a different way, as illustrated in the next proof.

Corollary 7.6. Consider an ergodic diffusion on $[0, \infty)$ with operator $L = a(x)\mathrm{d}^2/\mathrm{d}x^2 + b(x)\mathrm{d}/\mathrm{d}x$ and reflecting boundary. Suppose that the corresponding Dirichlet form $(D, \mathscr{D}(E))$ is regular, having core $\mathbb{C}_d[0, \infty)$: the set of all continuous functions with piecewise continuous derivatives and having compact support. Define $C(x) = \int_0^x b/a$ for $x \geqslant 0$. Then for Dirichlet boundary at 0, we have

$$B_{\mathbb{B}} = \sup_{x \geqslant 0} \|I_{(x,\infty)}\|_{\mathbb{B}} \int_0^x e^{-C}.$$

Proof. In view of (b) in the above proof, to compute the isoperimetric constant, we need only consider the compact $K = [n, m]$, $m > n$, $m, n \in \mathbb{R}_+$. Define

$$g(x) = \begin{cases} \int_0^x e^{-C} / \int_0^n e^{-C}, & \text{if } 0 \leqslant x \leqslant n, \\ 1, & \text{if } n \leqslant x \leqslant m, \\ 1 - \int_m^{x \wedge N} e^{-C} / \int_m^N e^{-C}, & \text{if } x \geqslant m. \end{cases}$$

We now show that $\mathrm{Cap}(K)$ can be computed in terms of this $g \in \mathbb{C}_d[0, \infty)$. Note that
$$\mathrm{Cap}(K) = \inf\{D(f) : f \in \mathbb{C}_d[0, \infty) : f|_K = 1\}.$$

Next, let $f_1 \in \mathbb{C}_d[0, n]$ with $f_1(0) = f_1(n) = 0$, $f_2 \in \mathbb{C}_d[m, N]$ with $f_2(m) = f_2(N) = 0$, and study the following variational problem with respect to ε_1 and ε_2:
$$H(\varepsilon_1, \varepsilon_2) = \int_0^n (g' + \varepsilon_1 f_1')^2 e^C + \int_m^N (g' + \varepsilon_2 f_2')^2 e^C.$$

If necessary, one may regard $\int_0^n$ as $\int_{0+}^{n-}$ and similarly for $\int_m^N$. Without loss of generality, assume that $f_k' \neq 0$. Otherwise, we can set $\varepsilon_k = 0$. Clearly, H should have a minimum in a bounded region. From $\partial H / \partial \varepsilon_k = 0$, it follows that

$$\varepsilon_1 = -\frac{\int_0^n g' f_1' e^C}{\int_0^n f_1'^2 e^C} = -\frac{\int_0^n f_1'}{\left(\int_0^n f_1'^2 e^C\right)\left(\int_0^n e^{-C}\right)} = -\frac{f_1(n) - f_1(0)}{\left(\int_0^n f_1'^2 e^C\right)\left(\int_0^n e^{-C}\right)} = 0,$$

$$\varepsilon_2 = -\frac{\int_m^N g' f_2' e^C}{\int_m^N f_2'^2 e^C} = -\frac{\int_m^N f_2'}{\left(\int_m^N f_2'^2 e^C\right)\left(\int_m^N e^{-C}\right)} = -\frac{f_2(N) - f_2(m)}{\left(\int_m^N f_2'^2 e^C\right)\left(\int_m^N e^{-C}\right)} = 0.$$

More precisely, if f' is discontinuous at $n_1, \ldots, n_k$, then
$$\int_0^n f' = \int_0^{n_1} f' + \cdots + \int_{n_k}^n f' = \left(f(n_1) - f(0)\right) + \cdots + \left(f(n) - f(n_k)\right)$$
$$= f(n) - f(0) = 0,$$

since f is continuous. Thus, $H(\varepsilon_1, \varepsilon_2)$ attains its minimum
$$D(g) = \left(\int_0^n e^{-C}\right)^{-1} + \left(\int_m^N e^{-C}\right)^{-1}$$

at $\varepsilon_1 = \varepsilon_2 = 0$. Moreover, due to the recurrence, we have $\int_m^\infty e^{-C} = \infty$. Combining these facts, we obtain $\mathrm{Cap}(K) = \left(\int_0^n e^{-C}\right)^{-1}$. The assertion now follows immediately. $\square$

For higher dimensions, the geometric aspect of estimating the isoperimetric constant or capacity has been developed extensively. Refer to V.G. Maz'ya (1985), I. Chavel (2001), D.R. Adams and L.I. Hedberg (1996), and references within. However, in view of the generality of Theorems 7.1–7.4, our knowledge about the isoperimetric constants is still rather limited (*open problem*).

7.2 Sketch of the proofs

The key to proving Theorem 7.1 is the following result.

Theorem 7.7. For a regular transient Dirichlet form $(D, \mathscr{D}(D))$, we have

$$\int_0^\infty \mathrm{Cap}(\{x \in E : |f(x)| \geqslant t\}) \mathrm{d}(t^2) \leqslant 4D(f), \qquad f \in \mathscr{D}(D) \cap C_0(E).$$

Proof. The simplified proof given here is due to M. Fukushima and T. Uemura (2003, Theorem 2.1). In this proof, one needs more knowledge about Dirichlet forms. Refer to the books by Fukushima et al. (1994), and by Ma and Röckner (1992).

Let $f \in \mathscr{D}(D) \cap C_0(E)$ and set $N_t = \{|f| \geqslant t\}$. Then there exist $e(t) \in \mathscr{D}_e(D)$, $e(t) = 1$, quasi everywhere (q.e.) on N_t, and a measure μ_t such that

$$\mathrm{Cap}(N_t) = \mu_t(N_t) = D(e(t)), \tag{7.19}$$

$$D(e(t), g) = \int_{N_t} g \mathrm{d}\mu_t, \qquad g \in \mathscr{D}_e(D). \tag{7.20}$$

Since $e(s) = 1$, $0 \leqslant s \leqslant t$, q.e. on N_t, by (7.20), we have

$$D(e(t), e(s)) = \mu_t(N_t) = D(e(t)). \tag{7.21}$$

Next, define $\|g\|_D^2 = D(g)$. By (7.21), we have

$$\|e(s) - e(t)\|_D^2 = \mathrm{Cap}(N_s) - \mathrm{Cap}(N_t).$$

Thus, $\|e(t)\|_D$ is measurable in t, since Cap is right-continuous. On the other hand, by (7.19), we have

$$\int_0^\infty \|e(t)\|_D \mathrm{d}t = \int_0^\infty \sqrt{\mathrm{Cap}(N_t)}\, \mathrm{d}t \leqslant \int_0^{\|f\|_\infty} \sqrt{\mathrm{Cap}(\mathrm{Supp}(f))}\, \mathrm{d}t$$

$$= \|f\|_\infty \sqrt{\mathrm{Cap}(\mathrm{Supp}(f))} < \infty.$$

We can define the Bochner integral $\psi = \int_0^\infty e(t)\mathrm{d}t$. Moreover,

$$D(\psi, g) = \int_0^\infty D(e(t), g)\mathrm{d}t, \qquad g \in \mathscr{D}_e(D). \tag{7.22}$$

With these preparations, we can now prove our assertion:

$$\int_0^\infty \mathrm{Cap}(N_t)\mathrm{d}t^2 = 2\int_0^\infty t\mathrm{Cap}(N_t)\mathrm{d}t = 2\int_0^\infty t\mu_t(N_t)\mathrm{d}t \quad \text{(by (7.19))}$$

$$\leqslant 2\int_0^\infty t \cdot \frac{1}{t}\int_{N_t} |f|\mathrm{d}\mu_t\mathrm{d}t \quad \text{(since } |f|/t \geqslant 1 \text{ on } N_t\text{)}$$

$$= 2\int_0^\infty D(e(t), |f|)\mathrm{d}t \quad \text{(by (7.20))}$$

$$= 2D(\psi, |f|) \quad \text{(by (7.22))}$$

$$\leqslant 2\sqrt{D(\psi)D(|f|)} \quad \text{(by the Schwarz inequality)}$$

$$\leqslant 2\sqrt{D(\psi)D(f)}.$$

But

$$D(\psi) = D\left(\int_0^\infty e(t)\mathrm{d}t, \int_0^\infty e(s)\mathrm{d}s \right)$$

$$= \int_0^\infty \int_0^\infty D(e(t), e(s))\mathrm{d}t\mathrm{d}s = 2 \int_0^\infty \mathrm{d}t \int_0^t D(e(t), e(s))\mathrm{d}s$$

$$= 2 \int_0^\infty \mathrm{d}t \int_0^t D(e(t))\mathrm{d}s \quad \text{(by (7.21))}$$

$$= 2 \int_0^\infty tD(e(t))\mathrm{d}t = 2 \int_0^\infty t\mathrm{Cap}(N_t)\mathrm{d}t \quad \text{(by (7.19))}$$

$$= \int_0^\infty \mathrm{Cap}(N_t)\mathrm{d}t^2,$$

and so the required assertion follows. □

Having Theorem 7.7 in mind, the proof of Theorem 7.2 (which is more general than Theorem 7.1) is quite standard. Here we follow the proof of V.A. Kaimanovich (1992, Theorem 3.1).

Proof of Theorem 7.2. Let $f \in \mathscr{D}(D) \cap C_0(E)$ and set $N_t = \{|f| \geqslant t\}$. Since N_t is compact, by (H_1), $I_{N_t} \in \mathbb{B}$. Next, since $|f| \leqslant \|f\|_\infty I_{\{\mathrm{Supp}(f)\}}$, by (H_1) and (H_2), $f^2 \in \mathbb{B}$. Note that

$$\int_0^\infty I_{N_t}\mathrm{d}(t^2) = 2 \int_0^\infty tI_{\{|f|\geqslant t\}}\mathrm{d}t = 2 \int_0^{|f|} t\,\mathrm{d}t = f^2 \quad \text{(coarea formula)}.$$

Since N_t is compact, by the definition of $B_\mathbb{B}$ and Theorem 7.7, we obtain

$$\|f^2\|_\mathbb{B} = \sup_{g \in \mathscr{G}} \int_E f^2 g\mathrm{d}\mu = \sup_{g \in \mathscr{G}} \int_E \left(\int_0^\infty I_{N_t}\mathrm{d}(t^2) \right) g\mathrm{d}\mu$$

$$= \sup_{g \in \mathscr{G}} \int_0^\infty \left(\int_E I_{N_t} g\mathrm{d}\mu \right) \mathrm{d}(t^2) \leqslant \int_0^\infty \|I_{N_t}\|_\mathbb{B}\mathrm{d}(t^2)$$

$$\leqslant B_\mathbb{B} \int_0^\infty \mathrm{Cap}(N_t)\mathrm{d}(t^2) \leqslant 4B_\mathbb{B}D(f).$$

This implies that $A_\mathbb{B} \leqslant 4B_\mathbb{B}$.

Next, for every compact K and any function f with $f|_K \geqslant 1$, we have

$$\|I_K\|_\mathbb{B} \leqslant \|f^2\|_\mathbb{B} \leqslant A_\mathbb{B}D(f).$$

Thus,

$$\|I_K\|_\mathbb{B} \leqslant A_\mathbb{B} \inf\{D(f) : f \in \mathscr{D}(D) \cap C_0(E), f|_K \geqslant 1\} = A_\mathbb{B}\mathrm{Cap}(K).$$

Dividing both sides by $\mathrm{Cap}(K)$ and taking the supremum with respect to K, it follows that $B_\mathbb{B} \leqslant A_\mathbb{B}$, and the proof is completed. □

The proof of Theorem 7.3 is based on the splitting technique and the following result.

Proposition 7.8. Let $(E, \mathscr{E}, \pi)$ be a probability space and $(\mathbb{B}, \|\cdot\|_{\mathbb{B}})$ a normed linear space, satisfying (H1) and (H2), of Borel measurable functions on $(E, \mathscr{E}, \pi)$.

(1) Let c_1 be given by (7.6). Then

$$\|\bar{f}^2\|_{\mathbb{B}} \leqslant \left(1 + \sqrt{c_1 \|1\|_{\mathbb{B}}}\,\right)^2 \|f^2\|_{\mathbb{B}}.$$

(2) Let $c_2(G)$ be given by (7.7). If $c_2(G)\pi(G)\|1\|_{\mathbb{B}} < 1$, then for every f with $f|_{G^c} = 0$, we have

$$\|f^2\|_{\mathbb{B}} \leqslant \|\bar{f}^2\|_{\mathbb{B}} \Big/ \Big[1 - \sqrt{c_2(G)\pi(G)\,\|1\|_{\mathbb{B}}}\,\Big]^2.$$

Proof. Note that $\pi(f)^2 \leqslant \pi(f^2) \leqslant c_1 \|f^2\|_{\mathbb{B}}$. For all $p, q > 1$ with $(p-1)(q-1) = 1$, we have $(x+y)^2 \leqslant px^2 + qy^2$, and so

$$\|\bar{f}^2\|_{\mathbb{B}} \leqslant p\|f^2\|_{\mathbb{B}} + q\pi(f)^2\|1\|_{\mathbb{B}} \leqslant \big(p + c_1 q\|1\|_{\mathbb{B}}\big)\|f^2\|_{\mathbb{B}}.$$

Minimizing the right-hand side with respect to p and q, we get the first assertion. The proof of the second one is similar. $\quad\square$

The proof of Theorem 7.4 uses $\mathscr{G}_1 = \{g \geqslant 0 : \int e^g \mathrm{d}\pi \leqslant e^2 + 1\}$ and $\mathscr{G}_2 = \{g \geqslant 0 : \int e^g \mathrm{d}\pi \leqslant 1\}$, respectively, for the upper and lower estimates.

7.3   Comparison with Cheeger's method

A typical case for which one needs the general form of Poincaré-type inequality is the *F-Sobolev inequality* [cf. F.Y. Wang (2000a), F.Z. Gong and F.Y. Wang (2002)]:

$$\int_E f^2 F(f^2)\mathrm{d}\mu \leqslant A_F D(f), \qquad f \in \mathscr{D}(D) \cap C_0(E). \qquad (7.23)$$

Theorem 7.9. Let $F : \mathbb{R}_+ \to \mathbb{R}_+$ satisfy the following conditions:

(1) $2F' + xF'' \geqslant 0$ on $[0, \infty)$.
(2) $\lim_{x \to 0} F(x) = 0$ and $\lim_{x \to \infty} F(x) = \infty$.
(3) $\sup_{x \gg 1} xF'(x)/F(x) < \infty$.

Then Theorem 7.2 is valid for the Orlicz space with N-function $\Phi(x) = |x|F(|x|)$. Furthermore, the *isoperimetric constant* is given by

$$B_\Phi = \sup_{\text{compact } K} \frac{\alpha_*(K)^{-1} + \mu(K)F(\alpha_*(K))}{\mathrm{Cap}(K)}, \qquad (7.24)$$

where $\alpha_*(K)$ is the minimal positive root of $\alpha^2 F'(\alpha) = \mu(K)$.

To compare this result with the generalized Cheeger's method, let us recall the symmetric form

$$D^{(\alpha)}(f) = \frac{1}{2} \int J^{(\alpha)}(\mathrm{d}x, \mathrm{d}y)[f(y) - f(x)]^2 + \int K^{(\alpha)}(\mathrm{d}x)f(x)^2, \qquad \alpha \geqslant 0,$$

as defined in Section 4.5, satisfying the normalizing condition

$$[J^{(1)}(\mathrm{d}x, E) + K^{(1)}(\mathrm{d}x)]/\pi(\mathrm{d}x) \leqslant 1.$$

Next, define

$$\lambda_0^{(\alpha)} = \inf \left\{ D^{(\alpha)}(f) : \|f\| = 1 \right\}, \qquad c_1 = \sup_{x \geqslant 0} \frac{x F'_{\pm}(x)}{F(x)} < \infty,$$

where $F'_{\pm}$ denotes the right and left derivatives of F.

Theorem 7.10 (Chen, 2000a). Suppose that F is a continuous increasing function on $[0, \infty)$ with $F(0) = 1$ such that F' is piecewise continuous and $c_1 < \infty$. Then the optimal A_F in (7.23) satisfies

$$A_F \geqslant \sup_{\pi(G) > 0} \frac{\pi(G) F\big(\pi(G)^{-1}\big)}{J(G \times G^c) + K(G)}, \tag{7.25}$$

$$A_F \leqslant \sup_{\pi(G) > 0} \frac{4(1 + c_1^2)\big(2 - \lambda_0^{(1)}\big)\pi(G)^2 F\big(\pi(G)^{-1}\big)}{\big[J^{(1/2)}(G \times G^c) + K^{(1/2)}(G)\big]^2}. \tag{7.26}$$

Comparing (7.26) with (7.24) and (7.4), it is clear that Cheeger's method is more explicit (without using capacity) but less precise qualitatively than the capacitary method. See also Example 4.19. In fact, as pointed out by A. Grigor'yan (1999), for the Laplacian on a Riemannian manifold, the capacitary result Theorem 7.1 implies Cheeger's lower bound for the first Dirichlet eigenvalue.

Moreover, it should also be clear that these two methods are much less explicit than the one-dimensional results studied in Chapter 6. The main reason is that in the latter case, our starting point given in Section 6.2 is much more precise than Theorem 7.1.

Comments on F-Sobolev inequalities

The F-Sobolev inequalities are important in the following sense.

(1) It was proved by F.Y. Wang (2000a), and extended by F.Z. Gong and F.Y. Wang (2002), that if the essential spectrum of the generator of the process is empty, then F-Sobolev inequality holds for a suitable function F. Here and in what follows, we are talking about the ergodic case only. The converse assertion holds once there exists a transition probability density with respect to the reversible probability measure.

(2) The F-Sobolev inequalities were used by F.Y. Wang (2000b) to estimate the higher eigenvalues λ_j, $j \geqslant 1$, not only the first one.

(3) Recently, F.Y. Wang (2004a; 2004b) has proved that the inequalities for suitable F are equivalent to the Bochner-type inequalities, which are additive and hence are useful in the infinite-dimensional situation to study the perturbation of independent systems.

(4) Clearly, one may regard some F-Sobolev inequalities as interpolations between the logarithmic Sobolev inequality and the Poincaré inequality (cf. F. Wang (2003a)). It is meaningful, especially for Markov jump processes, to give some sufficient conditions for exponential convergence in entropy.

7.4 General convergence speed

The aim of this section is to derive an inequality for general convergence speed for reversible Markov processes.

Let $\xi(t) \downarrow 0$ as $t \uparrow \infty$. Consider the general decay

$$\|P_t \bar{f}\|^2 = \|P_t f - \pi(f)\|^2 \leqslant CV(f)\xi(t), \tag{7.27}$$

where C is a constant and V is a suitable functional to be discussed below more carefully. On the right-hand side of (7.27), the variables t and f are separated. Of course, we are mainly looking for such a simple control, rather than a complicated expression. Now, a question arises. What is an analogue of the Poincaré inequality for such decay?

Note that if we define

$$V(f) = \sup_{t>0} \xi(t)^{-1} \|P_t \bar{f}\|^2,$$

then it is clear that the functional V should be homogeneous in degree two,

$$V(\alpha f + \beta) = \alpha^2 V(f), \tag{7.28}$$

which is the main condition we need for the functional V. Next, if we define $V(f) = \|\bar{f}\|^2$, then we will return to the exponential convergence $\xi(t) \sim e^{-\varepsilon t}$.

We now continue our search for an analogue of the Poincaré inequality. It is a simple fact that

$$0 \leqslant \frac{1}{t}(f - P_t f, f) = \int_{[0,\infty)} \frac{1 - e^{-\alpha t}}{t} d(E_\alpha f, f) \uparrow D(f) \qquad \text{as } t \downarrow 0,$$

where $\{E_\alpha\}_{\alpha \geqslant 0}$ is the spectral family of the semigroup, since $(1 - e^{-\alpha t})/t \uparrow$ as $t \downarrow$. Because $(P_t f, f)$ is decreasing in t, there exists a nonnegative, increasing function η ($\eta(r) = r$, for example) such that $\eta(r)/r \uparrow 1$ as $r \downarrow 0$, and then for each f with $\pi(f) = 0$, we have

$$\|f\|^2 - \eta(t)D(f) \leqslant (P_t f, f) \leqslant \|P_t f\| \, \|f\| \leqslant \sqrt{CV(f)\xi(t)} \, \|f\|,$$

by assumption. Solving this inequality in $\|f\|$, we get

$$\|f\| \leqslant \frac{1}{2}\left[\sqrt{CV(f)\xi(t)} + \sqrt{CV(f)\xi(t) + 4\eta(t)D(f)}\right].$$

Therefore, we obtain the required inequality

$$\|\bar{f}\|^2 \leqslant 2\eta(t)D(f) + C'V(f)\xi(t), \tag{7.29}$$

where C' is a constant.

In particular, if we set $\xi(t) = t^{1-q}\,(q > 1)$ and $\eta(t) = t$, then by optimizing the right-hand side of (7.29) with respect to t, we obtain the following *Liggett–Stroock inequality*:

$$\|\bar{f}\|^2 \leqslant C''D(f)^{1/p}V(f)^{1/q}, \tag{7.30}$$

where $1/p + 1/q = 1$. Refer to the proof of Theorem 5.10.

7.5 Two functional inequalities

Let us return to (7.29). By a transform if necessary, without loss of generality, we may assume that $V(f) = 1$. Then the right-hand side of (7.29) becomes $2\eta(t)D(f) + C'\xi(t)$. Define $\Phi(x) = \inf_{r>0}[2\eta(r)x + C'\xi(r)]$, $x \geqslant 0$. Then inequality (7.29) takes the following more compact form:

$$\|\bar{f}\|^2 \leqslant \Phi(D(f)), \qquad V(f) = 1.$$

However, this inequality is not practical, since Φ is not explicit. The trick now is to regard t as a new parameter r. Then we can rewrite (7.29) as

$$\|\bar{f}\|^2 \leqslant 2\eta(r)D(f) + C'V(f)\xi(r), \qquad r > 0 \tag{7.31}$$

Before moving further, we show that it is easy to go back to (7.27) from (7.31). Let $\pi(f) = 0$ and set $f_t = P_t f$, $F_t = \pi(f_t^2)$. Assuming that the semigroup is V-contractive in the sense that $V(f_t) \leqslant V(f)$ for all f, then by (7.31), we have

$$F_t' = -2D(f_t) \leqslant \frac{C'V(f_t)\xi(r)}{\eta(r)} - \frac{1}{\eta(r)}F_t$$
$$\leqslant \frac{C'V(f)\xi(r)}{\eta(r)} - \frac{1}{\eta(r)}F_t, \qquad t \geqslant 0,\ r > 0.$$

By Corollary A.2,

$$F_t \leqslant F_0 e^{-t/\eta(r)} + \frac{C'V(f)\xi(r)}{\eta(r)}\int_0^t e^{-(t-s)/\eta(r)}\mathrm{d}s$$
$$\leqslant \|f\|^2 e^{-t/\eta(r)} + C'V(f)\xi(r), \qquad t \geqslant 0,\ r > 0.$$

Suppose that $\eta(r) \uparrow$ as $r \uparrow$ and define

$$r(t) = \inf \left\{ r > 0 : -\eta(r) \log \xi(r) \leqslant t \right\},$$
$$\tilde{\xi}(t) = \xi(r(t)), \qquad \tilde{V}(f) = C'V(f) + \|\bar{f}\|^2.$$

Then as $t \uparrow \infty$, $r(t) \uparrow \infty$ and so $\tilde{\xi}(t) \downarrow 0$. Moreover,

$$\|P_t\bar{f}\|^2 = F_t \leqslant \tilde{V}(f)\tilde{\xi}(t), \qquad t > 0,$$

which gives us the required decay (7.27).

As an application of (7.31), by setting $\eta(r) = r/2$ and $V(f) = \pi(|f|)^2$, F.Y. Wang (2000a) introduced the so-called *super-Poincaré inequality*

$$\|f\|^2 \leqslant rD(f) + \beta(r)\pi(|f|)^2, \qquad \forall r > 0, \tag{7.32}$$

where $\beta(r) \downarrow$ as $r \uparrow$. The reason for choosing $V(f) = \pi(|f|)^2$ comes from the fact that the ordinary Poincaré inequality is equivalent to

$$\|f\|^2 \leqslant CD(f) + \pi(|f|)^2,$$

for some constant $C > 0$. It was also proved in F.Y. Wang (2000a), F.Z. Gong and F.Y. Wang (2002) that (7.32) is equivalent to the *F-Sobolev inequality* (ergodic case)

$$\int_E f^2 F(f^2)\mathrm{d}\pi \leqslant CD(f), \qquad \|f\| = 1, \tag{7.33}$$

where F satisfies $\sup_{r \in (0,1]} |rF(r)| < \infty$ and $\lim_{r \to \infty} F(r) = \infty$. The equivalence of (7.32) and (7.33) provides us not only a more intrinsic understanding of these two inequalities but also that we can use either of them according to our convenience. For instance, as shown in the previous paragraph, (7.32) describes the decay of the semigroup, and by Theorem 7.9 (and Corollaries 7.5 and 7.6), we obtain a criterion for (7.33).

Next, we are going to look for a slower convergence. Again, we start from (7.31). Note that the use of the new parameter r is mainly for our convenience. Actually, the right-hand side of (7.31) plays a role only at a point r_0 at which the right-hand side of (7.31) achieves the infimum. Note also that when $V = \|\cdot\|_\infty^2$ and $\eta(r) = r$, for instance, we have $\lim_{r \to 0} \beta(r) = \infty$. This singularity is sometimes reasonable (in the case of algebraic convergence for instance), and makes (7.32) stronger for smaller r. It may cost some difficulty in the applications and may exclude some slower convergence. For this, we need further consideration. It is clear that using a different pair (η, ξ) on the right-hand side of (7.31), one may obtain the same (or equivalent) inequality. Based on these observations, by exchanging the position of the functions r and $\beta(r)$, M. Röckner and F.Y. Wang (2001) introduced the so-called *weaker-Poincaré inequality* (WPI) as follows:

$$\|\bar{f}\|^2 \leqslant \alpha(r)D(f) + rV(f), \qquad \forall r > 0, \tag{7.34}$$

where $\alpha(r) > 0$, $\alpha(r) \downarrow$ as $r \uparrow$ on $(0, \infty)$. It was proved in the quoted paper that when $V = \|\cdot\|_\infty^2$, (7.34) is equivalent to the Kusuoka–Aida weak spectral gap property [cf. S. Aida (1998)]:

> For every sequence $\{f_n\} \subset \mathscr{D}(D)$ with $\pi(f_n) = 0$, $\|f_n\| \leqslant 1$, and $\lim_{n\to\infty} D(f_n) = 0$, we have $f_n \to 0$ in $\mathbb{P}$.

Here is one of the main results about WPI.

Theorem 7.11 (M. Röckner and F.Y. Wang, 2001).

(1) If $\|P_t\bar{f}\|^2 \leqslant \xi(t)V(f)$ for all $t > 0$ and $\xi(t) \downarrow 0$ as $t \uparrow \infty$, then WPI holds with the same V and

$$\alpha(r) = 2r \inf_{s>0} \frac{1}{s} \xi^{-1}\big(se^{1-s/r}\big), \qquad \xi^{-1}(t) := \inf\{r > 0 : \xi(r) \leqslant t\}.$$

(2) If WPI holds and $V(P_t f) \leqslant V(f)$ for all $t \geqslant 0$, then

$$\|P_t\bar{f}\|^2 \leqslant \xi(t)\big[V(f) + \|\bar{f}\|^2\big], \qquad \forall t > 0,$$

where $\xi(t) = \inf\{r > 0 : -\alpha(r)\log r/2 \leqslant t\}$.

To establish WPI, the generalized Cheeger's (isoperimetric) method is also very powerful [cf. Röckner and Wang (2001)].

The functional inequalities and much more material are explored in a monograph by F.Y. Wang (2004b).

7.6 Algebraic convergence

We now return to the Liggett–Stroock inequality (7.30). Instead of stating some general but technical theorems, we introduce only two examples, from which one can see the role played by different functionals V. Examples are the leading light of our study. Every meaningful theorem should be supported by a good example.

Example 7.12 (Chen and Y.Z. Wang, 2003). Consider the birth–death process with rates $a_i = b_i = i^\gamma$ for large i ($i \gg 1$) and $\gamma > 0$. The process is ergodic iff $\gamma > 1$.

(1) Let $\gamma > 1$. Then $\lambda_1 > 0$ iff $\gamma \geqslant 2$. In other words, with respect to $V(f) = \|\bar{f}\|^2$, the process has L^2-algebraic decay iff $\gamma \geqslant 2$.

(2) Let $\gamma \in (1, 2)$. Then with respect to V^s: $V^s(f) = \sup_{k\geqslant 0}\big[(k+1)^s|f_{k+1} - f_k|\big]^2$, where $0 < s \leqslant \gamma - 1$, the process has L^2-algebraic decay iff $\gamma \in (5/3, 2)$.

(3) Let $\gamma \in (1, 2)$. Then with respect to V_0: $V_0(f) = \sup_{i\neq j}(f_i - f_j)^2$, the process has L^2-algebraic decay for all $\gamma \in (1, 2)$.

Example 7.13 (Chen and Y.Z. Wang, 2003). Consider the birth–death process with rates $a_i = 1$, $b_i = 1 - \gamma/i$ for $i \gg 1$ and $\gamma > 0$. The process is ergodic iff $\gamma > 1$. We now let $\gamma > 1$.

(1) In general, we have $\lambda_1 = 0$ for all $\gamma > 1$.
(2) With respect to V^0: $V^0(f) = \sup_{k \geqslant 0} |f_{k+1} - f_k|^2$, the process has L^2-algebraic decay iff $\gamma > 3$.
(3) With respect to V_0: $V_0(f) = \sup_{i \neq j} (f_i - f_j)^2$, the process has L^2-algebraic decay for all $\gamma > 1$.

The beginning step of the proof

Note that the functionals V^s and V_0 are all of Lipschitz type with respect to some distance ρ: $\mathrm{Lip}_\rho(f)^2 = \sup_{x \neq y} \left| \frac{f(x) - f(y)}{\rho(x,y)} \right|^2$. As we have shown before, the distances play a very important role in the study of the spectral gap. The same happens in the present situation. Here we show a few of the lines in the original proof. First, recall the Liggett–Stroock inequality:

$$\|\bar{f}\|^2 \leqslant CD(f)^{1/p} V(f)^{1/q}.$$

Clearly, one has to use Hölder's inequality:

$$\mathrm{Var}(f) = \frac{1}{2} \sum_{i,j} \pi_i \pi_j (f_j - f_i)^2 = \sum_{\{i,j\}} \pi_i \pi_j (f_j - f_i)^2$$

$$\leqslant \left\{ \sum_{\{i,j\}} \pi_i \pi_j \left(\frac{f_j - f_i}{\phi_{ij}} \right)^2 \right\}^{1/p} \left\{ \sum_{\{i,j\}} \pi_i \pi_j \left(\frac{f_j - f_i}{\phi_{ij}^\delta} \right)^2 \phi_{ij}^{2(q+\delta-1)} \right\}^{1/q}.$$

Roughly speaking, in the last line, ϕ_{ij} represents a distance between i and j. The last inequality indicates a good use of the Hölder inequality. One may continue the proof by estimating the right-hand side.

General ergodic Markov chains

Finally, we consider a general ergodic Markov chain with transition probability matrix $(p_{ij}(t))$ on a countable set, $\pi_j := \lim_{t \to \infty} p_{ij}(t) > 0$, but looking for polynomial convergence only. Define

$$d_{ij}^{(n)} = \int_0^\infty t^n \big(p_{ij}(t) - \pi_j \big) \mathrm{d}t, \qquad n \in \mathbb{Z}_+,$$

$$m_{ij}^{(n)} = \mathbb{E}_i \sigma_j^n, \qquad \sigma_j = \inf\{t \geqslant \tau_1 : X_t = j\},$$

where τ_1 is the first jumping time of the chain.

Theorem 7.14 (Y.H. Mao, 2003). For an irreducible and ergodic Markov chain, the following assertions hold:

(1) $\left|d_{ij}^{(n)}\right| < \infty$ for all i, j iff $m_{jj}^{(n)} < \infty$ for some (equivalently, all) j.

(2) If $m_{jj}^{(n)} < \infty$, then $p_{ij}(t) - \pi_j = o\!\left(t^{-n+1}\right)$ as $t \to \infty$.

(3) $m_{jj}^{(n)} < \infty$ iff the inequalities

$$\begin{cases} \sum_k q_{ik} y_k \leqslant -n\, m_{ij}^{(n-1)}, & i \neq j, \\ \sum_{k \neq j} q_{jk} y_k < \infty, \end{cases} \tag{7.35}$$

have a finite nonnegative solution (y_i).

It is remarkable to note that equality in (7.35) indeed holds for $\big(y_i = m_{ij}^{(n)}\big)$. In other words, the nth moments are expressed in terms of the $(n-1)$th moments, and hence depend on all of the mth moments ($m \leqslant n - 1$). This indicates the complexity of a criterion, in general, for algebraic convergence, especially for irreversible processes. Recall that in the previous sections, general convergence speed depends heavily on the Dirichlet form (reversibility) and so is not available for the general irreversible situation.

7.7 General (irreversible) case

The discussion at the end of the last section leads to the following open problem.

Open Problem 7.15. What should be a criterion for slower convergence of a general time-continuous Markov process in terms of its operator?

As a reference, here we consider the time-discrete case. Let $(E, \mathscr{E})$ be a general measurable space and $(X_n)_{n \geqslant 0}$ a Markov process with state space $(E, \mathscr{E})$. Define the return time $\sigma_B = \inf\{n \geqslant 1 : X_n \in B\}$. Next, define

$$\mathscr{R}_0 = \left\{ r(n)_{n \in \mathbb{Z}_+} : 2 \leqslant r(n) \uparrow,\ \frac{\log r(n)}{n} \downarrow 0 \text{ as } n \uparrow \infty \right\},$$

$$\mathscr{R} = \left\{ r(n)_{n \in \mathbb{Z}_+} : \exists r_0 \in \mathscr{R}_0 \text{ such that } \varliminf_{n \to \infty} \frac{r(n)}{r_0(n)} > 0 \text{ and } \varlimsup_{n \to \infty} \frac{r(n)}{r_0(n)} < \infty \right\}.$$

Roughly speaking, $\mathscr{R}_0$ is the set of monotone speeds, and $\mathscr{R}$ is the perturbations of the elements in $\mathscr{R}_0$. Here is the general answer to Problem 7.15 in the context of time-discrete Markov processes.

For simplicity, one may think of the petite set and $\mathscr{E}^+$ below, respectively, as the compact set K and the Borel sets in $\mathbb{R}^d$ having positive Lebesgue measure.

Theorem 7.16 (P. Tuominen and R.L. Tweedie, 1994). Let a Markov process be irreducible and aperiodic. Fix $r \in \mathscr{R}$. Then

$$\lim_{n \to \infty} r(n) \| P_n(x, \cdot) - \pi \|_{\mathrm{Var}} = 0$$

for all x in the set

$$\left\{ x : \mathbb{E}_x \sum_{k=0}^{\sigma_B - 1} r(k) < \infty, \ \forall B \in \mathscr{E}^+ \right\},$$

provided one of the following equivalent conditions holds:

(1) There exists a petite set K such that $\mathbb{E}_x \sum_{k=0}^{\sigma_K - 1} r(k) < \infty$ for all $x \in K$.

(2) There exist $(f_n)_{n \in \mathbb{Z}_+} \colon E \to [1, \infty]$, a petite set K, and a constant b such that $\sup_{x \in K} f_0 < \infty$, $\{f_1 < \infty\} \subset \{f_0 < \infty\}$, and furthermore,

$$Pf_{n+1} \leqslant f_n - r(n) + b\, r(n) I_K, \qquad n \in \mathbb{Z}_+.$$

(3) There exists $A \in \mathscr{E}^+$ such that

$$\sup_{x \in A} \mathbb{E}_x \sum_{k=0}^{\sigma_B - 1} r(k) < \infty, \qquad \forall B \in \mathscr{E}^+.$$

It is proved by S.F. Jarner and G.O. Roberts (2002) that in the polynomial case, one can use a single f instead of the sequence (f_n) used above.

Finally, we mention that there is quite a number of publications devoted to the study on the ergodic convergence rates for time-discrete Markov processes, not included in this book. A natural extension of the classical types of ergodicity, largely developed in S.P. Meyn and R.L. Tweedie (1993b, Chapter 16), is the V-uniform ergodicity: $\|P^n - \pi\|_V \to 0$ as $n \to \infty$, where

$$\|P_1(x, \cdot) - P_2(x, \cdot)\|_V = \sup_{x \in E} V(x)^{-1} \sup_{|f| \leqslant V} |(P_1 - P_2)f(x)|$$

for a given function $V \in \mathscr{E} \colon 1 \leqslant V < \infty$. Intuitively, the function V comes from the equivalent "*draft condition*"

$$PV_0 \leqslant -\alpha V_0 + \beta I_B$$

for some petite set B and constants $\alpha > 0$ and $\beta < \infty$, where $V_0 \geqslant 1$ is equivalent to $V \colon c^{-1}V \leqslant V_0 \leqslant cV$ for some constant $c > 0$. The exponential ergodicity corresponds to a type of V-uniform ergodicity for a suitable V. In view of the drift condition, for smaller V, the V-uniform ergodicity becomes stronger. In particular, when $V = 1$ (the smallest one), we come back to the uniform (strong) ergodicity (cf. Section 1.4). Refer to I. Kontoyiannis and S.P. Meyn (2003), G.O. Roberts and J.S. Rosenthal (1997), J.S. Rosenthal (2002), L.M. Wu (2004), and references within for recent progress. Two results of the relation between the V-uniform ergodicity and the inequalities studied in the book are given in Figure 1.1 or Theorems 8.6, 8.8, and 8.13 in the next chapter. In general, they are not comparable, as shown by Examples 8.2–8.4. This is clear, since the inequalities deal with the norm of mappings from one normed linear space to another, not necessarily from one to itself.

Chapter 8

A Diagram of Nine Types of Ergodicity

This chapter consists of three sections. In the first section, we recall three basic inequalities and their ergodic meaning. Then we recall three traditional types of ergodicity. We compute the exact optimal constants of the inequalities or exact ergodic rates for the traditional ergodicity in the simplest case, that the state space consists of two points only. Next, we turn to study the diagram of nine types of ergodicity presented in Theorem 1.9. In the second section, we discuss the value of the diagram, consider its powerful applications, and make some comments on its completeness. The comparison of the different types of ergodicity are shown by some simple examples. The last section is devoted to the proof of the diagram with some addition.

8.1 Statements of results

Ergodicity by means of three inequalities

Three basic inequalities. Let $(E, \mathscr{E}, \pi)$ be a probability space and $(D, \mathscr{D}(D))$ a Dirichlet form. Denote by Var the variational norm: $\mathrm{Var}(f) = \|f\|^2 - \pi(f)^2$. The three inequalities mentioned several times before are as follows:

Poincaré inequality : $\quad \mathrm{Var}(f) \leqslant \lambda_1^{-1} D(f)$,

Nash inequality : $\quad \mathrm{Var}(f)^{1+2/\nu} \leqslant \eta^{-1} D(f) \|f\|_1^{4/\nu}, \quad \nu > 0$,

Logarithmic Sobolev inequality : $\quad \int f^2 \log\left(f^2/\|f\|^2\right) \mathrm{d}\pi \leqslant 2\sigma^{-1} D(f)$,

where $\| \cdot \|_p$ is the L^p-norm and $\| \cdot \| = \| \cdot \|_2$ We remark that for the Nash inequality, it holds as well if $\|f\|_1$ is replaced by $\|f\|_r$ for all $r \in (1, 2)$. To save on notation, here λ_1, η, and σ denote the optimal constants in the inequalities.

Ergodicity by means of the inequalities.   Let (P_t) be the semigroup determined by the Dirichlet form $(D, \mathscr{D}(D))$: $P_t = e^{tL}$ formally. Then

- Poincaré inequality $\Longleftrightarrow \text{Var}(P_t f) \leqslant \text{Var}(f)e^{-2\lambda_1 t}$.
- Logarithmic Sobolev inequality $\Longrightarrow$ exponential convergence in entropy:

$$\text{Ent}(P_t f) \leqslant \text{Ent}(f)e^{-2\sigma t},$$

  where $\text{Ent}(f) = \pi(f \log f) - \pi(f) \log \|f\|_1$. Actually, one can replace "$\Longrightarrow$" with "$\Longleftrightarrow$" in the context of diffusions.
- Nash inequality $\Longleftrightarrow \text{Var}(P_t f) \leqslant \left(\dfrac{\nu}{4\eta t}\right)^{\nu/2} \|f\|_1^2$.

At first glance, one may think that the Nash inequality is the weakest one, since the convergence speed is slower. However, this is incorrect, as shown in Figure 1.1. A result due to L. Gross (1976) says that

Logarithmic Sobolev inequality $\Longrightarrow$ Exponential convergence in entropy
$$\Longrightarrow \text{Poincaré inequality.}$$

Three traditional types of ergodicity

Recall that
$$\|\mu - \nu\|_{\text{Var}} = 2 \sup_{A \in \mathscr{E}} |\mu(A) - \nu(A)|.$$

Here are the three traditional types of ergodicity.

$$\textit{Ordinary ergodicity}: \qquad \lim_{t\to\infty} \|P_t(x, \cdot) - \pi\|_{\text{Var}} = 0,$$

$$\textit{Exponential ergodicity}: \qquad \lim_{t\to\infty} e^{\hat\alpha t} \|P_t(x, \cdot) - \pi\|_{\text{Var}} = 0,$$

$$\textit{Strong ergodicity}: \qquad \lim_{t\to\infty} \sup_x \|P_t(x, \cdot) - \pi\|_{\text{Var}} = 0$$

$$\Longleftrightarrow \lim_{t\to\infty} e^{\hat\beta t} \sup_x \|P_t(x, \cdot) - \pi\|_{\text{Var}} = 0,$$

where $\hat\alpha$ and $\hat\beta$ denote the largest positive constants in the corresponding equality to save on notation. For these types of ergodicity, there is a classical theorem, which will be proved in the last section of this chapter:

Strong ergodicity $\Longrightarrow$ Exponential ergodicity $\Longrightarrow$ Ordinary ergodicity.

As an illustration, we compute the optimal constants in the inequalities and the exact rates of ergodic convergence for the simplest example.

Example 8.1. Let $E = \{0, 1\}$ and consider the Q-matrix

$$Q = \begin{pmatrix} -b & b \\ a & -a \end{pmatrix}.$$

Then the Nash constant and the logarithmic Sobolev constant are given by

$$\eta = (a + b)\left(\frac{a \wedge b}{a \vee b}\right)^{1/q}, \qquad \sigma = \frac{2(a \vee b - a \wedge b)}{\log[(a \vee b)/(a \wedge b)]},$$

respectively. The rates of the L^2-exponential convergence, the exponentially ergodic convergence, and the strongly ergodic convergence (must be exponential) are all equal to $\lambda_1 = a + b$. The results are summarized in Table 8.1.

Table 8.1 The optimal constants of the inequalities for two points

$\lambda_1 = \hat{\alpha} = \hat{\beta}$	Log Sobolev σ	Nash η
$a + b$	$\dfrac{2(a \vee b - a \wedge b)}{\log a \vee b - \log a \wedge b}$	$(a + b)\left(\dfrac{a \wedge b}{a \vee b}\right)^{1 + 2/\nu}$

In general, we have $\lambda_1 \geqslant \sigma$ (Theorem 8.7), but $\lambda_1 = \sigma$ for this example iff $a = b$ (where σ is regarded as the limit as $a \to b$).

Proof. (a) Note that

$$P(t) = (p_{ij}(t)) = e^{tQ} = \frac{1}{a+b}\begin{pmatrix} a + be^{-\lambda_1 t} & b(1 - e^{-\lambda_1 t}) \\ a(1 - e^{-\lambda_1 t}) & b + ae^{-\lambda_1 t} \end{pmatrix}$$

and $\pi_0 = a/(a + b)$, $\pi_1 = b/(a + b)$. Hence

$$|p_{ij}(t) - \pi_j| \leqslant \frac{a \vee b}{a + b}e^{-\lambda_1 t}.$$

This proves the last assertion.

(b) Write

$$Q = (a + b)\begin{pmatrix} -\theta & \theta \\ 1 - \theta & \theta - 1 \end{pmatrix},$$

where $\theta = b/(a + b)$. Therefore, it suffices to consider the Q-matrix

$$Q_1 = \begin{pmatrix} -\theta & \theta \\ 1 - \theta & \theta - 1 \end{pmatrix}.$$

Without loss of generality, one may assume that $\theta \leqslant 1/2$, i.e., $b \leqslant a$.

(c) By P. Diaconis and L. Saloff-Coste (1996) or Chen (1997a), for Q_1, we have

$$\sigma = \frac{2(1 - 2\theta)}{\log(1/\theta - 1)}.$$

The computation of this constant is nontrivial. From this and (b), it is easy to obtain the second assertion.

(d) We now show that the Nash inequality is equivalent to

$$\|f - \pi(f)\|^2 \leqslant \eta^{-1} D(f)^{1/p} \|f - c\|_1^{2/q}, \qquad f \in L^2(\pi), \qquad (8.1)$$

where c is a median of f. To see this, replacing f with $f - c$ in the original Nash inequality, we get (8.1). The inverse implication follows from

$$\|f - c\|_1 = \inf_\alpha \|f - \alpha\|_1 \leqslant \|f\|_1.$$

Next, consider Q_1. Given a function f on $\{0, 1\}$, without loss of generality, assume that $f_0 > f_1$. Since $\theta \leqslant 1/2$, the median of f is f_0. Set $g = f - f_0$. Then

$$\|g\|_1 = \theta |g_1| = \theta(f_0 - f_1),$$
$$\mathrm{Var}(g) = \pi_1 g_1^2 + (\pi_1 g_1)^2 = \theta(1 - \theta)(f_0 - f_1)^2,$$
$$D(g) = \pi_0 q_{01}(g_1 - g_0)^2 = (1 - \theta)\theta(f_0 - f_1)^2.$$

Hence for Q_1,

$$\eta = \inf_g \frac{D(g)^{1/p} \|g\|_1^{2/q}}{\mathrm{Var}(g)} = \left(\frac{\theta}{1 - \theta}\right)^{1/q}.$$

Applying (b) again, we obtain the first assertion. □

Even in this simplest situation $E = \{0, 1\}$, the exact rate of the exponential convergence in entropy is still unknown.

A diagram of nine types of ergodicity

The main topic of this chapter is the diagram of the different types of ergodicity presented in Theorem 1.9.

8.2 Applications and comments

Here are some remarks about Figure 1.1.

The importance of the diagram is obvious. For instance, by using the estimates obtained from the study of the Poincaré inequality, based on the advantage of the analytic approach, the L^2-theory and the equivalence in the diagram, one can estimate exponentially ergodic convergence rates, for which current knowledge is still very limited. Actually, these two convergence rates often coincide (cf. the proofs given in Section 8.3). In particular, one obtains a criterion for exponential ergodicity in dimension one, which was open for a long period (cf. Tables 1.4 and 5.1). Conversely, from the well-known criteria for exponential ergodicity, one obtains immediately some criteria, which are indeed new, for the Poincaré inequality. Here is a criterion for exponential ergodicity. A ψ-irreducible, aperiodic Markov process with operator L is exponentially ergodic if there exist a probability measure ν, some functions $h : E \to (0, 1]$, $V : E \to [1, \infty]$, and some constants $\delta > 0$, $c < \infty$ such that

$$LV \leqslant (-\delta + ch)V, \qquad R_1 \geqslant h \otimes \nu,$$

where $R_\lambda = \int_0^\infty e^{-\lambda t} P_t \mathrm{d}t$, $\lambda > 0$, is the resolvent of the semigroup $\{P_t\}_{t \geqslant 0}$. Refer to S.P. Meyn and R.L. Tweedie (1993a), I. Kontoyiannis and S.P. Meyn (2003) for more details. Next, there is still very limited knowledge about the L^1-spectrum, due to the structure of the L^1-space, which is only a Banach but not a Hilbert space. Based on the probabilistic advantage and the identity in the diagram, from a study of the strong ergodicity, one learns a great deal about the L^1-spectral gap of the generator. Refer to L.M. Wu (2004) for a comprehensive study of related topics for time-discrete Markov processes.

As explained in Section 7.6, L^2-*algebraic convergence* means that $\mathrm{Var}(P_t f)$ $\leqslant CV(f) t^{1-q}$ for all $t > 0$ and for some V having the properties that V is homogeneous of degree two (in the sense that $V(cf + d) = c^2 V(f)$ for any constants c and d) and $V(f) < \infty$ for a class of functions f [continuous functions with compact support, for instance; cf. T.M. Liggett (1991)]. Refer also to J.D. Deuschel (1994), Chen and Y.Z. Wang (2003), Y.Z. Wang (2004; 2003b), M. Röckner and F.Y. Wang (2001) for a study of L^2-algebraic convergence.

Reversibility is used in both the identity and the equivalence. Without reversibility, L^2-exponential convergence still implies π-a.s. exponentially ergodic convergence.

An important fact is that the condition "having densities" is used only in the identity of L^1-exponential convergence and π-a.s. strongly ergodic convergence. Without this condition, L^1-exponential convergence still implies π-a.s. strongly ergodic convergence, and so the diagram needs only a little change (however, reversibility is still required here). Thus, it is a natural *open problem* to remove or relax this "density condition."

Except for the identity and the equivalence, all the implications in the diagram are suitable for general Markov processes, not necessarily reversible, even though the inequalities are mainly valuable in the reversible situation, or when the stationary distribution is known in advance. Clearly, the diagram extends the ergodic theory of Markov processes.

The diagram is complete in the following sense: each single implication cannot be replaced by a double one. Moreover, the L^1-exponential convergence (respectively, strong ergodicity) and the logarithmic Sobolev inequality (respectively, exponential convergence in entropy) are not comparable.

The differences among these types of ergodicity are illustrated by the following examples.

Examples 8.2. Comparisons of the different types of ergodicity for diffusions on the half-line with reflecting boundary at the origin. See Table 8.2. Here "$\sqrt{}$" means "always holds" and "$\times$" means "never holds."

Table 8.2 Comparisons of diffusions on $[0, \infty)$

	Erg.	Exp. erg.	LogS	Strong erg.	Nash
$b(x) = 0$ $a(x) = x^\gamma$	$\gamma > 1$	$\gamma \geqslant 2$	$\gamma > 2$	$\gamma > 2$	$\gamma > 2$
$b(x) = 0$ $a(x) = x^2 \log^\gamma x$	✓	$\gamma \geqslant 0$	$\gamma \geqslant 1$	$\gamma > 1$	✗
$a(x) = 1$ $b(x) = -b$	✓	✓	✗	✗	✗

Examples 8.3. Comparisons of the different types of ergodicity for birth–death processes. See Table 8.3.

Table 8.3 Comparisons of birth–death processes

	Ergodicity	Exp. erg.	LogS	Strong erg.	Nash
$a_i = b_i$ $= i^\gamma$	$\gamma > 1$	$\gamma \geqslant 2$	$\gamma > 2$	$\gamma > 2$	$\gamma > 2$
$a_i = b_i$ $= i^2 \log^\gamma i$	✓	$\gamma \geqslant 0$	$\gamma \geqslant 1$	$\gamma > 1$	✗
$a_i = a >$ $b_i = b$	✓	✓	✗	✗	✗

We have seen from the above tables that strong ergodicity is usually stronger than the logarithmic Sobolev inequality. The next example goes in the opposite way.

Example 8.4 (Chen, 1999b). Let $(\pi_i > 0)$ and take $q_{ij} = \pi_j (j \neq i)$. Then the process is strongly ergodic but the logarithmic Sobolev inequality does not hold.

Proof. (a) Since the Q-matrix is bounded, the logarithmic Sobolev inequality cannot hold (cf. Theorem 4.12).

(b) For strong ergodicity, note that the sequence $y_0 = 0$, $y_i = 1/\pi_0$ $(i \neq 0)$ satisfies the following criterion [cf. Theorem 5.1 (3)]:

$$\begin{cases} \sum_j q_{ij}(y_j - y_i) \leqslant -1, & i \neq 0, \\ \sum_{j \neq 0} q_{0j} y_j < \infty, \\ (y_i) \text{ is nonnegative and bounded.} \end{cases}$$

Hence the process is strongly ergodic. □

For one-dimensional diffusions, a counterexample was constructed by F.Y. Wang (2001) to show that strong ergodicity does not imply exponential convergence in entropy (equivalently, the logarithmic Sobolev inequality).

The diagram was presented in Chen (1999c; 2002b), originally stated mainly for Markov chains. Recently, the identity of L^1-exponential convergence and the π-a.s. strong ergodicity were proven by Y.H. Mao (2002c).

8.3 Proof of Theorem 1.9

The detailed proofs and some necessary counterexamples were presented in Chen (1999c; 2002b) for reversible Markov processes, except for the identity of the L^1-exponential convergence and π-a.s. strong ergodicity. Note that for a discrete state space, one can rule out "a.s." used in the diagram. Here, we collect the complete proofs of the diagram, with some more careful estimates for the general state spaces. The author would like to acknowledge Y.H. Mao for his nice ideas, which are included in this section. The steps of the proofs are listed as follows:

(a) Nash inequality $\Longrightarrow$ L^1-exponential convergence and π-a.s. strongly ergodic convergence.

(b) L^1-exponential convergence $\Longleftrightarrow$ π-a.s. strongly ergodic convergence.

(c) Strong ergodicity $\Longrightarrow$ exponential ergodicity $\Longrightarrow$ ordinary ergodicity.

(d) L^2-exponential convergence $\Longrightarrow$ L^2-algebraic convergence.

(e) L^2-algebraic convergence $\Longrightarrow$ ordinary ergodic convergence.

(f) Nash inequality $\Longrightarrow$ logarithmic Sobolev inequality.

(g) Logarithmic Sobolev inequality $\Longrightarrow$ Poincaré inequality.

(h) L^2-exponential convergence $\Longrightarrow$ π-a.s. exponentially ergodic convergence.

(i) exponentially ergodic convergence $\Longrightarrow$ L^2-exponential convergence.

Moreover, at each step, once we have a more general result, it will be stated more precisely.

(a) Nash inequality $\Longrightarrow$ L^1-exponential convergence and π-a.s. strong ergodicity [Chen, 1999b]

Theorem 8.5.

(1) In general, Nash inequality $\Longrightarrow$ L^1-exponential convergence.

(2) In the reversible case, Nash inequality $\Longrightarrow$ π-a.s. strong ergodicity.

Proof. Denote by $\|\cdot\|_{p\to q}$ the operator's norm from $L^p(\pi)$ to $L^q(\pi)$. From proof (a) of Theorem 5.10, it follows that

$$\text{Nash inequality} \Longrightarrow \text{Var}(P_t(f)) = \|P_t f - \pi(f)\|_2^2 \leqslant C^2 \|f\|_1^2 / t^{q-1}$$
$$\Longleftrightarrow \|(P_t - \pi)f\|_2 \leqslant C\|f\|_1 / t^{(q-1)/2}.$$
$$\Longleftrightarrow \|P_t - \pi\|_{1\to 2} \leqslant C / t^{(q-1)/2} \quad (q := \nu/2 + 1).$$

Since $\|P_t - \pi\|_{1\to 1} \leqslant \|P_t - \pi\|_{1\to 2}$, we have

$$\text{Nash inequality} \Longrightarrow L^1\text{-algebraic convergence.}$$

Furthermore, because of the semigroup property, the convergence of $\|\cdot\|_{1\to 1}$ must be exponential, and we indeed have

$$\text{Nash inequality} \Longrightarrow L^1\text{-exponential convergence.}$$

That is the first assertion.

In the reversible case, $P_t - \pi = (P_t - \pi)^*$, and so

$$\|P_{2t} - \pi\|_{1\to\infty} \leqslant \|P_t - \pi\|_{1\to 2}\|P_t - \pi\|_{2\to\infty} = \|P_t - \pi\|_{1\to 2}^2.$$

Hence $\|P_t - \pi\|_{1\to\infty} \leqslant C/t^{q-1}$. Thus,

$$\text{ess sup}_x \|P_t(x,\cdot) - \pi\|_{\text{Var}} = \text{ess sup}_x \sup_{|f|\leqslant 1} |(P_t(x,\cdot) - \pi)f|$$
$$\leqslant \text{ess sup}_x \sup_{\|f\|_1\leqslant 1} |(P_t(x,\cdot) - \pi)f|$$
$$= \sup_{\|f\|_1\leqslant 1} \text{ess sup}_x |(P_t(x,\cdot) - \pi)f|$$
$$= \|P_t - \pi\|_{1\to\infty}$$
$$\leqslant C/t^{q-1} \to 0 \qquad \text{as } t \to \infty.$$

This gives us the π-a.s. strong ergodicity. $\square$

(b) L^1-exponential convergence $\Longleftrightarrow$ π-a.s. strong ergodicity [Y. H. Mao, 2002c]

Theorem 8.6.

(1) For reversible Markov process, we have L^1-exponential convergence $\Longrightarrow$ π-a.s. strongly ergodic convergence. Moreover, the exponential convergence rate of the latter one is bounded below by the former one.

(2) If additionally, $P_t^*(x,\cdot) \ll \pi$, then the two convergence rates coincide with each other.

Proof. Since $(L^1)^* = L^\infty \implies \|P_t - \pi\|_{1\to 1} = \|P_t^* - \pi\|_{\infty \to \infty}$, we have

$$\|P_t^* - \pi\|_{\infty \to \infty} = \operatorname{ess\,sup}_x \sup_{\|f\|_\infty = 1} |(P_t^* - \pi)f(x)|$$

$$\geqslant \operatorname{ess\,sup}_x \sup_{\sup |f| = 1} |(P_t^* - \pi)f(x)|$$

$$= \operatorname{ess\,sup}_x \|P_t^*(x, \cdot) - \pi\|_{\mathrm{Var}}.$$

This proves the first assertion. If, moreover, $P_t^*(x, \cdot) \ll \pi$, then the sign of the second equality holds, and so π-a.s. strong ergodicity is exactly the same as L^1-exponential convergence. $\square$

(c) Strong ergodicity$\implies$exponential ergodicity$\implies$ordinary ergodicity

If the Markov process corresponding to the semigroup (P_t) is irreducible and aperiodic in the Harris sense, then the implications hold. To see this, noting that by [Chen (1992a, Section 4.4); D. Down, S.P. Meyn, and R.L. Tweedie (1995)], the continuous-time case can be reduced to the discrete-time one, and then the conclusion follows from S.P. Meyn and R L. Tweedie (1993b, Chapter 16).

(d) L^2-exponential convergence $\implies$ L^2-algebraic convergence

Simply take $V(f) = \|f\|^2$ in (5.20) and apply Theorem 5.10 (1).

(e) L^2-algebraic convergence $\implies$ ordinary ergodic convergence

The proof is very much the same as proof (a) of Theorem 8.8 below.

(f) Nash inequality $\implies$ logarithmic Sobolev inequality [Chen, 1999b]

Because $\|f\|_1 \leqslant \|f\|_p$ for all $p \geqslant 1$, we have

$$\|\cdot\|_{2\to 2} \leqslant \|\cdot\|_{1\to 2} \leqslant C/t^{(q-1)/2},$$

and so

Nash inequality $\implies$ Poincaré inequality $\iff$ $\lambda_1 > 0$,

$$\|P_t\|_{p\to 2} \leqslant \|P_t\|_{1\to 2} \leqslant \|P_t - \pi\|_{1\to 2} + \|\pi\|_{1\to 2} < \infty, \qquad p \in (1, 2).$$

By D. Bakry (1992, Theorem 3.6), this implies (4.15). The assertion now follows from (4.16).

(g) Logarithmic Sobolev inequality $\Longrightarrow$ Poincaré inequality [O.S. Rothaus, 1981]

Theorem 8.7. $\lambda_1 = \mathrm{gap}(L) \geqslant \sigma$.

Proof. Consider $f = 1 + \varepsilon g$ for sufficiently small ε. Then $D(f) = \varepsilon^2 D(g)$. Next, expand $f^2 \log f^2$ and $f^2 \log \|f\|^2$ in ε up to order 2:

$$f^2 \log f^2 = 2\varepsilon g + 3\varepsilon^2 g^2 + O(\varepsilon^3),$$
$$f^2 \log \|f\|^2 = 2\varepsilon \pi(g) + \varepsilon^2 \big(-2\pi(g)^2 + \pi(g^2) + 4g\pi(g) \big) + O(\varepsilon^3).$$

Then we get $\int f^2 \log f^2 / \|f\|^2 \mathrm{d}\pi = 2\varepsilon^2 \mathrm{Var}(g) + O(\varepsilon^3)$. The proof can be done using the definitions of λ_1 and σ and letting $\varepsilon \to 0$. $\square$

The remainder of the section is devoted to the proof of the following assertion:

L^2-exponential convergence $\Longleftrightarrow$ π-a.s. exponentially ergodic convergence.

$$(8.2)$$

This was done by Chen (2000a). Because by assumption, the process is reversible and $P_t(x, \cdot) \ll \pi$, set

$$p_t(x, y) = \frac{\mathrm{d}P_t(x, \cdot)}{\mathrm{d}\pi}(y).$$

Then we have $p_t(x, y) = p_t(y, x)$, $\pi \times \pi$-a.s. (x, y). Hence

$$\int p_s(x, y)^2 \pi(\mathrm{d}y) = \int p_s(x, y) p_s(y, x) \pi(\mathrm{d}y) = p_{2s}(x, x) < \infty$$

$$(8.3)$$

$$(\text{E.A. Carlen et al., 1987}).$$

This means that $p_t(x, \cdot) \in L^2(\pi)$ for all $t > 0$ and π-a.s. $x \in E$. Thus, by Chen (2000a, Theorem 1.2) and the remarks right after the theorem, (8.2) holds.

The proof above is mainly based on the time-discrete analogue result by G.O. Roberts and J.S. Rosenthal (1997). Here, we present a more direct proof of (8.3) as follows.

(h) L^2-exponential convergence $\Longrightarrow$ π-a.s. exponentially ergodic convergence [Chen, 1991b; 1998b; 2000a]

Theorem 8.8.

(1) In general, we have $\|\mu P_t - \pi\|_{\mathrm{Var}} \leqslant \|\mathrm{d}\mu/\mathrm{d}\pi - 1\|_2 e^{-t\,\mathrm{gap}(L)}$ provided $\mathrm{d}\mu/\mathrm{d}\pi \in L^2(\pi)$.

(2) In the reversible case, we have $\varepsilon_1 \geqslant \mathrm{gap}(L) = \lambda_1$, where ε_1 is the largest ε such that

$$\|P_t(x,\cdot) - \pi\|_{\mathrm{Var}} \leqslant C(x)e^{-\varepsilon t} \qquad (8.4)$$

for some $C(x)$ and all t. Hence L^2-exponential convergence $\Longrightarrow \pi$-a.s. exponentially ergodic convergence.

(3) In the φ-irreducible case, L^2-exponential convergence $\Longrightarrow \pi$-a.s. exponentially ergodic convergence.

Proof. (a) Let $\mu \ll \pi$. Then

$$
\begin{aligned}
\|\mu P_t - \pi\|_{\mathrm{Var}} &= \sup_{|f|\leqslant 1} |(\mu P_t - \pi)f| = \sup_{|f|\leqslant 1} \left|\pi\left(\frac{\mathrm{d}\mu}{\mathrm{d}\pi}P_t f - f\right)\right| \\
&= \sup_{|f|\leqslant 1} \left|\pi\left(f P_t^*\left(\frac{\mathrm{d}\mu}{\mathrm{d}\pi}\right) - f\right)\right| = \sup_{|f|\leqslant 1} \left|\pi\left[f\left(P_t^*\left(\frac{\mathrm{d}\mu}{\mathrm{d}\pi} - 1\right)\right)\right]\right| \\
&\leqslant \left\|P_t^*\left(\frac{\mathrm{d}\mu}{\mathrm{d}\pi} - 1\right)\right\|_1 \leqslant \left\|\frac{\mathrm{d}\mu}{\mathrm{d}\pi} - 1\right\|_2 e^{-t\,\mathrm{gap}(L^*)} \\
&= \left\|\frac{\mathrm{d}\mu}{\mathrm{d}\pi} - 1\right\|_2 e^{-t\,\mathrm{gap}(L)}.
\end{aligned}
\qquad (8.5)
$$

This gives us the first assertion. We now consider two cases separately.

(b) In the reversible case with $P_t(x,\cdot) \ll \pi$, by (8.3), we have

$$
\begin{aligned}
\|P_t(x,\cdot) - \pi\|_{\mathrm{Var}} &\leqslant \left\|P_{t-s}\left(\frac{\mathrm{d}P_s(x,\cdot)}{\mathrm{d}\pi} - 1\right)\right\|_1 \\
&\leqslant \|p_s(x,\cdot) - 1\|_2 e^{-(t-s)\,\mathrm{gap}(L)} \\
&= \left[\sqrt{p_{2s}(x,x) - 1}\, e^{s\,\mathrm{gap}(L)}\right] e^{-t\,\mathrm{gap}(L)}, \qquad t \geqslant s.
\end{aligned}
\qquad (8.6)
$$

Therefore, there exists $C(x) < \infty$ such that

$$\|P_t(x,\cdot) - \pi\|_{\mathrm{Var}} \leqslant C(x)e^{-t\,\mathrm{gap}(L)}, \qquad t \geqslant 0, \quad \pi\text{-a.s. } (x).$$

This proves the second assertion.

(c) In the φ-irreducible case, without using the reversibility and transition density, from (8.5), one can still derive π-a.s. exponential ergodicity (but may have different rates). Refer to G.O. Roberts and R.L. Tweedie (2001) for a proof in the time-discrete situation (the title of the quoted paper is confused, where the term "L^1-convergence" is used for the π-a.s. exponentially ergodic convergence, rather than the standard meaning of L^1-exponential convergence used in this book. These two types of convergence are essentially different, as shown in Theorem 1.9 and Section 8.2). In other words, the reversibility and the existence of the transition density are not completely necessary in this implication. $\square$

(i) π-a.s. exponentially ergodic convergence $\Longrightarrow$ L^2-exponential convergence [Chen (2000a), Y.H. Mao (2002c)]

In the time-discrete case, a similar assertion was proved by G.O. Roberts and J.S. Rosenthal (1997) and so can be extended to the time-continuous case by the standard technique [cf. Chen (1992a, Section 4.4)]. The proof given below provides more precise estimates. To begin with, we prove some short lemmas.

Lemma 8.9. Let $\mathscr{E}$ be countably generated. Then π-a.s. exponentially ergodic convergence and exponential convergence of $\| \, \|P_t(\bullet, \cdot) - \pi\|_{\mathrm{Var}}\|_1$ are equivalent. Here and in what follows, the L^1-norm is taken with respect to the variable "$\bullet$".

Proof. By E. Numemelin and P. Tuominen (1982) or E. Numemelin (1984, Theorem 6.14 (iii)), we have in the time-discrete case that

$$\pi\text{-a.s. geometrically ergodic convergence}$$
$$\Longleftrightarrow \| \, \|P^n(\bullet, \cdot) - \pi\|_{\mathrm{Var}}\|_1 \text{ geometric convergence.}$$

This implies the time-continuous case as stated in the lemma. $\square$

From now on, assume that $\| \, \|P_t(\bullet, \cdot) - \pi\|_{\mathrm{Var}}\|_1 \leqslant Ce^{-\varepsilon_2 t}$ with largest ε_2.

Lemma 8.10. If $P_t(x, \cdot) \ll \pi$ or the process is reversible, then we have $\| \, \|P_t(\bullet, \cdot) - \pi\|_{\mathrm{Var}}\|_1 \geqslant \|P_t - \pi\|_{\infty \to 1}$.

Proof. Let $\|f\|_\infty = 1$. Then

$$\|(P_t - \pi)f\|_1 = \int \pi(\mathrm{d}x)\big|(P_t(x, \cdot) - \pi)f\big|$$
$$\leqslant \int \pi(\mathrm{d}x) \sup_{\|g\|_\infty \leqslant 1} \big|(P_t(x, \cdot) - \pi)g\big|$$
$$= \int \pi(\mathrm{d}x) \sup_{|g| \leqslant 1} \big|(P_t(x, \cdot) - \pi)g\big|$$
$$= \| \, \|P_t(\bullet, \cdot) - \pi\|_{\mathrm{Var}}\|_1.$$

The second-to-last equality comes from the following fact. If $P_t(x, \cdot) \ll \pi$, then every π-zero set is a $P_t(x, \cdot)$-zero set. Next, if P_t is reversible, then for every π-zero set B, we have $P_t(x, B) = 0$, π-a.s.(x). $\square$

Lemma 8.11. (1) In general, we have $\|P_t - \pi\|_{\infty \to 1} \geqslant \|P_t - \pi\|_{\infty \to 2}^2/2$.
  (2) In the reversible case, we have $\|P_{2t} - \pi\|_{\infty \to 1} = \|P_t - \pi\|_{\infty \to 2}^2$.

Proof. The first assertion comes from

$$\|(P_t - \pi)f\|_2^2 \leqslant \int |(P_t - \pi)f|^2 \mathrm{d}\pi \leqslant 2\|f\|_\infty \int |(P_t - \pi)f| \mathrm{d}\pi$$
$$\leqslant 2\|f\|_\infty^2 \|P_t - \pi\|_{\infty \to 1}, \qquad f \in L^\infty(\pi).$$

To prove the second assertion, note that

$$\|(P_t - \pi)f\|_2^2 = ((P_t - \pi)f, (P_t - \pi)f) = (f, (P_t - \pi)^2 f)$$
$$= (f, (P_{2t} - \pi)f) \leqslant \|f\|_\infty \|(P_{2t} - \pi)f\|_1 \qquad (8.7)$$
$$\leqslant \|f\|_\infty^2 \|P_{2t} - \pi\|_{\infty \to 1}.$$

Hence $\|P_{2t} - \pi\|_{\infty \to 1} \geqslant \|P_t - \pi\|_{\infty \to 2}^2$. The inverse inequality is obvious by using the semigroup property and symmetry:

$$\|P_{2t} - \pi\|_{\infty \to 1} \leqslant \|P_t - \pi\|_{\infty \to 2}\|P_t - \pi\|_{2 \to 1} = \|P_t - \pi\|_{\infty \to 2}^2. \qquad \square$$

Lemma 8.12. In the reversible case, $\lambda_1 = \mathrm{gap}(L) \geqslant \varepsilon_2$.

Proof. By Lemmas 8.10 and 8.11 (2), for every f with $\pi(f) = 0$ and $\|f\|_2 = 1$,

$$\|P_t f\|_2^2 \leqslant C\|f\|_\infty^2 e^{-2\varepsilon_2 t}.$$

Following F.Y. Wang (2000a, Lemma 2.2), or M. Röckner and F.Y. Wang (2001), by the spectral representation theorem, we have

$$\|P_t f\|_2^2 = \int_0^\infty e^{-2\lambda t} \mathrm{d}(E_\lambda f, f)$$
$$\geqslant \left[\int_0^\infty e^{-2\lambda s} \mathrm{d}(E_\lambda f, f) \right]^{t/s} \qquad \text{(by Jensen's inequality)}$$
$$= \|P_s f\|_2^{2t/s}, \qquad t \geqslant s.$$

Thus, $\|P_s f\|_2^2 \leqslant \left[C\|f\|_\infty^2 \right]^{s/t} e^{-2\varepsilon_2 s}$. Letting $t \to \infty$, we get

$$\|P_s f\|_2^2 \leqslant e^{-2\varepsilon_2 s}, \qquad \pi(f) = 0, \ \|f\|_2 = 1, \ f \in L^\infty(\pi).$$

Since $L^\infty(\pi)$ is dense in $L^2(\pi)$, we have

$$\|P_s f\|_2^2 \leqslant e^{-2\varepsilon_2 s}, \qquad s \geqslant 0, \quad \pi(f) = 0, \ \|f\|_2 = 1.$$

Therefore, $\lambda_1 \geqslant \varepsilon_2$. $\square$

Theorem 8.13.

(1) In general, if $P_t(x, \cdot) \ll \pi$, then $\| \|P_t(\bullet, \cdot) - \pi\|_{\mathrm{Var}}\|_1 \geqslant \|P_t - \pi\|_{\infty \to 2}^2 / 2$.

(2) In the reversible case, $\lambda_1 = \mathrm{gap}(L) \geqslant \varepsilon_2$. In particular, if $\mathscr{E}$ is countably generated, then π-a.s. exponentially ergodic convergence $\Longrightarrow L^2$-exponential convergence.

(3) In the reversible case, if $p_{2s}(\cdot, \cdot) \in L^{1/2}(\pi)$, then $\lambda_1 = \varepsilon_2$.

(4) In the reversible case, if $p_{2s}(\cdot, \cdot) \in L_{\mathrm{loc}}^{1/2}(\pi)$ and the set $\mathscr{K}$ of bounded functions with compact support is dense in $L^2(\pi)$, then $\lambda_1 = \varepsilon_1$.

Proof. The first assertion comes from Lemmas 8.10 and 8.11 (1). The second one comes from Lemma 8.12. Its particular consequence then follows from Lemma 8.9.

To prove the third assertion, by (8.6), it follows that there exists a constant C such that $\| \, \|P_t(\bullet, \cdot) - \pi\|_{\mathrm{Var}}\|_1 \leqslant Ce^{-\lambda_1 t}$. Hence, $\varepsilon_2 \geqslant \lambda_1$. Combining this with Lemma 8.12, we indeed have $\lambda_1 = \varepsilon_2$.

To prove the last assertion, we follow C.R. Hwang et al. (2002). From (8.6), we have seen that if the process has L^2-exponential convergence, then (8.4) holds with $C \in L_{\mathrm{loc}}^{1/2}(\pi)$ by assumption. Under this condition, as in (8.7), we have

$$
\begin{aligned}
\|(P_t - \pi)f\|_2^2 &\leqslant (f, (P_{2t} - \pi)f) \\
&\leqslant \|f\|_\infty \int \pi(\mathrm{d}x)|f(x)| \|P_{2t} - \pi\|_{\mathrm{Var}} \\
&\leqslant \|f\|_\infty^2 \int_{\mathrm{supp}\,(f)} \pi(\mathrm{d}x)C(x)e^{-2\varepsilon_1 t} \\
&=: C_f e^{-2\varepsilon_1 t}, \qquad f \in \mathscr{K}.
\end{aligned}
$$

The constant C_f can be removed, as we did in the proof of Lemma 8.12, by using the denseness of $\mathscr{K}$. Therefore, we have $\lambda_1 \geqslant \varepsilon_1$. Combining this with Theorem 8.8 (2), we obtain the last assertion. $\square$

Refer also to F.Y. Wang (2002) for related estimates.

Chapter 9

Reaction–Diffusion Processes

This chapter surveys the main progress made in the past twenty years or so in the study of reaction–diffusion (RD) processes. The processes are motivated from some typical models in modern nonequilibrium statistical physics and are an important class of interacting particle systems, which is currently an active research field in probability theory, mathematical physics, and chemistry. The models are concrete, but as a part of infinite-dimensional mathematics, the topic is quite hard. It is explained how new problems arise and how some new ideas and new mathematical tools are introduced. Surprisingly, the mathematical tools produced from studying these simple models then turn out to have a number of powerful applications not only in probability theory but also in other branches of mathematics. Nevertheless, the story is still far from finished, and some important open problems are proposed for further study.

The chapter consists of five sections. We begin with an introduction of the models (Section 9.1). Then we turn to the finite-dimensional case, in which the processes are indeed Markov chains (Section 9.2). This study leads to a powerful criterion for the uniqueness of Markov chains. The infinite-dimensional processes are constructed in Section 9.3. The main tool for the construction is the coupling methods discussed in Chapter 2. The existence of the stationary distribution, the ergodicity of the processes, and the phase transitions for several models are discussed in Section 9.4. Again, the coupling methods play a key role in our study of ergodicity. In the last section, the relation between the RD processes and RD equations (hydrodynamic limits) and their interaction are studied.

A more complete exploration of RD processes can be found in Part IV of the book by Chen (1992a).

9.1   The models

Let $S = \mathbb{Z}^d$, the d-dimensional lattice. Consider a chemical reaction in a container. Divide the container into small vessels, imagining each $u \in S$ as a small vessel in which there is a reaction. The reaction is described by some Markov chains (MCs) with Q-matrices $Q_u = (q_u(i,j) : i, j \in \mathbb{Z}_+)$, $u \in S$. That is, the rate of the MC in u jumping from i to $j \neq i$ is given by $q_u(i,j)$. Throughout the chapter, we consider only a totally stable and conservative Q-matrix: $-q_u(i,i) = \sum_{j \neq i} q_u(i,j) < \infty$ for all $i \in \mathbb{Z}_+$. Thus, the reaction part of the formal generator of the process is as follows:

$$\Omega_r f(x) = \sum_{u \in S} \sum_{k \in \mathbb{Z} \setminus \{0\}} q_u(x_u, x_u + k) \big[f(x + ke_u) - f(x) \big],$$

where e_u is the element in $E := \mathbb{Z}_+^S$ whose value at site u is equal to one, and the values at other sites are zero. Moreover, we have used the following convention: $q_u(i,j) = 0$ for $i \in \mathbb{Z}_+$, $j \notin \mathbb{Z}_+$, and $u \in S$. Mathematically, one may regard x_u as the uth component of x in the product space $\mathbb{Z}_+^S$. The other part of the generator of the process consists of diffusions between the vessels, which are described by a transition probability matrix $(p(u,v) : u, v \in S)$ and a function c_u ($u \in S$) on $\mathbb{Z}_+$. For instance, if there are k particles in u, then the rate function of the diffusion from u to v is $c_u(k)p(u,v)$, where c_u satisfies

$$c_u \geqslant 0, \qquad c_u(0) = 0, \qquad u \in S. \tag{9.1}$$

Thus, the diffusion part of the formal generator becomes

$$\Omega_d f(x) = \sum_{u,v \in S} c_u(x_u)p(u,v)[f(x - e_u + e_v) - f(x)].$$

Finally, the whole formal generator of the process is $\Omega = \Omega_r + \Omega_d$. A simple description of the models is given by Figure 9.1.

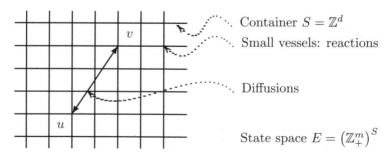

Figure 9.1   The models of reaction–diffusion processes

Sometimes, it is more convenient to lift the spin spaces (regarded as "fibers"). Then we have Figure 9.2.

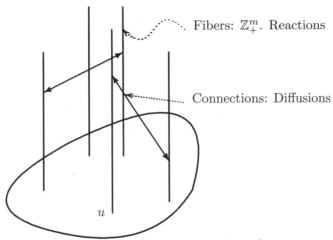

Figure 9.2 Informal interpretation: dynamics of infinite-dimensional fiber bundle

Example 9.1 (*Polynomial model*). The diffusion rates are described by $c_u(k) = k$ and $p(u, v)$, which is the simple random walk on $\mathbb{Z}^d$. The reaction rates are of birth–death type:

$$q_u(k, k + 1) = b_k = \sum_{j=0}^{m_0} \beta_j k^{(j)}, \qquad q_u(k, k - 1) = a_k = \sum_{j=1}^{m_0+1} \delta_j k^{(j)},$$

where $k^{(j)} = k(k - 1) \cdots (k - j + 1)$, $\beta_j, \delta_j \geqslant 0$ and $\beta_0, \beta_{m_0}, \delta_1, \delta_{m_0+1} > 0$.

In particular, we have the following example.

Example 9.2. (1). *Schlögl's first model*: $m_0 = 1$.
(2). *Schlögl's second model*: $m_0 = 2$ but $\beta_1 = \delta_2 = 0$.

All these examples have a single type of particle and so the number of particles is valued in $\mathbb{Z}_+$. If we consider two types of particles, then the reaction part becomes a Markov chain valued in $\mathbb{Z}_+^2$. Here is a typical example.

Example 9.3 (*Brusselator model*). For each type of particle, the diffusion part of the formal generator is the same as in Example 9.1. As for the reaction part, the MC has the following transition behavior:

$$\begin{aligned}
\mathbb{Z}_+^2 \ni (i, j) &\to (i + 1, j) & \text{at rate} \quad & \lambda_1 \\
&\to (i - 1, j) & \text{at rate} \quad & \lambda_4 i \\
&\to (i - 1, j + 1) & \text{at rate} \quad & \lambda_2 i \\
&\to (i + 1, j - 1) & \text{at rate} \quad & \lambda_3 i(i - 1)j/2,
\end{aligned}$$

where the λ_k's are positive constants.

These examples are typical models in nonequilibrium statistical physics. Refer to Chen (1986b) or Chen (1992a, Part IV) for more information about the background and references. Fifteen models are treated in these books. The author learned about the models from Prof. S.J. Yan in 1980.

9.2 Finite-dimensional case

Replacing $S = \mathbb{Z}^d$ with a finite set S (which is fixed in this section) in the above definitions of Ω_r and Ω_d, the corresponding processes are simply MCs, since the state space $E = \mathbb{Z}_+^S$ (or $(\mathbb{Z}_+^2)^S$) is countable. At the beginning, one may think this step can be ignored, because there already is a well-developed theory of MCs. However, the object is not so easy as it stands. Indeed, we did not know how to prove the uniqueness of the MCs for several years. The usual criterion for the uniqueness says that the equations

$$(\lambda - \Omega)u(x) = 0, \qquad 0 \leqslant u(x) \leqslant 1, \qquad x \in \mathbb{Z}_+^S,$$

have only the trivial solution zero for some (equivalently, for all) $\lambda > 0$. It should be clear that the equations are quite hard to handle, especially in higher dimensions. The criterion does not take the geometry of the MC into account.

To overcome this difficulty, we regard the set $\{x : \sum_{u \in S} x_u = n\}$ as a single point n $(n \geqslant 0)$. Construct a single birth process (i.e., when $k > 0$, $q_{i,i+k} > 0$ iff $k = 1$ but there is no restriction on the death rates q_{ij} for all $j < i$) on $\mathbb{Z}_+$ that dominates the original process. See Figure 9.3.

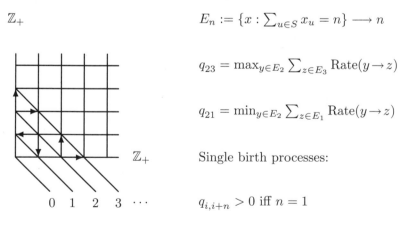

$\mathbb{Z}_+$

$$E_n := \{x : \textstyle\sum_{u \in S} x_u = n\} \longrightarrow n$$

$$q_{23} = \max_{y \in E_2} \textstyle\sum_{z \in E_3} \mathrm{Rate}(y \to z)$$

$$q_{21} = \min_{y \in E_2} \textstyle\sum_{z \in E_1} \mathrm{Rate}(y \to z)$$

$\mathbb{Z}_+$

Single birth processes:

$$0 \quad 1 \quad 2 \quad 3 \quad \cdots$$

$$q_{i,i+n} > 0 \text{ iff } n = 1$$

Figure 9.3   Reduce higher dimensions to dimension one

Since for single birth processes we do have a computable criterion for uniqueness (cf. Section 5.5), then we can prove the uniqueness of the original processes by a comparison argument. This and related results are presented in

S.J. Yan and Chen (1986). See also Chen (1999d) for some improvements and Sections 5.5 and 5.6 for additional results. By using an approximation of the processes with bounded rates (in this case, the process is always unique), a more general uniqueness result (even for Markov jump processes on general state spaces) was proved in Chen (1986a). The following result is also included in Chen (1986b; 1991c; 1992a).

Theorem 9.4. Let $Q = (q_{ij})$ be a Q-matrix on a countable set E. Suppose that there exist a sequence $\{E_n\}_1^\infty$, a constant $c \in \mathbb{R}$, and a nonnegative function φ such that

(1) $E_n \uparrow E$, $\sup_{i \in E_n}(-q_{ii}) < \infty$ for all $n \geqslant 1$,
(2) $\lim_{n \to \infty} \inf_{i \notin E_n} \varphi_i = \infty$, and
(3) $\sum_j q_{ij}(\varphi_j - \varphi_i) \leqslant c \varphi_i$ for all $i \in E$.

Then the process (MC) is unique (i.e., the minimal process is nonexplosive).

To justify the power of the theorem, for Examples 9.1 and 9.2, simply take $\varphi(x) = c[1 + \sum_{u \in S} x_u]$ and $E_n = \{x : \sum_{u \in S} x_u \leqslant n\}$ for some suitable constant c. A slight modification works also for Example 9.3. Indeed, it can be proved that the conditions of the theorem are also necessary for the single birth processes [see Chen (1986b) or Chen (1992a, Remark 3.20)], and up to now we do not know any counterexample for which the process is unique but the conditions of Theorem 9.4 fail. The theorem now has a very wide range of applications. For instance, it is a basic result used in the study of RD processes. Here is a long list of publications:

M. Bebbington et al. (1995), C. Boldrighini et al. (1987), Chen (1985b; 1987; 1989c; 1991d; 1994c), Chen, W.D. Ding and D.J. Zhu (1994), Chen, L.P. Huang and X.J. Xu (1991) A. DeMasi and E. Presutti (1992), W.D. Ding, R. Durrett and T.M. Liggett (1990), W.D. Ding and X.G. Zheng (1989), D. Han (1990; 1991; 1992; 1995), L.P. Huang (1987), Y. Li (1991; 1995), J.S. Lü (1997), T.S. Mountford (1992), C. Neuhauser (1990), A. Perrut (2000), T. Shiga (1988), X.G. Zheng and W.D. Ding (1987), S.Z. Tang (1985).

For the mean field models, see

D.A. Dawson and X.G. Zheng (1991), S. Feng (1994a; 1994b; 1995), S. Feng and X.G. Zheng (1992).

These will be discussed later. The theorem was used in R.R. Chen (1997b) to study an extended class of branching processes, and it was actually a key in J.S. Song (1988) to study Markov decision programming in the unbounded case. The theorem is also included in the book by W.J. Anderson (1991, Corollary 2.2.16) and is followed with some extension by the following:

K. Hamza and F.C. Klebaner (1995), G. Kersting and F.C. Klebaner (1995), S.P. Meyn and R.L. Tweedie (1993a).

The generalization of Theorem 9.4 to general state spaces given in Chen (1986a) is also meaningful in quantum mechanics, refer to A. Konstantinov et al. (1990) and references within.

Sketch of the proof of Theorem 9.4. Instead of $(p_{ij}(t))$, we use its Laplace transform: $p_{ij}(\lambda) = \int_0^\infty p_{ij}(t)e^{-\lambda t}dt$. The advantage of this is to reduce an integral equation to an algebraic one.

(a) Let $q_{ij}^{(n)} = q_{ij}I_{E_n}(i)$ for $i \neq j$ and $q_i^{(n)} = \sum_{j \neq i} q_{ij}^{(n)}$. Then, from condition (1), it follows that $\sup_i q_i^{(n)} < \infty$, and so there is uniquely a Q-process $P_n(\lambda) = (p_{ij}^{(n)}(\lambda))$. Next, replacing c with $c_+ = c \vee 0$, condition (3) also holds for $Q_n = (q_{ij}^{(n)})$. Because $(p_{ij}^{(n)}(\lambda) : i \in E)$ is the minimal solution to the backward Kolmogorov equation

$$x_i = \sum_{k \neq i} \frac{q_{ik}^{(n)}}{\lambda + q_i^{(n)}}x_k + \frac{\delta_{ij}}{\lambda + q_i^{(n)}}, \qquad i \in E,$$

by the linear combination theorem (Theorem 5.17), $\left(\sum_j p_{ij}^{(n)}(\lambda)\varphi_j : i \in E\right)$ is the minimal solution to the equation

$$x_i = \sum_{k \neq i} \frac{q_{ik}^{(n)}}{\lambda + q_i^{(n)}}x_k + \frac{\varphi_i}{\lambda + q_i}.$$

By condition (3), we have

$$\frac{\varphi_i}{\lambda - c_+} \geqslant \sum_{k \neq i} \frac{q_{ik}^{(n)}}{\lambda + q_i^{(n)}} \cdot \frac{\varphi_k}{\lambda - c_+} + \frac{\varphi_i}{\lambda + q_i^{(n)}}, \qquad \lambda > c_+$$

$$\left[\iff (\lambda + q_i^{(n)})\varphi_i \geqslant \sum_{k \neq i} q_{ik}^{(n)}\varphi_k + \varphi_i(\lambda - c_+)\right].$$

Then, the comparison theorem (Theorem 5.16) gives us

$$\sum_j p_{ij}^{(n)}(\lambda)\varphi_j \leqslant \frac{\varphi_i}{\lambda - c_+} < \infty, \qquad \lambda > c_+.$$

(b) Denote by $\left(p_{ij}^{\min}(\lambda) : i \in E\right)$ the minimal solution to the backward Kolmogorov equation

$$x_i = \sum_{k \neq i} \frac{q_{ik}}{\lambda + q_i}x_k + \frac{\delta_{ij}}{\lambda + q_i}, \qquad i \in E.$$

By the linear combination theorem, $\left(p_{iA}^{\min}(\lambda) := \sum_{j \in A} p_{ij}^{\min}(\lambda) : i \in E\right)$ is the minimal solution to the equation

$$x_i = \sum_{k \neq i} \frac{q_{ik}}{\lambda + q_i}x_k + \frac{\delta_{iA}}{\lambda + q_i},$$

where $\delta_{iA} = 1$ if $i \in A$ and $\delta_{iA} = 0$ otherwise.

When $i \in E_n$, for all $A \subset E_n$, we have

$$p_{iA}^{\min}(\lambda) = \sum_{k \neq i} \frac{q_{ik}}{\lambda + q_i} p_{kA}^{\min}(\lambda) + \frac{\delta_{iA}}{\lambda + q_i} = \sum_{k \neq i} \frac{q_{ik}^{(n)}}{\lambda + q_i^{(n)}} p_{kA}^{\min}(\lambda) + \frac{\delta_{iA}}{\lambda + q_i^{(n)}}.$$

On the other hand, for $i \notin E_n$, we have

$$P_{iA}^{\min}(\lambda) \geqslant 0 = \sum_{k \neq i} \frac{q_{ik}^{(n)}}{\lambda + q_i^{(n)}} p_{kA}^{\min}(\lambda) + \frac{\delta_{iA}}{\lambda + q_i^{(n)}}, \qquad A \subset E_n.$$

By the comparison theorem again, we get

$$p_{iA}^{\min}(\lambda) \geqslant p_{iA}^{(n)}(\lambda), \qquad i \in E, \ A \subset E_n.$$

(c) Finally, by (b) and (a), we have

$$\lambda p_{iE_n}^{\min}(\lambda) \geqslant \lambda p_{iE_n}^{(n)}(\lambda) = 1 - \lambda p_{iE_n^c}^{(n)}(\lambda) \quad \left(\text{since } \left(p_{ij}^{(n)}(\lambda)\right) \text{ is nonexplosive}\right)$$

$$\geqslant 1 - \lambda \sum_{j \neq E_n} p_{ij}^{(n)}(\lambda)\varphi_j \bigg/ \inf_{i \notin E_n} \varphi_i$$

$$\geqslant 1 - \frac{\lambda \varphi_i}{\inf_{i \notin E_n} \varphi_i} \cdot \frac{1}{\lambda - c_+}, \qquad \lambda > c_+.$$

Letting $n \to \infty$, by condition (2), it follows that $\lambda p_{iE}^{\min}(\lambda) \geqslant 1$. This implies the uniqueness as required. $\square$

We have seen how the models led us to resolve one of the classical problems for MCs and produce some effective results. Some new solutions to the recurrence and positive recurrence problems are also given in S.J. Yan and Chen (1986), Chen (1986b), and Chen (1992a, Chapter 4). However, the positive recurrence for the Brusselator model was proved only in 1991 by D. Han in the case of $d = 1$, and by J.W. Chen (1995) for the general finite-dimensional situation [cf. Chen (1992a, Example 4.50)]. From the papers listed above, one can see again a number of applications of these results, but we are not going into details here. In conclusion, the finite-dimensional Schlögl's and Brusselator models are all ergodic and so have no phase transitions. Thus, in order to study the phase transition phenomena for these systems, we have to go to the infinite-dimensional setup.

Before going further, let us compare the above models with the famous Ising model (cf. Section B.1).

- The state space $E = \{-1, +1\}^{\mathbb{Z}^d}$ for the Ising model is compact, but for Schlögl's models, the state space $E = \mathbb{Z}_+^{\mathbb{Z}^d}$ is neither compact nor locally compact.

- The Ising model is reversible, its local Gibbs distributions are explicit. But the Schlögl's models has no such advantages except a very special case.

- The Ising model has at least one stationary distribution, since every Feller process with compact state space does. But for noncompact case, the conclusion may not be true.

- The generator of the Ising model is locally bounded but it is not so for the Schlögl's models.

In summary, we have Table 9.1.

Table 9.1   Comparison of Ising model and Schlögl models

Comparison	Ising model	Schlögl model
Space	$\{-1, +1\}^{\mathbb{Z}^d}$ compact	$\mathbb{Z}_+^{\mathbb{Z}^d}$: not locally compact
System	equilibrium reversible	nonequilibrium irreversible
Operator	locally bounded	not locally bounded and nonlinear
Stationary distribution	always exists and locally explicit	? locally no expression

From these facts, it should be clear that the Ising model and the Schlögl's models are very different.

9.3   Construction of the processes

The diffusion part of the operator for RD processes cannot be ignored; otherwise, there is no interaction, and then the processes are simply the independent product of the classical MCs. If we forget the reaction part, then the processes are reduced to the well-known zero range processes, for which the construction was completed step by step by several authors. In a special case, the process was first constructed by R. Holley (1970), and the general case was done by T.M. Liggett (1973). Then, E.D. Andjel (1982), T.M. Liggett, and F. Spitzer (1981) simplified the construction. For all the models considered in the last quoted paper, the coefficients of the operator are assumed to be locally bounded and linear. Thus, even in this simpler case, the construction is still not simple.

A standard tool in constructing Markov processes is semigroup theory, as was used by T.M. Liggett (1985) to construct a large class of interacting particle systems. However, the theory is not suitable in the present situation. Even if one has a semigroup at hand, it is still quite a distance to construct the process, since in our case we do not have the Riesz representation theorem for constructing the transition probability kernel. Moreover, from the author's knowledge, since the state space is so poor, the usual weak convergence (even on the path space) is not effective for the construction. What we adopt is a stronger convergence.

Recall that for two given probability measures P_1 and P_2 on a measurable state space $(E, \mathscr{E})$, a *coupling* of P_1 and P_2 is a probability measure $\widetilde{P}$ on the product space $(E \times E, \mathscr{E} \times \mathscr{E})$ having the marginality $\widetilde{P}(A \times E) = P_1(A)$ and $\widetilde{P}(E \times A) = P_2(A)$ for all $A \in \mathscr{E}$. Next, assume that $(E, \rho, \mathscr{E})$ is a metric space with distance ρ. The *Wasserstein distance* $W(P_1, P_2)$ of P_1 and P_2 is defined by

$$W(P_1, P_2) = \inf_{\widetilde{P}} \int_{E^2} \rho(x_1, x_2) \widetilde{P}(\mathrm{d}x_1, \mathrm{d}x_2), \tag{9.2}$$

where $\widetilde{P}$ varies over all couplings of P_1 and P_2. Refer to Section 2.2 and Chen (1992a, Chapter 5) for further properties of the Wasserstein distance.

Now our strategy goes as follows. Take a sequence of finite subsets $\{\Lambda_n\}$ of $S = \mathbb{Z}^d$, $\Lambda_n \uparrow S$. Using Λ_n instead of S, we obtain an MC $P_n(t, x, \cdot)$ as mentioned in the last section. For each $n < m$, one may regard $P_n(t, x, \cdot)$ as an MC on the larger space $E_m := \mathbb{Z}_+^{\Lambda_m}$ and hence for fixed $t \geqslant 0$ and $x \in E_m$, the distance $W(P_n(t, x, \cdot), P_m(t, x, \cdot))$ of $P_n(t, x, \cdot)$ and $P_m(t, x, \cdot)$ is well defined. Clearly, one key step in our construction is to prove that

$$W(P_n(t, x, \cdot), P_m(t, x, \cdot)) \longrightarrow 0 \qquad \text{as } m, n \to \infty. \tag{9.3}$$

Certainly, there is no hope of computing exactly the W-distance, since $P_n(t, x, \cdot)$ is not explicitly known. In virtue of (9.3), we need only an upper bound of the distance, and moreover, it follows from (9.2) that every coupling gives us such a bound. The problem is that a coupling measure of $P_n(t, x, \cdot)$ and $P_m(t, x, \cdot)$ for fixed t and x is still not easy to construct, again due to the fact that these marginal measures are not known explicitly. What we know is mainly the operators Ω_n obtained from Ω but replacing $\mathbb{Z}^d$ with Λ_n. Thus, in order to get some practical coupling, it is natural to restrict ourselves to the Markovian coupling; i.e., the coupling process itself is again an MC. This analysis leads us to explore a theory of couplings for time-continuous Markov processes, which dates back to Chen (1984).

Since then, we have gone a long way in the field: from MC to general jump processes [Chen (1986a)], from discrete state spaces to continuous spaces [Chen and S.F. Li (1989)], from Markovian couplings to optimal Markovian couplings [Chen (1994a; 1994b)], from exponential convergence to the estimation of spectral gap [Chen and F.Y. Wang (1993b), Chen (1994a)], from compact manifolds to noncompact ones [Chen and F.Y. Wang (1995; 1997a; 1997b)], and from finite dimensions to infinite ones [Chen (1987; 1989c; 1991d; 1994c), F.Y. Wang (1994c; 1995; 1996)]. No doubt, the coupling methods are now a powerful tool and have many applications. The story of our study of couplings is presented in Chapter 2.

We now return to our main construction. We will restrict ourselves to a single reactant for a while. Let (k_u) be a positive summable sequence and set

$$E_0 = \left\{ x \in E : \|x\| := \sum_{u \in S} x_u k_u < \infty \right\},$$

i.e., an L^1-subspace of E with respect to (k_u). Roughly speaking, the key to our construction (which is rather lengthy and technical) is to get the following estimates:

(E.1) $P_n(t)\| \cdot \|(x) \leqslant (1 + \|x\|)e^{ct}$, $x \in E_0$, and
(E.2) $W_{\Lambda_n}(P_n(t, x, \cdot), P_m(t, x, \cdot)) \leqslant c(t, \Lambda_n, x; n, m)$, $x \in E_0$,

where c is a constant, independent of n, $c(t, \Lambda_n, x; n, m) \in \mathbb{R}_+$ satisfy

$$\lim_{m \geqslant n \to \infty} c(t, \Lambda_n, x; n, m) = 0,$$

and W_V is the Wasserstein distance restricted to $\mathbb{Z}_+^V$, with respect to the underlying distance $\sum_{u \in V} |x_u - y_u| k_u$. The second condition (E.2) shows that $\{P_n(t, x, \cdot) : n \geqslant 1\}$ is a Cauchy sequence in the W_V-distance (for each fixed finite V). Noting that our operators are not locally bounded and the particles from infinite sites may move to a single site, the process may explode at some single site. This explains why we use E_0 instead of E. Then, the first moment condition (E.1) ensures that E_0 is a closed set of the process. Finally, in order to prove that the limiting process satisfies the Chapman–Kolmogorov equation, some kind of uniform control in the second condition is also needed.

To state our main result, we need some assumptions:

$$\sup_v \sum_u p(u, v) < \infty, \tag{9.4}$$

$$\sum_{k \neq 0} q_u(i, i + k)|k| < \infty, \qquad u \in S, \tag{9.5}$$

$$\sup_{k, u} |c_u(k) - c_u(k + 1)| < \infty, \tag{9.6}$$

$$\sup \left\{ g_u(j_1, j_2) + h_u(j_1, j_2) : u \in S, \; j_2 > j_1 \geqslant 0 \right\} < \infty, \tag{9.7}$$

where

$$g_u(j_1, j_2) = \frac{1}{j_2 - j_1} \sum_{k \neq 0} (q_u(j_2, j_2 + k) - q_u(j_1, j_1 + k))k, \qquad j_2 > j_1 \geqslant 0,$$

$$h_u(j_1, j_2) = \frac{2}{j_2 - j_1} \sum_{k=0}^{\infty} \Big[(q_u(j_2, j_1 - k) - q_u(j_1, 2j_1 - j_2 - k))^+$$

$$+ (q_u(j_1, j_2 + k) - q_u(j_2, 2j_2 - j_1 + k))^+ \Big] k, \qquad j_2 > j_1 \geqslant 0.$$

Conditions (9.1), (9.4), and (9.5) are natural. For instance, when $p(u, v)$ is the simple random walk, (9.4) becomes trivial. However, conditions (9.6) and (9.7) are essential in this construction; they are keys to the estimates (E.1) and (E.2) mentioned above and also to the study of mean field models discussed below. To get a feeling for condition (9.7), let us explain the coupling adopted to deduce (E.2). For the diffusion part, in the box Λ_n, we use the coupling of marching soldiers. For each pair $\{u, v\}, u, v \in \Lambda_n$, let

$$
\begin{aligned}
(x, y) &\to (x - e_u + e_v, \, y - e_u + e_v) && \text{at rate } p(u, v)\big(c_u(x_u) \wedge c_u(y_u)\big) \\
&\to (x - e_u + e_v, \, y) && \text{at rate } p(u, v)\big(c_u(x_u) - c_u(y_u)\big)^+ \\
&\to (x, \, y - e_u + e_v) && \text{at rate } p(u, v)\big(c_u(y_u) - c_u(x_u)\big)^+.
\end{aligned}
$$

In the box $\Lambda_m \setminus \Lambda_n$, since $(x_u : u \in \Lambda_m \setminus \Lambda_n)$ is absorbed, the second process (y_u) evolves alone. That is, for each $u \in \Lambda_n$ and $v \in \Lambda_m \setminus \Lambda_n$, take

$$
(x, y) \to (x, \, y - e_u + e_v) \quad \text{at rate } p(u, v) c_u(y_u).
$$

Conversely, for $u \in \Lambda_m \setminus \Lambda_n$ and $v \in \Lambda_m$, we have the same evolution as in the last line. Moreover, for different pairs, the couplings are taken to be independent. In other words, we have the coupling operator $\widetilde{\Omega}^d_{n,m}$ for the diffusion part as follows:

$$
\begin{aligned}
\widetilde{\Omega}^d_{n,m} & f(x, y) \\
= & \sum_{u, v \in \Lambda_n} p(u, v)\big(c_u(x_u) \wedge c_u(y_u)\big)[f(x - e_u + e_v, \, y - e_u + e_v) - f(x, y)] \\
& + \sum_{u, v \in \Lambda_n} p(u, v)\big(c_u(x_u) - c_u(y_u)\big)^+ [f(x - e_u + e_v, \, y) - f(x, y)] \\
& + \sum_{u, v \in \Lambda_n} p(u, v)\big(c_u(y_u) - c_u(x_u)\big)^+ [f(x, \, y - e_u + e_v) - f(x, y)] \\
& + \sum_{u \in \Lambda_n, \, v \in \Lambda_m \setminus \Lambda_n} p(u, v) c_u(y_u)[f(x, \, y - e_u + e_v) - f(x, y)] \\
& + \sum_{u \in \Lambda_m \setminus \Lambda_n, \, v \in \Lambda_m} p(u, v) c_u(y_u)[f(x, \, y - e_u + e_v) - f(x, y)].
\end{aligned}
$$

For the reaction part, in the box Λ_n, we also use the coupling of marching soldiers. For each $u \in \Lambda_n$, let

$$
\begin{aligned}
(x, y) &\to (x + ke_u, \, y + ke_u) && \text{at rate } q_u(x_u, x_u + k) \wedge q_u(y_u, y_u + k) \\
&\to (x + ke_u, \, y) && \text{at rate } \big(q_u(x_u, x_u + k) - q_u(y_u, y_u + k)\big)^+ \\
&\to (x, \, y + ke_u) && \text{at rate } \big(q_u(y_u, y_u + k) - q_u(x_u, x_u + k)\big)^+.
\end{aligned}
$$

Again, for each $u \in \Lambda_m \setminus \Lambda_n$, let the second process (y_u) evolve alone:

$$
(x, y) \to (x, \, y + ke_u) \quad \text{at rate } q_u(y_u, y_u + k).
$$

Finally, for the reaction part, let each component evolve independently. Thus, we have defined a coupling operator $\widetilde{\Omega}^r_{n,m}$ for the reaction part. Then the whole coupling of the operators Ω_m and Ω_n is defined by $\widetilde{\Omega}_{n,m} = \widetilde{\Omega}^d_{n,m} + \widetilde{\Omega}^r_{n,m}$. Computing the action of the coupling operator on the distance in $\mathbb{Z}^{\Lambda_m}_+$, we get condition (9.7) and an estimate for (E.2). The computations are rather long and technical; only a part of them are illustrated at the end of this chapter. Refer to Chen (1992a, Chapter 13) for details.

The next result is due to Chen (1985b), first reported at the Second International Conference on Random Fields, Hungary, 1984. See also Chen (1987, 1986b, 1992a) for more general results.

Theorem 9.5. Denote by $\mathscr{E}_0$ the Borel σ-algebra generated by the distance $\| \cdot \|$ on E_0. Under (9.1) and (9.4)–(9.7), there exists a Markov process on $(E_0, \mathscr{E}_0)$, and the corresponding semigroup (P_t) maps the set of Lipschitz functions on E_0 with respect to $\| \cdot \|$ into itself. Moreover, for every Lipschitz function f on E_0, the derivative of $P_t f$ at the origin coincides with Ωf in a dense set of E_0.

It is now a simple matter to justify the assumptions of Theorem 9.5 for Examples 9.1 and 9.2. However, up to now, we do not know how to choose a distance so that our general theorem [Chen (1987; 1986b; 1992a)] can be applied to obtain a Lipschitz semigroup for Example 9.3. In the case where the diffusion rates are bounded or growing at most as fast as $\log x_u$, an infinite-dimensional process corresponding to Example 9.3 was constructed by S.Z. Tang (1985) [see also Chen (1992a, Example 13.38)] and D. Han (1990; 1992; 1995). For the mean field models, the problem was solved by S. Feng (1995). In the latter papers, the martingale approach was adopted but not the analytic one used here.

Open Problem 9.6. Construct a Markov process for the Brusselator model.

The next result is due to Y. Li (1991), which improves the author's one in (1991d).

Theorem 9.7. Under the same assumptions as in Theorem 9.5, if additionally,

$$\sup_u \sum_{k \neq 0} q_u(i, i+k)\big[(i+k)^m - i^m\big] \leqslant \text{constant } (1 + i^m), \qquad i \in \mathbb{Z}_+ \quad (9.8)$$

for some $m > 1$, then the process constructed by Theorem 9.5 is also unique.

The proof of Theorem 9.7 is also nontrivial. It uses an infinite-dimensional version of the maximum principle, due to S.Z. Tang (1985) and Y. Li (1991). This is the third mathematical tool developed from the study of RD processes.

9.4 Ergodicity and phase transitions

Existence of stationary distributions

When the state space is compact, it is known that every Feller process has a stationary distribution. But for the noncompact case, there is no such general theorem, and so one needs to work things out case by case. The next result is a particular case of Chen (1986b; 1992a). See also L.P. Huang (1987).

Theorem 9.8. There always exists at least one stationary distribution for the polynomial model.

The intuition for the result is quite clear. Since the order of the death rate is higher than the birth one, the number of particles at each site is kept to be nearly bounded, and then we may return to the compact situation. However, the proof depends heavily on the construction of the process. We will not go into the details here.

Ergodicity

There are two cases: the general case and the reversible case.

(a) *General case.* By using the coupling methods again, some general sufficient conditions for the ergodicity of the processes were presented in Chen (1986b; 1989c). The result was then improved in C. Neuhauser (1990) and further improved in Chen (1990). In the case that the coefficients of the operator are translation-invariant and with an absorbing state, some refined results are given in Y. Li (1995). A particular result from Chen (1990) can be stated as follows.

Theorem 9.9. For the polynomial model, when $\beta_1, \ldots, \beta_{m_0}$ and $\delta_1, \ldots, \delta_{m_0+1}$ are fixed, the processes are exponentially ergodic, uniformly in the initial points, for all large enough β_0.

We will come back to this topic at the end of this chapter (Theorems 9.18 and 9.19).

(b) *Reversible case.* When the reaction part is a birth–death process with birth rates $b(k)$ and death rates $a(k)$, the RD process is reversible iff $p(u, v) = p(v, u)$ and $(k + 1)b(k)/a(k) = $ constant, independent of k [cf. Chen, W.D. Ding, and D.J. Zhu (1994)].

The next result is due to W.D. Ding, R. Durrett, and T.M. Liggett (1990).

Theorem 9.10. For the reversible polynomial model, the processes are always ergodic.

The proof of the result is a nice illustration of the application of the free energy method. It also uses the power of the monotonicity of the processes. The result was then extended by Chen, W.D. Ding, and D.J. Zhu (1994) to the nonpolynomial case.

If we replace β_0, $\delta_1 > 0$ with $\beta_0 = \delta_1 = 0$, then we obtain two stationary distributions; one is trivial and the other one is nontrivial. The question is whether starting from a nontrivial initial distribution, the process converges to the nontrivial stationary distribution (ergodic). The affirmative assertion is called *Shiga's conjecture* (T. Shiga (1988)), which was solved by T.S. Mountford (1992).

Theorem 9.11. For the reversible polynomial model with $\beta_0 = \delta_1 = 0$, under mild assumptions, Shiga's conjecture is correct.

Phase transitions

(a) *RD processes with absorbing state.* The following result was first proved by Y. Li and X.G. Zheng (1988) using a colored graphic representation, and then simplified by R. Durrett (1988) using oriented percolation [see Chen (1992a, Theorem 15.8)].

Theorem 9.12. Take $S = \mathbb{Z}$. Consider the RD process with birth rates $b(k) = \lambda k$, arbitrary death rates $a(k) > 0$ ($k \geqslant 1$), and diffusion coefficient $x_u p(u, v)$, where $p(u, v)$ is the simple random walk. Then for the process $X^0(t)$ starting from x^0: $x_0^0 = 1$ and $x_u^0 = 0$ for all $u \neq 0$, we have

$$\inf\{\lambda : \mathbb{P}[X^0(t) \not\equiv 0 \text{ for all } t > 0] > 0\} < \infty.$$

In other words, for some $\lambda > 0$, there exists a nontrivial stationary distribution.

(b) *Mean field models.* In statistical physics, one often studies the mean field models as simplified approximations of the original ones. It is usually a common phenomenon that with the mean field models it is easier to exhibit phase transitions. Roughly speaking, the mean field model of an RD process is the time-inhomogeneous birth–death process on $\mathbb{Z}_+$ with death rates $a(k)$ as usual but with birth rates $b(k) + \mathbb{E}X(t)$, where $(X(t))_{t \geqslant 0}$ denotes the process. The term $\mathbb{E}X(t)$ represents the interaction of the particle at the present site with the particles at the other sites in the original models. The next result is due to S. Feng and X.G. Zheng (1992).

Theorem 9.13. For the mean field of the second Schlögl model, there always exists at least one stationary distribution. There is precise one if $\delta_1, \delta_3 \gg 1$ and there are more than two if $1 < \delta_1 < \sqrt{1/2 + (2\beta_2 + 1)/(3\delta_1 + 6\delta_3)}$ and β_0 is small enough.

For more information about mean field models, refer to D.A. Dawson and X.G. Zheng (1991), S. Feng (1994a; 1994b; 1995), S. Feng and X.G. Zheng (1992). In B. Djehiche and I. Kaj (1995), the models are treated as a measure-valued process.

Finally, we mention another model, the linear growth model that exhibits phase transitions; refer to W.D. Ding and X.G. Zheng (1989). However, we are still unable to solve the following problem.

Open Problem 9.14. Does there exist more than one stationary distribution for the polynomial model with no absorbing states?

The last phrase means that $\beta_0 > 0$. In physics, this represents an exchange of energy between inside and outside (nonequilibrium). From the mathematical point of view, there is an essential difference between $\beta_0 = 0$ and $\beta_0 > 0$. For instance, when $\beta_0 = 0$, the process restricted to $\{x : \sum_u x_u < \infty\}$ is simply an MC, but this is no longer true when $\beta_0 > 0$.

Because the RD processes are quite involved, partially due to the non-compactness of the state space, one may construct some similar models with finite spin space to simplify the problem. There are many publications in this direction. Refer to R. Durrett and S. Levin (1994), R. Durrett (1995), R. Durrett and C. Neuhauser (1994), and references therein.

9.5 Hydrodynamic limits

Consider again the polynomial model. However, we now study the process with the rescaled operator $\Omega^\varepsilon = \varepsilon^{-2}\Omega_d + \Omega_r$. Our main purpose is to look for the limiting behavior of the scaling processes as $\varepsilon \to 0$. To do so, let $\mu^\varepsilon (\varepsilon > 0)$ be the independent product of the Poisson measures for which $\mu^\varepsilon(x_u) = \rho(\varepsilon u)$, $u \in \mathbb{Z}^d$, where ρ is a nonnegative, bounded $C^2(\mathbb{R}^d)$ function with bounded first derivative.

Denote by $\mathbb{E}^\varepsilon_{\mu^\varepsilon}$ the expectation of the process with formal generator Ω^ε and initial distribution μ^ε. The next result is due to C. Boldrighini, A. DeMasi, A. Pellegrinotti, and E. Presutti (1987) [See Chen (1992a, Theorem 16.1)]. Refer also to T. Funaki (1997; 1999), J.F. Feng (1996), and A. Perrut (2000).

Theorem 9.15. For all $r = (r^1, \ldots, r^d) \in \mathbb{R}^d$ and $t \geqslant 0$, the limit $f(t,r) := \lim_{\varepsilon \to 0} \mathbb{E}^\varepsilon_{\mu^\varepsilon} X_{[r/\varepsilon]}(t)$, where $[r/\varepsilon] = ([r^1/\varepsilon], \ldots, [r^d/\varepsilon]) \in \mathbb{Z}^d$, exists and satisfies the RD equation

$$\begin{cases} \dfrac{\partial f}{\partial t} = \dfrac{1}{2} \sum_{i=1}^{d} \dfrac{\partial^2 f}{\partial (r^i)^2} + \sum_{j=0}^{m_0} \beta_j f^j - \sum_{j=1}^{m_0+1} \delta_j f^j, \\ f(0,r) = \rho(r). \end{cases} \tag{9.9}$$

This result explains the relation between the RD process and RD equation, and it is indeed the original reason why the processes were named RD processes in Chen (1985b). Certainly, at that time, a result like Theorem 9.15 did not exist, we had only a rough impression that the RD equations describe the macroscopic behavior of the physical systems, and our aim was to introduce the processes as a microscopic description of the same systems.

To give some insight into the relationships between these two subjects, we need some notation. Let $\lambda \geqslant 0$ satisfy the algebraic equation

$$\sum_{j=0}^{m} \beta_j \lambda^j - \sum_{j=1}^{m+1} \delta_j \lambda^j = 0, \tag{9.10}$$

which is the simplest solution to the first equation of (9.9). A (constant equilibrium) solution λ is called *asymptotically stable* if there exists a $\delta > 0$ such that for any solution $f(t, r)$ to (9.9), whenever $|f(0, r) - \lambda| < \delta$, we have $\lim_{t \to \infty} |f(t, r) - \lambda| = 0$.

The following result is due to X.J. Xu (1991) [see Chen (1992a, Theorem 16.2)].

Theorem 9.16. Denote by $\lambda_1 > \lambda_2 > \cdots > \lambda_k$ the nonnegative roots of (9.10), where λ_j has multiplicity m_j. Then, $f(t, r) \equiv \lambda_i$ is asymptotically stable iff m_i is odd and $\sum_{j \leqslant i-1} m_j$ is even.

All known results are consistent with the assertion that a model has no phase transition iff every λ_j is asymptotically stable and it is a case of Schlögl's first model. This leads to the following conjecture.

Conjecture 9.17. (1). Schlögl's first model has no phase transition.
(2). Schlögl's second model has phase transitions.

To conclude this chapter, we want to show a use of the RD equation. Note that for Schlögl's second model, the role played by the parameters β_k and δ_k is not clear at all. It seems too hard and may not be necessary to consider all of the parameters. Based on the above observation and to keep the physical meaning, we fix $\beta_2 = 6\alpha \, (\alpha > 0)$, $\delta_1 = 9\alpha$, and $\delta_3 = \alpha$. Then, when $\beta_0 \in (0, 4\alpha)$, there are three roots $\lambda_1 > \lambda_2 > \lambda_3 \geqslant 0$; λ_1 and λ_3 are asymptotically stable, but not λ_2. We have thus reduced the four parameters to only one. Now we want to know for which region of α the process can be ergodic. The following result is based on recent progress on couplings [cf. Chapter 2 or Chen (1994a)]; it is complementary to Theorem 9.9, and is also the most precise information we have so far.

Theorem 9.18 (Chen, 1994c). Consider the second Schlögl model with $\beta_0 = 2\alpha$, $\beta_2 = 6\alpha$, $\delta_1 = 9\alpha$, and $\delta_3 = \alpha$. Then the processes are exponentially ergodic, uniformly in the initial points, for all $\alpha \geqslant 0.7303$.

We now sketch the proof of the previous theorem. Actually, we have a general result as follows.

Theorem 9.19. Consider the polynomial model. Let (u_k) be a positive sequence on $\mathbb{Z}_+$ with $u_0 = 1$ and $\bar{u} := \sup_{k \geqslant 0} u_k < \infty$. Set $u^* = \sup_{j > i \geqslant 0}(u_j - u_i) \vee 0$. Suppose that there exists an $\varepsilon > 0$ such that

$$b_{k+1} u_{k+1} - (b_k + a_{k+1} + k + 1 - \varepsilon) u_k + (a_k + k) u_{k-1} + \bar{u} + k u^* \leqslant 0, \qquad k \geqslant 0,$$

where $a_0 = 0$ and $u_{-1} = 1$. Then the reaction–diffusion processes are exponentially ergodic, uniformly in the initial points.

Sketch of the proof. (a) Define a distance in $\mathbb{Z}_+$ as follows:

$$\rho(k, \ell) = \left| \sum_{j < k} u_j - \sum_{j < \ell} u_j \right|, \qquad k, \ell \in \mathbb{Z}_+.$$

By Theorem 2.28, for birth–death processes, the couplings mentioned in Chapter 2, except the independent one, are all ρ-optimal. Thus, we now adopt the simplest classical coupling. Denote by $\widetilde{\Omega}_c$ the coupling operator of the reaction–diffusion processes, that is, using the classical coupling for each component of the reaction part, but for the diffusion part still using the coupling of marching soldiers mentioned in Section 9.3. Fix $x \leqslant y$ and $u \in S$, and write $x_u = i \leqslant j = y_u$. We have

$$\widetilde{\Omega}_c\rho(i,j) = \big\{ -b_iu_i + a_iu_{i-1} + b_ju_j - a_ju_{j-1}\big\}I_{\{j-i\geqslant 1\}} - (j-i)u_{j-1}$$
$$- i(u_{j-1} - u_{i-1}) + \sum_v (y_v - x_v)p(v,u)u_j + \sum_v x_vp(v,u)(u_j - u_i)$$
$$= \big\{b_ju_j - b_iu_i - (a_j + j)u_{j-1} + (a_i + i)u_{i-1}\big\}I_{\{j-i\geqslant 1\}}$$
$$+ \sum_v (y_v - x_v)p(v,u)u_j + \sum_v x_vp(v,u)(u_j - u_i).$$

The last term on the right-hand side appears because ρ is not translation-invariant. Now, by assumption, we have

$$\big\{b_ju_j - b_iu_i - (a_j + j)u_{j-1} + (a_i + i)u_{i-1}\big\}I_{\{j-i\geqslant 1\}}$$
$$= \sum_{\ell=i}^{j-1} \big\{(b_{\ell+1}u_{\ell+1} - b_\ell u_\ell) - [(a_{\ell+1} + \ell + 1)u_\ell - (a_\ell + \ell)u_{\ell-1}]\big\}$$
$$\leqslant -\varepsilon \sum_{\ell=i}^{j-1} u_\ell - (j-i)\bar{u} - (j-i)iu^*$$
$$\leqslant -\varepsilon\rho(i,j) - (j-i)\bar{u} - iu^*.$$

On the other hand, by the order preservation of the coupling and the translation invariance of the processes, for every translation-invariant x and y with $x \leqslant y$, we have

$$\sum_v \widetilde{\mathbb{E}}^{x,y}\big(Y_v(t) - X_v(t)\big)p(v,u)u_{Y_u(t)} + \sum_v \widetilde{\mathbb{E}}^{x,y}X_v(t)p(v,u)\big(u_{Y_u(t)} - u_{X_u(t)}\big)$$
$$\leqslant \bar{u}\widetilde{\mathbb{E}}^{x,y}\big(Y_u(t) - X_u(t)\big) + u^*\widetilde{\mathbb{E}}^{x,y}X_u(t).$$

Collecting the above estimates, replacing i and j by X_u and Y_u, respectively, we arrive at

$$\widetilde{\mathbb{E}}^{x,y}\widetilde{\Omega}_c\rho\big(X_u(t), Y_u(t)\big) \leqslant -\varepsilon\widetilde{\mathbb{E}}^{x,y}\rho\big(X_u(t), Y_u(t)\big), \qquad t \geqslant 0.$$

By Gronwall's lemma or Lemma A.6, this gives us

$$\widetilde{\mathbb{E}}^{x,y}\rho\big(X_u(t), Y_u(t)\big) \leqslant \widetilde{\mathbb{E}}^{x,y}\rho\big(X_u(1), Y_u(1)\big)e^{-\varepsilon t}, \qquad t \geqslant 0,$$

for every translation-invariant x and y.

(b) The reason we use the time $t = 1$ as initial value rather than $t = 0$ is the first moment estimate

$$\mathbb{E}^x\big[X_u(t)^m\big] \leqslant \varphi_m(t) < \infty, \qquad t > 0,\ m \in \mathbb{N}$$

[cf. Chen (1992a, Lemma 14.12)]. Thus, we can extend the initial state to be ∞ everywhere. Let (X_t^n) be the process starting from $(x_u = n, u \in \mathbb{Z}^d)$. Then, by (a), we obtain

$$\widetilde{\mathbb{E}}\rho\big(X_u^0(t), Y_u^\infty(t)\big) \leqslant \widetilde{\mathbb{E}}\rho\big(X_u^0(1), Y_u^\infty(1)\big)e^{-\varepsilon t}, \qquad t \geqslant 0.$$

This certainly implies the ergodicity of the process, because of the translation invariance and monotonicity. Clearly, the convergence is exponential, uniformly in the initial points (x, y). $\quad\square$

To complete the proof of Theorem 9.18, by Theorem 9.19, it remains to choose a suitable positive sequence (u_i). Regarding the reaction–diffusion processes at a site as perturbation of the birth–death processes, it seems natural to choose the sequence from the mimic eigenfunction that produces the explicit criterion for exponential convergence (or equivalently, spectral gap). More precisely, take $u_i = (g_{i+1} - g_i)/(g_1 - g_0)$, $i \geqslant 0$, where

$$g_i = \sum_{j=0}^{i-1} \frac{1}{\mu_j b_j} \sum_{k=j+1}^{\infty} \mu_k \sqrt{\varphi_k}, \qquad \varphi_i = \sum_{j=0}^{i-1} \frac{1}{\mu_j b_j}, \qquad i \geqslant 0.$$

Due to the diffusion part, one may replace the original b_k and a_k by $b_k + k$ and $a_k + k$, respectively. Intuitively, this means that the interaction is ignored. However, the resulting sequence (u_i) is indeed not good enough for Theorem 9.18. Practically, we adopt a more direct and economic way to define the sequence. Take $\varepsilon \leqslant 10^{-5}$ and define

$$u_0 = 1, \qquad u_1 = u_2 = 3/2 + \varepsilon \quad \text{(trick!)},$$

$$u_{k+1} = \frac{(a_{k+1} + b_k + k + 1 - \varepsilon)u_k - (a_k + k)u_{k-1} - (k+1)u_1 + k}{b_{k+1}}, \qquad k \geqslant 2.$$

Refer to Chen (1994c) for more details.

A large number of publications of the study on hydrodynamic limits are collected in the book by C. Kipnis and C. Landim (1999), from which one sees that the spectral gap and the logarithmic Sobolev inequalities play a crucial role. The spectral gap for the Ising model in dimension one was computed explicitly by R.A. Minlos and A.G. Trishch (1994). For higher-dimensional results, refer to A.D. Sokal and L.E. Thomas (1988), L.E. Thomas (1989), R.H. Schonmann (1994), S.L. Lu and H.T. Yau (1993), and R.A. Minlos (1996). For other equilibrium particle systems, here are some recent excellent explorations: A. Guionnet and B. Zegarlinski (2003), M. Ledoux (1999; 2001), F. Martinelli (1999), and so on. Some remarkable approaches were created or developed in these quoted papers.

Chapter 10

Stochastic Models of
Economic Optimization

This chapter deals with some stochastic models of economic optimization. Due to their value in practice, the models are quite attractive. But our knowledge on them is still very limited, and some fundamental problems remain open.

We begin with a short review of some global economic models (or economy on a large scale), the well-known input–output method, and L.K. Hua's fundamental theorem for the stability of an economy. Then we show that it is necessary to study the stochastic models. A collapse theorem for a non-controlling stochastic economic system is introduced. In the analysis of the system, the products of random matrices play a crucial role. In particular, the first eigenvalue, the corresponding eigenfunctions, and an ergodic theorem of Markov chains play a nice role here. Partial proofs are included. Some challenging open problems are also mentioned.

10.1　Input–output method

First, we fix the units of each product: kilogram, kilovolt, and so on. Denote by $x = \left(x^{(1)}, x^{(2)}, \ldots, x^{(d)}\right)$ the quantity of the main products in which we are interested; it is called the *vector of products*. Throughout this chapter, all vectors are row vectors.

To understand the present economy, we need to examine three things: the input, the output, and the structure matrix. Suppose that the starting vector of products last year was

$$x_0 = \left(x_0^{(1)}, x_0^{(2)}, \ldots, x_0^{(d)}\right).$$

For production, assume that the jth product distributed amount $x_{ij}^{(0)}$ to the

ith product, and the vector of the products this year becomes

$$x_1 = \left(x_1^{(1)}, x_1^{(2)}, \ldots, x_1^{(d)} \right).$$

Next, set

$$a_{ij}^{(0)} = x_{ij}^{(0)} / x_1^{(i)}, \qquad 1 \leqslant i,\, j \leqslant d.$$

The matrix $A_0 = \left(a_{ij}^{(0)} \right)$ is called a *structure matrix* (or matrix of *expending coefficients*). This matrix is essential, since it describes the efficiency of the current economy: to produce one unit of the ith product, one needs $a_{ij}^{(0)}$ units of the jth product. Suppose for a moment that all the products are used for reproduction (*idealized model*). Then

$$x_0^{(j)} = \sum_i x_{ij}^{(0)} = \sum_i x_1^{(i)} a_{ij}^{(0)}.$$

That is, $x_0 = x_1 A_0$. Similarly, we have $x_{n-1} = x_n A_{n-1}$ for all $n \geqslant 1$. Suppose that the structure matrices are time-homogeneous: $A_n = A$ for all $n \geqslant 0$ (this is reasonable if one considers a short time unit). Then we have a simple expression for the nth output:

$$x_n = x_0 A^{-n}, \qquad n \geqslant 1. \tag{10.1}$$

Thus, once the structure matrix and the input x_0 are known, we may predict the future output. This is called the *input–output method* or *Leontief's method* [cf. Leontief (1936; 1951; 1986)]. It is a well known method. As far as I know, up to the 1960s, more than 100 countries had used this method in their national economies.

10.2 L.K. Hua's fundamental theorem

Let us return to the original equation

$$x_1 = x_0 A^{-1}.$$

We now fix A. Then x_1 is determined by x_0 only. The question is, which choice of x_0 is the optimal one? Furthermore, what sense of optimality are we talking about? The first choice would be "average." If someone tells you that the average of the members' ages in a group is twenty, you may think that everyone in the group is a vigorous young adult, and that the group might be a team of volleyball players. However, the group could be a nursery school consisting of six babies and two older women who are over seventy. The average of the ages in this group is still twenty. The misleading point is that the variance is too big in this situation, and so the average is not a good tool. To avoid this, we adopt the *minimax principle*, i.e., finding out the best solution among the worst cases. It is the safest strategy, and is used widely in optimization and game theory. In other words, we want to find x_0 such that $\min_{1 \leqslant j \leqslant d} x_1^{(j)} / x_0^{(j)}$ attains the maximum below

$$\max_{x_1 > 0,\, x_0 = x_1 A}\, \min_{1 \leqslant j \leqslant d}\, x_1^{(j)} / x_0^{(j)}.$$

By using the classical Frobenius theorem, L.K. Hua (1984b, Part III) proved the following result.

Theorem 10.1 (L.K. Hua (1984b, Part III)). Given an irreducible nonnegative matrix A, let u be the (positive) left eigenvector of A, corresponding to the largest eigenvalue $\rho(A)$ of A. Then, up to a constant, the solution to the above problem is $x_0 = u$. In this case, we have

$$x_1^{(j)} / x_0^{(j)} = \rho(A)^{-1} \qquad \text{for all } j.$$

In what follows, we call the above technique (i.e., setting $x_0 = u$) the *eigenvector method*.

Next, we are going to study further the stability of economies. From (10.1), we obtain a simple expression,

$$x_n = x_0 \rho(A)^{-n},$$

whenever $x_0 = u$. What happens if we take $x_0 \neq u$ (up to a constant)?

Stability of an economy

For convenience, set

$$T^x = \inf \left\{ n \geqslant 1 : x_0 = x \text{ and there is some } j \text{ such that } x_n^{(j)} \leqslant 0 \right\},$$

which is called the *collapse time* of the economic system.

We can now state Hua's important result as follows.

Theorem 10.2 (L.K. Hua (1984b, Part III; 1985, Part IX)). Let A be nonnegative, irreducible, invertible, and not of the form that every row and column has one and only one positive element. Then for every $x_0 \neq u$, we have $T^{x_0} < \infty$.

In the case that the collapse time is bigger than 150 years, we do not need to worry about the stability of the economy, since none of us will be still alive. However, the next example shows that we are not in this situation.

Example 10.3 (L.K. Hua (1984b, Part I)). Consider two products only: industry and agriculture. Let

$$A = \frac{1}{100} \begin{pmatrix} 20 & 14 \\ 40 & 12 \end{pmatrix}.$$

Then $u = \left(5\left(\sqrt{2400} + 13\right)/7,\, 20\right) \approx (44.34397483,\, 20)$. For different input x_0, the collapse time T^{x_0} is listed in Table 10.1.

Table 10.1   Input and collapse time

x_0	T^{x_0}
$(44, 20)$	3
$(44.344, 20)$	8
$(44.34397483, 20)$	13

This shows that the economy is very sensitive!

We point out that Theorem 10.2 is essential. Recall that the Frobenius theorem and Brouwer fixed point theorem, often used in the study of economics, do not provide any information about the collapse phenomena.

To understand Hua's theorem, for probabilists, it is very natural to consider the particular case that $A = P$. That is, A is a transition probability matrix. Then, from an ergodic theorem for Markov chains (irreducible and aperiodic), it follows that

$$P^n \to \Pi \qquad \text{as } n \to \infty,$$

where Π is the matrix having the same row $\left(\pi^{(1)}, \pi^{(2)}, \ldots, \pi^{(d)}\right)$, which is just the stationary distribution of the corresponding Markov chain. Since the distribution is the only stable solution for the chain, it should have some meaning in economics even though the economic model goes in a converse way:

$$x_n = x_0 P^{-n}, \qquad n \geq 1.$$

From the above facts, it is not difficult to prove, as shown in the next paragraph, that if

$$x_0 \neq u = \left(\pi^{(1)}, \pi^{(2)}, \ldots, \pi^{(d)}\right)$$

up to a positive constant, then $T^{x_0} < \infty$. Next, since the general case can be reduced to the above particular case, we think that this is a very natural way to understand Hua's theorem.

Proof of Theorem 10.2 [L.K. Hua (1984b, Part IX) and Chen (1992b, Part I)].

(a) First, consider the special case that $A = P$. We need to show that if $x_0 \neq \pi$ up to a positive constant and $d \geq 2$, we must have $x_n \not> 0$ for some n. In other words, if $x_n > 0$ for all n, then $x_0 = \pi$.

Let $x_0 > 0$ be normalized such that $x_0 \mathbb{1}^* = 1$, where $\mathbb{1}^*$ is the column vector having components 1 everywhere. Because $x_0 = x_n A^n = x_n P^n$ and $P\mathbb{1}^* = \mathbb{1}^*$, we have

$$1 = x_0 \mathbb{1}^* = x_n P^n \mathbb{1}^* = x_n \mathbb{1}^*, \qquad n \geq 1.$$

Since the set $\{x : x \geqslant 0, \ x\mathbb{1}^* = 1\}$ is compact, there exists a subsequence $\{x_{n_k}\}_{k \geqslant 1}$ and a vector $\bar{x}$ such that

$$\lim_{k \to \infty} x_{n_k} = \bar{x}, \qquad \bar{x} \geqslant 0, \qquad \bar{x}\mathbb{1}^* = 1.$$

Therefore

$$x_0 = (x_0 P^{-n_k}) P^{n_k} = x_{n_k} P^{n_k} \to \bar{x}\Pi = \bar{x}\mathbb{1}^* \pi = \pi.$$

Thus, we must have $x_0 = \pi$.

(b) For general nonnegative primitive (irreducible and aperiodic) A, by F.R. Gantmacher (1989, Chapter 13, Section 6) or L.K. Hua (1984a, Chapter 9, Sections 2–4), there exists a diagonal matrix D with positive diagonals such that $A = \rho(A) D^{-1} P D$, and so we are done.

(c) Finally, assume that A has periodic r. That is, there are r eigenvalues $\{\lambda_j\}$ with the same modulus $|\lambda_j| = \rho(A)$. Then A can be represented as (cyclic decomposition)

$$\begin{pmatrix} 0 & A_{12} & 0 & \cdots & 0 \\ 0 & 0 & A_{23} & \cdots & 0 \\ \multicolumn{5}{c}{\dotfill} \\ A_{r1} & 0 & 0 & \cdots & 0 \end{pmatrix},$$

where the A_{ij}'s are $s \times s$ $(d = rs)$ matrices, since A is invertible. The case that $s = 1$ is exceptional, since then $A^{-1} \geqslant 0$. This is ruled out by the last assumption of the theorem. Thus $s > 1$. Then, we have $A^r = \mathrm{diag}\,[B_1, B_2, \ldots, B_r]$, where each B_j is primitive. Without loss of generality, assume that $\rho(A) = 1$. Then $\rho(B_j) = 1$ for all j. According to (b), there exist a diagonal matrix D_j, transition probability matrix P_j, and stationary distribution $\pi^{(j)}$ such that $B_j = D_j^{-1} P_j D_j$, and

$$B_j^n = D_j^{-1} P_j^n D_j \to D_j^{-1} \mathbb{1}_s^* \pi^{(j)} D_j, \qquad n \to \infty,$$

where $\mathbb{1}_s^*$ is the column vector with s elements 1. Replacing A with A^r and renormalizing the vector $x := (x^{(1)}, x^{(2)}, \ldots, x^{(r)})$ by $x^{(j)} D_j^{-1} \mathbb{1}_s^* = 1$ for all $1 \leqslant j \leqslant r$, the same argument in (a) shows that we must have $x_0 = (\pi^{(1)}, \ldots, \pi^{(r)}) D$, where $D = \mathrm{diag}\,[D_j]$, once $x_n > 0$ for all n. Because A is irreducible, the left eigenvector is unique. Moreover, if $u = uA$, then $u = uA^r$. It is clear that the eigenvector must coincide with $(\pi^{(1)}, \ldots, \pi^{(r)}) D$, up to a positive constant. $\square$

We have seen the crucial role played by the largest or the first eigenvalue and its eigenvectors, for which the computations are far from nontrivial, especially for large-scale matrices, as we have seen from the previous chapters. In the numerical computation of the largest eigenvalue, it is important to have good initial data; this is just an application of estimation of the eigenvalue. Having the known eigenvalue at hand, the computation of eigenvectors is easier, for which one needs only to solve a linear equation (in contrast, the equation of the eigenvalue is polynomial).

Economy in markets

In L.K. Hua's eleven reports (1984–1985), he also studied some more general economic models. But the above two theorems are the key to his idea. The title of the reports (written in that specific period) may cause some misunderstanding, since one may think that the theory works only for planned economies. Actually, market economies were also treated in (Hua, 1984b, Part VII). The only difference is that in the latter case one needs to replace the structure matrix A with $V^{-1}AV$, where V is the diagonal matrix diag(v_i/p_i), (p_i) is the vector of prices in the market, and (v_i) is the right eigenvector of A. Note that the eigenvalues of $V^{-1}AV$ are the same as those of A. Corresponding to the eigenvalue $\rho(V^{-1}AV) = \rho(A)$, the left eigenvector of $V^{-1}AV$ becomes uV. Therefore, for a market economy, we have a new structure matrix $V^{-1}AV$ and a new left eigenvector uV, which are all that we need in Hua's model. Thus, from a mathematical point of view, the consideration of markets makes no essential difference in Hua's model.

10.3   Stochastic model without consumption

In the case that randomness does not play a crucial role, one may simply ignore it and insist on a deterministic system. Thus, we started our study by examining the influence of a smaller random perturbation in Hua's example: Example 10.3.

Consider the perturbation

$$\widetilde{a}_{ij} = a_{ij} \qquad \text{with probability } 2/3,$$
$$= a_{ij}(1 \pm 0.01) \quad \text{with probability } 1/6.$$

Let the elements $\widetilde{a}_{ij}$ be independent. Taking $(\widetilde{a}_{ij})$ instead of (a_{ij}), we get a random matrix. Next, let $\{A_n;\, n \geqslant 1\}$ be a sequence of independent random matrices (regarding each A_n as a vector, the independence of a sequence of random vectors is standard) with the same distribution as above. Then $x_n = x_0 \prod_{k=1}^{n} A_k^{-1}$ gives us a stochastic model of an economy without consumption.

Again, starting from $x_0 = (44.344, 20)$ (recall that the collapse time is 8 in the deterministic case), then the collapse probability in the above stochastic model is the following:

$$\mathbb{P}[T^{x_0} = n] = \begin{cases} 0 & \text{for } n = 1, \\ 0.09 & \text{for } n = 2, \\ 0.65 & \text{for } n = 3. \end{cases}$$

Surprisingly, we have $\mathbb{P}[T \leqslant 3] \approx 0.74$. This observation tells us that randomness plays a crucial role in an economy. It also explains why the traditional input–output is not very practicable, as people often think, because randomness has been ignored and so the deterministic model is far away from actual practice.

Now, what is the analogue of Hua's theorem for the stochastic case?

Theorem 10.4 (Chen (1992b, Part II)).   Under some mild conditions, we have

$$\mathbb{P}[T^{x_0} < \infty] = 1, \qquad \forall x_0 > 0.$$

Note that the limit theory of products of random matrices is quite different from that in the deterministic case (cf. P. Bougerol and J. Lacroix (1985)); the problem is nontrivial. We have to deal with the product of random matrices

$$M_n = A_n A_{n-1} \cdots A_1.$$

As we have seen before, in the deterministic case, the leading order of M_n is $\prod_{j=1}^n \rho(A_j)$. Thus, in the random case, one may study the limiting behavior of $\prod_{j=1}^n A_j / \rho(A_j)$, as suggested by L.K. Hua (1984b, Part II). By Kolmogorov's strong law of large numbers, we have

$$\frac{1}{n} \log \prod_{j=1}^n \rho(A_j) \xrightarrow{\text{a.s.}} \mathbb{E} \log \rho(A_1), \qquad n \to \infty,$$

once $\mathbb{E}|\log \rho(A_1)| < \infty$. So we have a deterministic exponent $\mathbb{E} \log \rho(A_1)$. However, the leading order $\rho(M_n)$ of M_n is not $\prod_{j=1}^n \rho(A_j)$, as shown by the following result.

Theorem 10.5 (V.I. Oseledec, 1968).   Let $\mathbb{E} \log^+ \|A_1\| < \infty$. Then

$$\frac{1}{n} \log \|M_n\| \xrightarrow{\text{a.s.}} \gamma \in \{-\infty\} \cup \mathbb{R},$$

where

$$\gamma = \lim_{n \to \infty} \frac{1}{n} \mathbb{E} \log \|M_n\|.$$

Note that the *Lyapunov exponent* γ is independent of the norms of the matrices.

Oseledec's theorem is the most popular result we learned from the limit theory of products of random matrices, sometimes called the "strong law of large numbers." However, this result is still not enough for our purpose, since we need more, the limiting behavior of M_n under suitable scaling. What we adopted is a much stronger result. To state the result, we need the following assumptions, which are analogues of the irreducible and aperiodic conditions.

(H_1) $A_1 \geqslant 0$ a.s., and there exists an integer m such that

$$\mathbb{P}[M_m \text{ is positive}] > 0,$$

where $M_n = A_1 \cdots A_n$.

(H_2) $\mathbb{P}[A_1 \text{ has a zero row or column}] = 0$.

Theorem 10.6 (H. Kesten and F. Spitzer, 1984).   Under (H_1) and (H_2), $M_n > 0$ for large n with probability one and $M_n/\|M_n\|$ converges in distribution to a positive matrix $M = L^*R$ with rank one, where $\|A\| = \sup_i \sum_j |a_{ij}|$ denotes the norm of $A = (a_{ij})$, and L and R are independent, positive row vectors satisfying the normalizing condition:

$$\max_{1 \leqslant i \leqslant d} R(i) = 1, \qquad \sum_{j=1}^{d} L(j) = 1. \tag{10.2}$$

By a change of the probabilistic frame, one may replace "convergence in distribution" by "convergence almost surely" (Skorohod's theorem, cf. Section 2.2 or N. Ikeda and S. Watanabe (1988, page 9)). In this sense, the last result is really the strong law of large numbers. Having these remarks in mind, the proof of Theorem 10.4 is not difficult and is given in Section 10.5.

One may refer to A. Mukherjea (1991), H. Hennion (1997), and references within for more recent progress on the limit theory of products of random matrices.

10.4   Stochastic model with consumption

The model without consumption is idealized and so is not practical. For practicality one should have consumption, that is, allow a part of the production to turn into consumption and not be used for reproduction.

Suppose that in every year we take the $\theta^{(i)}$ times amount of the increment of the ith product to be consumed. Then in the first year, the vector of products that can be used for reproduction is

$$y_1 = x_0 + (x_1 - x_0)(I - \Theta),$$

where I is the $d \times d$ unit matrix and $\Theta = \text{diag}\left(\theta^{(1)}, \theta^{(2)}, \ldots, \theta^{(d)}\right)$, which is called a *consumption matrix*. Therefore

$$y_1 = y_0[A_0^{-1}(I - \Theta) + \Theta], \qquad y_0 = x_0.$$

Similarly, in the nth year, the vector of the products that can be used for reproduction is

$$y_n = y_0 \prod_{k=0}^{n-1} [A_{n-k-1}^{-1}(I - \Theta) + \Theta], \qquad n \geqslant 1.$$

Let

$$B_n = [A_{n-1}^{-1}(I - \Theta) + \Theta]^{-1}.$$

Then

$$y_n = y_0 \prod_{k=1}^{n} B_{n-k+1}^{-1}, \qquad n \geqslant 1.$$

We have thus obtained a *stochastic model with consumption*. In the deterministic case, a collapse theorem was obtained by L.K. Hua (1985, Part X), L.K. Hua and S. Hua (1985). The conclusion is that the system becomes more stable than the idealized model. More precisely, the dimension of (x_0), for which the economy will not collapse, can be greater than one. This is consistent with our practice.

To state our result in this general case, we need some notation. Denote by $Gl(d, \mathbb{R})$ the general linear group of real invertible $d \times d$ matrices and by $O(d, \mathbb{R})$ the orthogonal matrices in $Gl(d, \mathbb{R})$. Next, given a family of random matrices with distribution μ, denote by $\mathscr{G}_\mu$ the smallest closed semigroup of $Gl(d, \mathbb{R})$ containing the support of μ.

Definition 10.7.

- $\mathscr{G}$ is called strongly irreducible if there exist no proper linear subspaces of $\mathbb{R}^d$, $\mathscr{V}_1, \ldots, \mathscr{V}_k$, such that
$$(\cup_{i=1}^k \mathscr{V}_i)B = \cup_{i=1}^k \mathscr{V}_i, \qquad \forall B \in \mathscr{G}.$$

- $\mathscr{G}$ is said to be contractive if there exists $\{B_n\} \subset \mathscr{G}$ such that $\|B_n\|^{-1}B_n$ converges to a matrix with rank one.

- We call $B = K \mathrm{diag}(a_i)U$ a polar decomposition if $K, U \subset O(d, \mathbb{R})$ and $a_1 \geqslant a_2 \geqslant \cdots \geqslant a_d > 0$.

Theorem 10.8 (Chen and Y. Li, 1994). Let $\{B_n\}$ be an i.i.d. sequence of random matrices with common distribution μ. Suppose that $\mathscr{G}_\mu$ is strongly irreducible and contractive and that the sequence $\{K_n\}$ in the polar decomposition satisfies a "tightness condition." Then $\mathbb{P}[T^x < \infty] = 1$ for all x: $0 < x \in \mathbb{R}^d$.

Naturally, we have the following question.

Open Problem 10.9. How fast does the economy go to collapse?

As we have seen before, since the economy is very sensitive, one certainly expects the following large deviation result:
$$\mathbb{P}[T > n] \leqslant C e^{-\alpha n}.$$

Clearly, Theorem 10.8 is still a distance from being complete. Furthermore, in practice, a collapse result is not expected and less useful. Now another question arises.

Open Problem 10.10. How can one control the economy and what is the optimal one?

To date, we have no idea how to handle this problem; we even do not understand what kind of optimality should be adopted here.

Finally, we mention that a probabilistic exploration of Hua's model, closely related to the ergodic theorem used in the proof of Theorem 10.2, was investigated by K.L. Chung (1995). The topic of this chapter is now explored, with much more extension and recent references, in the book by D. Han and X.J. Hu (2003).

10.5   Proof of Theorem 10.4

Given i.i.d., nonnegative random matrices $\{A_n\}_{n=1}^{\infty}$, since we are working on the economic model

$$x_n = x_0 A_1^{-1} \cdots A_n^{-1},$$

it is natural to assume that

$$\mathbb{P}[\det A_1 = 0] = 0. \tag{10.3}$$

We mainly investigate the collapse probability $\mathbb{P}[T < \infty]$, where T is the same as before,

$$T = \{n \geqslant 1 : \text{ there exists some } 1 \leqslant j \leqslant d \text{ such that } x_n^{(j)} \leqslant 0\}.$$

The following result is a more precise statement of Theorem 10.4.

Theorem 10.11 (Chen (1992b, Part II)). Let (H_1), (H_2), and (10.3) hold. Given a deterministic $x_0 > 0$ with $\max_i x_0^{(i)} = 1$, we have

$$\mathbb{P}[T = \infty] \leqslant \mathbb{P}[R = x_0].$$

In particular, if $\mathbb{P}[R = x_0] = 0$, then $\mathbb{P}[T = \infty] = 0$.

Proof. (a) Write $M_n = A_n \cdots A_1$, $\|A\| = \sup_i \sum_j |a_{ij}|$ for $A = (a_{ij})$, and set $\overline{M}_n = M_n / \|M_n^*\|$. Note that the product M_n is in a different order from that in Theorem 10.6, from which we know that $\overline{M}_n$ converges in distribution to $R^* L$, where R and L are independent, positive row vectors satisfying (10.2).
   (b) By condition (10.3), we have $\|M_n^*\| > 0$, a.s., and so

$$x_n > 0 \iff x_0 M_n^{-1} > 0 \iff x_0 \overline{M}_n^{-1} > 0, \qquad n \geqslant 1.$$

Hence

$$\mathbb{P}[T = \infty] = \mathbb{P}[x_n > 0, \forall n \geqslant 1] = \mathbb{P}[x_0 \overline{M}_n^{-1} > 0, \forall n \geqslant 1].$$

Thus, we can use $\overline{M}_n$ instead of M_n.
   (c) By Skorohod's theorem (cf. Section 2.2 or N. Ikeda and S. Watanabe (1988, page 9)), there exists a probability space $(\widetilde{\Omega}, \widetilde{\mathscr{F}}, \widetilde{\mathbb{P}})$ on which there are $\widetilde{M}_n$ and $\widetilde{M}$ such that

$$\widetilde{M}_n \text{ and } \overline{M}_n \text{ have the same distribution for each } n \geqslant 1,$$
$$\widetilde{M} \text{ and } M := R^* L \text{ have the same distribution}, \tag{10.4}$$
$$\widetilde{M}_n \to \widetilde{M} \quad \text{as } n \to \infty, \qquad \widetilde{\mathbb{P}}\text{-a.s.}$$

In particular,

$$\widetilde{\mathbb{P}}\big[\widetilde{M} \text{ has rank } 1\big] = \mathbb{P}\big[M \text{ has rank } 1\big] = 1.$$

From these facts and the normalizing condition, it is easy to see that there exist positive $\widetilde{R}$ and $\widetilde{L}$, $\widetilde{\mathbb{P}}$-a.s. unique, such that $\widetilde{M} = \widetilde{R}^* \widetilde{L}$ and

$$\max_i \widetilde{R}(i) = 1, \qquad \sum_j \widetilde{L}(j) = 1, \qquad \widetilde{\mathbb{P}}\text{-a.s.}$$

Therefore we must have

$$\mathbb{P}[x_0 \overline{M}_n^{-1} > 0, \forall n \geqslant 1] = \widetilde{\mathbb{P}}[x_0 \widetilde{M}_n^{-1} > 0, \forall n \geqslant 1].$$

Thus we can ignore $\tilde{}$ and use the original $(\Omega, \mathscr{F}, \mathbb{P})$ instead of $(\widetilde{\Omega}, \widetilde{\mathscr{F}}, \widetilde{\mathbb{P}})$, and assume that $\overline{M}_n$ converges to $R^* L$ almost everywhere, rather than convergence in distribution.

(d) By (10.4), there exists a $\mathbb{P}$-zero set Λ such that

$$\overline{M}_n^* \to R^* L \qquad \text{as } n \to \infty \text{ on } \Lambda^c.$$

Write $\bar{x}_n = x_0 \overline{M}_n^{-1}$. Fix $\omega \in \Lambda^c$. If

$$\bar{x}_n(\omega) > 0, \qquad \forall n \geqslant 1,$$

because of the normalizing condition of x_0 and $\overline{M}_n$, there must exist a subsequence $\{n_k = n_k(\omega)\}$ such that

$$\lim_{k \to \infty} \bar{x}_{n_k}(\omega) =: \bar{x}(\omega) \in [0, \infty]^d.$$

But

$$\begin{aligned}
x_0 &= \lim_{k \to \infty} \left[x_0 \overline{M}_{n_k}(\omega)^{-1} \overline{M}_{n_k}(\omega) \right] \\
&= \lim_{k \to \infty} \left[\bar{x}_{n_k}(\omega) \overline{M}_{n_k}(\omega) \right] \\
&= \bar{x}(\omega) L^*(\omega) R(\omega).
\end{aligned}$$

Combining this with the positivity of x_0, L, and R, it follows that

$$c := \bar{x} L^* \in (0, \infty), \qquad \text{a.s.}$$

Furthermore, since $\max_i x_0(i) = \max_i R(i) = 1$, we know that $c = 1$, a.s. Therefore we have

$$[T = \infty] \subset [R = x_0] \qquad \text{a.s. on } \Lambda^c,$$

as required. □

Finally, we mention that the condition "$\mathbb{P}[R = x_0] = 0$" can be removed in some cases, as was proven in Chen and Y. Li (1994).

To conclude this chapter, let us make some remarks about the theory of random matrices. The theory is a traditional and important branch of mathematics and has a very wide range of applications including statistics, physics, number theory, and even the Riemann hypothesis. Refer to M. Mehta (1991), V.L. Girko (1990), J.B. Conrey (2003), and references within.

Mainly, there are two topics in the study of eigenvalues. The first one is the estimation of the first few eigenvalues, as dealt with in this book. The second one, omitted in the book, is the asymptotic behavior of the eigenvalues. In the context of random matrices, concerning the second topic there is the famous beautiful Wigner's semicircle law (1955). For its modern generalization to operator algebras, called free probability, see D. Voiculescu, K. Dykema, and A. Nica (1992) and E. Haagerup (2002), for instance.

Appendix A

Some Elementary Lemmas

This appendix is an extension of Chen (1999c). Lemma A.1 below with $m - 1$ generalizes a result that appeared in Y. Li (1995). It is essential in our applications that the functions φ and/or ψ below are not required to be nonnegative.

Lemma A.1. Let u, $\varphi: [a, b] \to \mathbb{R}^m$, $\psi : [a, b] \to \mathbb{R}^m \otimes \mathbb{R}^m$, and $G : U \to \mathbb{R}^m$ for some open $U \subset \mathbb{R}^m$. Denote by "$\leqslant$" the ordinary partial order in $\mathbb{R}^m$. Suppose that the following conditions hold:

(1) u is absolutely continuous.

(2) The derivative matrix DG is continuous, nonnegative, and invertible.

(3) Range $(u) \subset U$ and $H(a, t) \in \text{Range}(G)$ for all $t \in (a, b)$, where $H(a, t)$ is the unique solution to the linear equation $v'(t) = \varphi(t) + \psi(t)v(t)$, a.e. t. If $\psi(t)\psi(s) = \psi(s)\psi(t)$ for all t and s, then we have the expression

$$H(a, t) = \exp\left[\int_a^t \psi(s)\mathrm{d}s\right]\left(G(u(a)) + \int_a^t \exp\left[-\int_a^s \psi(\xi)\mathrm{d}\xi\right]\varphi(s)\mathrm{d}s\right).$$

(4) $u'(t) \leqslant (DG)^{-1}(u(t))(\varphi(t) + \psi(t)G(u(t)))$ for a.e. $t \in (a, b)$.

If (i) ψ is diagonal, or (ii) ψ is nonnegative and bounded on finite intervals, then we have $u(t) \leqslant G^{-1}(H(a, t))$ for all $t \in [a, b)$, where G^{-1} is the inverse function of G. Equality holds if it does in (4).

Proof. By the assumptions on G, the inverse function G^{-1} not only exists but also is continuously differentiable. Because DG is nonnegative, G is increasing with respect to "$\leqslant$". Thus, by (3), it suffices to show that

$$G(u(t)) \leqslant H(a, t), \qquad t \in [a, b).$$

The idea is to make a change of variables, reducing to the solvable linear case. Define $v(t) = G(u(t))$. In view of (1), $v'(t) = DG(u(t))u'(t)$ for a.e. t. Then, by (2) and (4), we obtain the linear form $v' \leqslant \varphi + \psi v$, a.e. t. Next, set $w = \exp[-\Psi]v$, $\Psi = \int_a^{\cdot} \psi$. In case (i), since ψ is diagonal, it follows that $\exp[-\Psi]$ is nonnegative, and hence $w' \leqslant \exp[-\Psi]\varphi$, a.e. t. By integration, we get an upper bound of w, and then assertion (i) follows, since $\exp[\Psi]$ is also nonnegative. At the same time, we obtain the expression for H whenever $\psi(t)\psi(s) = \psi(s)\psi(t)$. For case (ii), assume $b < \infty$ and let $Av = \int_a^{\cdot} \psi v$, $\Phi = \int_a^{\cdot} \varphi$, and $Bv = \Phi + Av$. Since $\psi \geqslant 0$, B is order-preserving, $v \leqslant B^n v$. Next, since ψ is bounded, the spectral radius $r(A) = 0$. Therefore $A^n \to 0$ and $B^n v = \sum_{k=0}^{n-1} A^k \Phi + A^n v \to (I - A)^{-1}\Phi = H$ as $n \to \infty$, since H is the unique fixed point of B. Actually, this is a consequence of the abstract Grownwall lemma (cf. E. Zeidler (1986, Proposition 7.17)). □

Corollary A.2 (Exponential form). Let u, φ, ψ, and H be the same as in Lemma A.1 with $G(u) = u$. If $u'(t) \leqslant$ (respectively, $=$) $\varphi(t) + \psi(t)u(t)$ for a.e. t, then we have for all $t \in [a, b)$, $u(t) \leqslant$ (respectively, $=$) $H(a, t)$.

Proof. For $G(u) = u$, we have $DG = I$, and so $(DG)^{-1} = I$.   □

From now on, we consider $m = 1$ only. We may and will assume that $G(u) = \int_{u_0}^u 1/g$ for some $g > 0$. Then condition (4) takes the following form:

$$u'(t) \leqslant g(u(t))(\varphi(t) + \psi(t)G(u(t)))   \text{for a.e. } t \in (a, b).$$

If moreover $\psi = 0$, then an alternative proof of Lemma A.1 goes as follows:

$$G(u(t)) = \int_{u_0}^{u(t)} \frac{ds}{g(s)} = G(u(a)) + \int_{u(a)}^{u(t)} \frac{ds}{g(s)} = G(u(a)) + \int_a^t \frac{u'(s)}{g(u(s))}ds.$$

Remark A.3. (1) When $m = 1$, $\psi = 0$, condition (4) is equivalent to the integral form $u(t_2) - u(t_1) \leqslant \int_{t_1}^{t_2} \varphi(s)g(u(s))ds$, $t_1, t_2 \in [a, b)$, $t_2 \geqslant t_1$. This means that condition (4) is stronger than the above integral form with fixed $t_1 = a$.

(2) A related result is due to I. Bihari (1956): Let $m = 1$. If $\varphi \geqslant 0$, g is nondecreasing, and $u(t) \leqslant C + \int_a^t \varphi(s)g(u(s))ds$, $t \in [a, b)$, for some constant C, then $u(t) \leqslant G^{-1}\left(G(C) + \int_a^t \varphi(s)ds\right)$ for all $t \in [a, b)$.

Denote by $U(t)$ the right-hand side. Replacing $u(t)$ by $\sup_{a \leqslant s \leqslant t} U(s)$, one can assume that $u(t)$ is continuous and nondecreasing. Then replacing a with $\inf\{t \geqslant a : u(t) > C\}$, one can also assume that $u(t) > C$ for $t > a$.

The proof of the result goes as follows. Set $w(t) = C + \int_a^t \varphi(s)g(u(s))ds$. Then $u(t) \leqslant w(t)$ and $w'(t) = \varphi(t)g(u(t)) \geqslant 0$. Furthermore,

$$G(u(t)) = \int_{u_0}^{u(t)} \frac{ds}{g(s)} \leqslant \int_{u_0}^{w(t)} \frac{ds}{g(s)} = G(C) + \int_{w(a)}^{w(t)} \frac{ds}{g(s)} = G(C) + \int_a^t \frac{w'(s)}{g(w(s))}ds$$

$$= G(C) + \int_a^t \frac{\varphi(s)g(u(s))}{g(w(s))} ds \leqslant G(C) + \int_a^t \frac{\varphi(s)g(u(s))}{g(u(s))} ds$$

$$= G(C) + \int_a^t \varphi(s)ds.$$

Corollary A.4 (Algebraic form). Let u be nonnegative, absolutely continuous with $u(0) > 0$, let $\varphi \geqslant 0$, φ and ψ be locally integrable. If

$$u'(t) \leqslant -\varphi(t)u(t)^p + \psi(t)u(t) \qquad \text{for a.e. } t \text{ on } (0, \infty)$$

and some $p > 1$, then we have for $t \geqslant 0$,

$$u(t) \leqslant \exp\left[\int_0^t \psi(s)ds\right]\left[u(0)^{1-p} + (p-1)\int_0^t \exp\left[(p-1)\int_0^s \psi(\xi)d\xi\right]\varphi(s)ds\right]^{1-q},$$

where $1/p + 1/q = 1$. Equality holds if it does so in the assumption.

Proof. Note that one can reduce to the case that $\psi = 0$. If $u' \leqslant -\varphi u^p + \psi u$ a.e. for some $p \neq 1$, then $w := \exp\left[-\int_0^{\bullet} \psi\right]u$ is nonnegative, absolutely continuous, $w(0) = u(0)$, and satisfies $w' \leqslant -\varphi \exp\left[(p-1)\int_0^{\bullet} \psi\right]w^p =: vw^p$, a.e.

If $t_0 := \inf\{t > 0 : w(t) = 0\} < \infty$, we claim that $w(t) = 0$, and so the estimate becomes trivial for all $t \geqslant t_0$. Otherwise, suppose that there is a $t_2 > 0$ such that $w(t_2) > 0$. Let $t_1 = \sup\{t \in [t_0, t_2) : w(t) = 0\}$. Then $w(t_1) = 0$ by the continuity of w and the initial condition $w(t_0) = 0$. Clearly, $t_0 \leqslant t_1 < t_2$ and $w(t) > 0$ for $t \in (t_1, t_2)$. Thus, by assumption, $0 < w(t_2) = \int_{t_1}^{t_2} w'(s)ds \leqslant \int_{t_1}^{t_2} v(s)w(s)^p ds \leqslant 0$, which is a contradiction.

For simplicity, assume that $t_0 = \infty$, and so $w(t) > 0$ for all t. Take $(a, b) = (0, \infty) = U$ and $g(x) = x^p$. Then

$$G(u) = \int_\varepsilon^u x^{-p}dx = \frac{1}{1-p}(u^{1-p} - \varepsilon^{1-p})$$

and

$$G^{-1}(u) = [\varepsilon^{1-p} + (1-p)u]^{1/(1-p)} = [\varepsilon^{1-p} + (1-p)u]^{1-q}.$$

Hence by Lemma A1 with $m = 1$, $\psi = 0$, and $\varphi = v$, for sufficiently small $\varepsilon > 0$, we have

$$w(t) \leqslant \left(\varepsilon^{1-p} + (1-p)\left[\frac{1}{1-p}(w(0)^{1-p} - \varepsilon^{1-p}) + \int_0^t v(s)ds\right]\right)^{1-q}$$

$$= \left(u(0)^{1-p} - (p-1)\int_0^t v(s)ds\right)^{1-q}. \qquad \square$$

We remark that the corollary is meaningful for general φ and $p \neq 1$ within the interval for which the estimate is positive, provided u is assumed to be positive. The conclusion can be deduced directly from Corollary A.2 using the transform $w = u^{1-p}/(1-p)$.

Corollary A.5. If a function $u \geqslant 1$ satisfies $u'(t) \leqslant \psi(t)u(t)\log u(t)$ for a.e. t on $[a, b)$, then

$$u(t) \leqslant u(a)^{\exp[\int_a^t \psi(s)ds]}, \qquad t \in [a, b).$$

In particular, when $\psi(t) = 1/(t(1 - bt))$ on $[a, b)$, we have

$$u(t) \leqslant u(a)^{t(1-ab)/a(1-bt)}, \qquad t \in [a, b).$$

Proof. The main assertion follows from Lemma A.1 with $m = 1$, $\varphi = 0$, and $g(x) = x$ ($u_0 = 1$) or from Corollary A.2 using the transform $u \to \log u$. The particular case then follows from

$$\psi(s) = \frac{1}{s(1 - bs)} = \frac{1}{s} + \frac{b}{1 - bs},$$

$$\int_a^t \psi(s)ds = \log t - \log a - \log(1 - bt) + \log(1 - ab) = \log \frac{t(1 - ab)}{a(1 - bt)}. \qquad \square$$

For the remainder of this part, we consider a Markov semigroup $\{P(t)\}_{t \geqslant 0}$ with weak operator Ω having domain

$$\mathscr{D}_w(\Omega) = \left\{ f : \frac{d}{dt}P(t)f(x) = P(t)\Omega f(x) \text{ for all } x \in E \text{ and } t \geqslant 0 \right\}.$$

The next two results describe the exponential or algebraic decay of the semigroup in terms of its operator.

Lemma A.6 (Exponential form). Let $f \in \mathscr{D}_w(\Omega)$ and $\alpha > 0$ be a constant. Then $P(t)f \leqslant e^{-\alpha t}f$ iff $\Omega f \leqslant -\alpha f$.

Proof. Let $f_t = P(t)f$. Then $f_t' = P(t)\Omega f \leqslant -\alpha P(t)f = -\alpha f_t$. The sufficiency now follows from Corollary A3. The necessity follows from

$$\Omega f = \lim_{t \to 0} \frac{P(t)f - f}{t} \leqslant \lim_{t \to 0} \frac{e^{-\alpha t} - 1}{t}f = -\alpha f. \qquad \square$$

Lemma A.7 (Algebraic form). Fix $p > 1$. Let $f \in \mathscr{D}_w(\Omega)$, $f > 0$, and $C > 0$ be a constant. Then $P(t)f \leqslant [f^{1-p} + (p - 1)Ct]^{1-q}$ iff $\Omega f \leqslant -Cf^p$, where $1/p + 1/q = 1$.

Proof. Again, let $f_t = P(t)f$. Then $f_t' = P(t)\Omega f \leqslant -CP(t)(f^p)$. However, by the Hölder inequality, $P(t)(f^p) \geqslant (P(t)f)^p$. Hence $f_t' \leqslant -Cf_t^p$. The sufficiency now follows from Corollary A4. Next, note that $p - 1 = p/q$ and $q - 1 = q/p$. The necessity follows from

$$\Omega f = \lim_{t \to 0} \frac{P(t)f - f}{t} \leqslant \lim_{t \to 0} \frac{\left[f^{1-p} + (p - 1)Ct\right]^{1-q} - f}{t}$$

$$= \lim_{t \to 0}(1 - q)(p - 1)C\left[f^{1-p} + (p - 1)Ct\right]^{-q} = -Cf^{q(p-1)}$$

$$= -Cf^p. \qquad \square$$

Appendix B

Examples of the Ising Model on Two to Four Sites

This appendix introduces some ideas for reducing the higher-dimensional or graphic cases to dimension one, based on the symmetry of the model, to estimate the spectral gap of Markov chains by couplings. As an illustration, we compute the spectral gap for the Ising model on two, three, or four sites. The discussions here show, as in Section 3.4, that an effective estimation depends on not only the couplings but also the choice of "distance." Such a choice becomes even more serious in higher dimensions. Clearly, the discrete world is still largely open.

Some idea of this appendix appeared in the review [MR: 1768241 (2002b: 60129)] of the paper by K. Burdzy and W.S. Kendall (2000).

For the study on the spectral gap and logarithmic Sobolev inequality for the original Ising model and other statistical models, refer to the publications listed at the end of Chapter 9.

B.1 The model

Let $S = \{1, 2, \ldots, N\}$, $N = |S| < \infty$. Regard S as a circle such that the distance between 0 and N is equal to one (nearest neighbor). Consider a time-continuous Markov chain, which jumps at each site $u \in S$ with flip rate from η_u to $-\eta_u$ as follows:

$$c(u, \eta) = \exp\left[-\beta \sum_{v:|v-u|=1} \eta(u)\eta(v) \right], \qquad u \in S, \ \eta \in \{-1, +1\}^S.$$

The jumps at different sites are assumed to be independent.

We are interested in estimating or computing the spectral gap (or the first nontrivial eigenvalue) λ_1 of the chain by coupling. Here we compute λ_1 for

$|S| = 2, 3, 4$. Note that in the last case, the state space consists of 16 points, and so a direct computation seems impossible.

In the next section, we explain the role played by the symmetry of the eigenfunction and compute λ_1 in the case of two sites. In Section B.3, we introduce a reduction procedure and compute λ_1 in the case of three sites. The model with four sites is treated in Section B.4, in which a modification technique is explained.

B.2   Distance based on symmetry: two sites

To illustrate the idea clearly, we begin with the simplest case, $|S| = 2$. Label the state space as in Figure B.1.

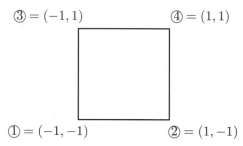

$$③ = (-1, 1) \qquad\qquad ④ = (1, 1)$$

$$① = (-1, -1) \qquad\qquad ② = (1, -1)$$

Figure B.1   The state space

Starting from ①, after one jump, we arrive at ② or ③. The next jump goes to ④. Half of the transition rates are given in Figure B.2. To get the transition rates for the opposite direction, simply exchange $e^{\pm\beta}$ and $e^{\mp\beta}$. This figure indicates the symmetry of the model.

However, we would like to consider a simplified model as shown by Figure B.3. In other words, we regard ② and ③ as one class in the sense of having distance zero. The transition rates are the same as those given in Figure B.2.

Figure B.2   The rates of transition   **Figure B.3**   A distance on the graph

As explained in Chapter 1, we need only to construct a coupling operator $\widetilde{\Omega}$ and a distance ρ. Our result says that if

$$\widetilde{\Omega}\rho(\eta_1, \eta_2) \leqslant -\alpha\rho(\eta_1, \eta_2), \qquad \eta_1 \neq \eta_2, \tag{B.1}$$

then we have $\lambda_1 \geqslant \alpha$.

We now define the distance between ① and ②, as well as ② and ④, to be the same $x > 0$ (by symmetry). Actually, one may simply set $x = 1$ because of the homogeneity of (B.1). Thus, the distance between ① and ④ equals 2. Next, we define a coupling as follows.

- When the two processes start from ① and ④, respectively, we adopt the basic coupling. In tabular form, we have

$$(①, ④) \quad \rightarrow \quad (③, ③) \quad \text{at rate} \ e^{-\beta} \wedge e^{-\beta} = e^{-\beta}$$
$$\rightarrow \quad (②, ②) \quad \text{at rate} \ e^{-\beta} \wedge e^{-\beta} = e^{-\beta}.$$

Thus,

$$\widetilde{\Omega}\rho(①, ④) = 2e^{-\beta}(0 - 2) = -4e^{-\beta}. \tag{B.2}$$

- When the two processes start from ① and ②, respectively, for the jumps ① → ② (the first component) and ② → ① (the second component), let the two components jump independently; for the jumps ① → ③ (the first component) and ② → ④ (the second component), we adopt the coupling by marching soldiers (that is, the term $\{\cdots\}$ in (B.3) below). In tabular form, we have

$$(①, ②) \quad \rightarrow \quad (①, ①) \quad \text{at rate} \ e^{\beta}$$
$$\rightarrow \quad (②, ②) \quad \text{at rate} \ e^{-\beta}$$
$$\rightarrow \quad (③, ④) \quad \text{at rate} \ e^{\beta} \wedge e^{-\beta} = e^{-\beta}$$
$$\rightarrow \quad (①, ④) \quad \text{at rate} \ \left(e^{\beta} - e^{-\beta}\right)^{+} = e^{\beta} - e^{-\beta}.$$

Then

$$\widetilde{\Omega}\rho(①, ②) = \left(e^{\beta} + e^{-\beta}\right)(0 - 1) + \left\{e^{-\beta}(1 - 1) + \left(e^{\beta} - e^{-\beta}\right)(2 - 1)\right\}$$
$$= -2e^{-\beta}. \tag{B.3}$$

- For the other cases, due to the symmetry again, one can define the coupling as above.

Inserting (B.2) and (B.3) into (B.1), it follows that the largest α is equal to $2e^{-\beta}$. Hence, $\lambda_1 \geqslant 2e^{-\beta}$ for this modified model.

We now return to our original model. The distance is chosen to be the same as in Figure B.3, except for the distance between ② and ③, which is set to be an arbitrarily small $\varepsilon > 0$. Besides, we need to modify the coupling by adding the case that the two processes start from ② and ③, respectively. For this, we adopt the basic coupling or the coupling of marching soldiers:

$$(②, ③) \quad \rightarrow \quad (①, ①) \quad \text{at rate} \ e^{\beta} \wedge e^{\beta} = e^{\beta}$$
$$\rightarrow \quad (④, ④) \quad \text{at rate} \ e^{\beta} \wedge e^{\beta} = e^{\beta}.$$

Thus,

$$\widetilde{\Omega}\rho(②, ③) = 2e^{\beta}(0 - \varepsilon) = -2e^{\beta}\varepsilon \leqslant -2e^{\beta}\rho(②, ③). \tag{B.4}$$

Combining this with (B.3), (B.2), and (B.1), we still get $\lambda_1 \geqslant 2e^{-\beta}$, which is indeed exact. Actually, for this example, the four eigenvalues are the following:

$$0, \quad 2e^{-2\beta}, \quad 2e^{2\beta}, \quad 2\left(e^{-2\beta} + e^{-2\beta}\right).$$

One may feel strange in first looking at the last distance $\rho(②,③) = \varepsilon$. It is clearly different from the geometric distance in the graph. However, this distance is quite natural, due to the symmetry of the eigenfunction of λ_1. More precisely, the eigenfunction of λ_1 has the same value at ② and ③. For this, one may say that the potential (eigenfunction) has three levels. We define the distance according to the levels. Note that we do not really need to prove this property of the eigenfunction for this specific situation, because of the "ε technique" used above. Unfortunately, this technique does not work in general. For this, we need more work, as explained in the next section.

B.3   Reduction: three sites

Consider a general Markov chain with finite state space $E = \{\eta, \zeta, \xi, \dots\}$. Let g be a fixed eigenfunction (eigenvector) of λ_1 (it may have multiplicity). Define a partial ordering on E as follows:

$$\eta \prec \eta \quad \text{for all } \eta \in E \text{ and} \tag{B.5}$$
$$\text{if } \eta \neq \zeta, \quad \text{then } \eta \prec \zeta \text{ iff } g(\eta) < g(\zeta). \tag{B.6}$$

Next, let $\varphi \geqslant 0$ be symmetric and satisfy

$$\varphi(\eta, \zeta) = 0 \quad \text{iff } g(\eta) = g(\zeta).$$

Since g is not a constant, we have $\varphi \neq 0$.

Proposition B.1.   (1) If there exists a coupling process $\widetilde{P}_t$ such that

$$\widetilde{P}_t \varphi(\eta, \zeta) \leqslant \varphi(\eta, \zeta) e^{-\alpha t}, \qquad t > 0, \tag{B.7}$$

for some $\eta \prec \zeta$, $\eta \neq \zeta$, then $\lambda_1 \geqslant \alpha$.
(2) If the Markov chain is monotone with respect to the partial ordering, then the assumption of part (1) holds with $\varphi(\eta, \zeta) = |g(\eta) - g(\zeta)|$ and $\alpha = \lambda_1$. Actually, (B.7) holds for all $\eta \prec \zeta$ and $\zeta \prec \eta$.
(3) If there are a coupling operator $\widetilde{\Omega}$ and a constant $\alpha \geqslant 0$ such that

$$\widetilde{\Omega} \varphi(\eta, \zeta) \leqslant -\alpha \, \varphi(\eta, \zeta) \quad \text{for all } \eta \text{ and } \zeta, \tag{B.8}$$

then the coupling process $\widetilde{P}_t$ determined by $\widetilde{\Omega}$ satisfies (B.7) for all η and ζ.

Proof. Part (3) of the proposition is standard. The proof of part (1) is very much the same as those given in Section 1.2. First, we have seen that

$$\mathbb{E}^\eta g(\eta_t) = g(\eta) e^{-\lambda_1 t}, \qquad t \geqslant 0, \, \eta \in E.$$

Fix $\eta \neq \zeta$, $\eta \prec \zeta$. Then

$$e^{-\lambda_1 t}|g(\eta) - g(\zeta)| \leqslant \widetilde{\mathbb{E}}^{\eta,\zeta}|g(\eta_t) - g(\zeta_t)|$$

$$\leqslant \left[\sup_{g(\xi_1) \neq g(\xi_2)} \frac{|g(\xi_1) - g(\xi_2)|}{\varphi(\xi_1, \xi_2)}\right] \widetilde{\mathbb{E}}^{\eta,\zeta} \varphi(\eta_t, \zeta_t)$$

$$\leqslant \left[\sup_{g(\xi_1) \neq g(\xi_2)} \frac{|g(\xi_1) - g(\xi_2)|}{\varphi(\xi_1, \xi_2)}\right] \varphi(\eta, \zeta) e^{-\alpha t}. \qquad (B.9)$$

Since $g(\eta) \neq g(\zeta)$, we have $\varphi(\eta, \zeta) \neq 0$. Letting $t \to \infty$, it follows that $\lambda_1 \geqslant \alpha$.

To prove part (2), by Theorem 2.35 (2), there exists an order-preserving coupling $\overline{P}_t^{\eta,\zeta}$ such that

$$\overline{P}_t^{\eta,\zeta}[\eta_t \prec \zeta_t, \forall t \geqslant 0] = 1, \qquad \eta \prec \zeta. \qquad (B.10)$$

We now fix $\eta \neq \zeta$, $\eta \prec \zeta$. Then for $\varphi(\eta, \zeta) = |g(\eta) - g(\zeta)|$, we have

$$0 \leqslant e^{-\lambda_1 t}\varphi(\eta, \zeta) = e^{-\lambda_1 t}\big(g(\zeta) - g(\eta)\big)$$

$$= \mathbb{E}^\zeta g(\zeta_t) - \mathbb{E}^\eta g(\eta_t) = \overline{\mathbb{E}}^{\eta,\zeta}\big[g(\zeta_t) - g(\eta_t)\big]$$

$$= \overline{\mathbb{E}}^{\eta,\zeta}\big[g(\zeta_t) - g(\eta_t) : \eta_t \prec \zeta_t\big]$$

$$= \overline{\mathbb{E}}^{\eta,\zeta}\varphi(\eta_t, \zeta_t). \qquad (B.11)$$

Thus, (B.7) holds for all $\eta \prec \zeta$. □

Before moving further, let us make some remarks on condition (B.8). For a monotone process, from (B.10), it follows that

$$\overline{\Omega}\varphi(\eta, \zeta) = \lim_{t \to 0} \frac{1}{t}\left[\overline{\mathbb{E}}^{\eta,\zeta}\varphi(\eta_t, \zeta_t) - \varphi(\eta, \zeta)\right]$$

$$= \lim_{t \to 0} \frac{1}{t}\left(e^{-\lambda_1 t} - 1\right)\varphi(\eta, \zeta)$$

$$= -\lambda_1 \varphi(\eta, \zeta), \qquad \eta \prec \zeta. \qquad (B.12)$$

Thus, (B.8) holds with $\alpha = \lambda_1$ for all $\eta \prec \zeta$. When $g(\eta) = g(\zeta)$, as in (B.11), we have

$$0 = e^{-\lambda_1 t}\varphi(\eta, \zeta) = \overline{\mathbb{E}}^{\eta,\zeta}\big[g(\zeta_t) - g(\eta_t)\big]$$

$$= \overline{\mathbb{E}}^{\eta,\zeta}\big[\varphi(\eta_t, \zeta_t)I_{[\eta_t \prec \zeta_t]}\big] - \overline{\mathbb{E}}^{\eta,\zeta}\big[\varphi(\eta_t, \zeta_t)I_{[\zeta_t \prec \eta_t]}\big]. \qquad (B.13)$$

The two terms on the right-hand side of (B.13) may not vanish, since an order-preserving coupling does not provide any information about the partial ordering of (η_t, ζ_t) when started from (η, ζ) with $g(\eta) = g(\zeta)$. What we can get from (B.13) is the following relation:

$$\overline{\Omega}(\varphi I_F)(\eta, \zeta) = \overline{\Omega}(\varphi I_{F'})(\eta, \zeta) \qquad \text{whenever } g(\eta) = g(\zeta), \qquad (B.14)$$

where $F = \{(\eta, \zeta) : \eta \prec \zeta\}$ and $F' = \{(\eta, \zeta) : \zeta \prec \eta\}$. On the other hand, in order to estimate $\widetilde{P}_t \varphi$, we need to estimate $\widetilde{\Omega}\varphi(\eta, \zeta)$ for all η and ζ. Note that condition (B.8) implies that

$$\widetilde{\Omega}\varphi(\eta, \zeta) = 0 \qquad \text{for all } \eta \text{ and } \zeta \text{ with } g(\eta) = g(\zeta). \tag{B.15}$$

Remark B.2. Let P_t be monotone. Suppose that there exists a coupling operator $\widetilde{\Omega}$ satisfying (B.15). Then there exists a coupling satisfying (B.8) with $\varphi(\eta, \zeta) = |g(\eta) - g(\zeta)|$ and $\alpha = \lambda_1$.

Proof. As mentioned before, there is an order-preserving coupling $\overline{\Omega}$ such that (B.12) holds for all $\eta \prec \zeta$ and symmetrically for all $\eta \prec \zeta$. Reconstruct a coupling operator $\overline{\Omega}_{\text{New}}$ as follows:

$$\overline{\Omega}_{\text{New}} f(\eta, \zeta) = \begin{cases} \overline{\Omega} f(\eta, \zeta) & \text{if } \eta \prec \zeta \text{ or } \zeta \prec \eta, \\ \widetilde{\Omega} f(\eta, \zeta) & \text{if } g(\eta) = g(\zeta). \end{cases}$$

By (B.12) and (B.15), it is clear that (B.8) holds. Furthermore, it is not difficult to check that $\overline{\Omega}_{\text{New}}$ is an order-preserving coupling operator. □

The proof given in Section B.2 with $\varepsilon = 0$ illustrates an application of Remark B.2 and Proposition B.1. Note that (B.2) and (B.3) give us the same solution. This is due to the freedom in the eigenequation

$$\Omega g = -\lambda_1 g, \tag{B.16}$$

again based on the symmetry of the model. This leads to a reduction procedure that we are going to discuss.

Let $(q(\eta, \zeta) : \eta, \zeta \in E)$ be an irreducible Q-matrix with finite state space E. Denote by $r_0 < \cdots < r_m$ the values of a fixed eigenfunction g of λ_1. Note that the values of g have different signs, the order of $\{r_0, \ldots, r_m\}$ is well defined, and $m \geqslant 1$. However, the values $\{r_0, \ldots, r_m\}$ are determined by g up to a positive constant only. We may renormalize g according to our convenience.

Proposition B.3. Given g as above, set $E_j = \{\eta : g(\eta) = r_j\}$, $0 \leqslant j \leqslant m$ and define

$$\bar{q}_{ij} = \min_{\eta \in E_i} \sum_{\zeta \in E_j} q(\eta, \zeta), \qquad j \neq i, \ 0 \leqslant i, j \leqslant m. \tag{B.17}$$

Suppose that $(\bar{q}_{ij})$ is reversible and denote by $\bar{\lambda}_1$ its spectral gap.

(1) If

$$\sum_{j \neq i} \left(\sum_{\zeta \in E_j} q(\eta, \zeta) - \bar{q}_{ij} \right) (g(\zeta) - g(\eta)) = 0, \qquad \eta \in E_i, \ 0 \leqslant i \leqslant m, \tag{B.18}$$

then $\lambda_1 \geqslant \bar{\lambda}_1$.

(2) If additionally, there exists a strictly increasing eigenfunction $(\bar{g}_i)$ of $\bar{\lambda}_1$ such that

$$\sum_{j\neq i}\left(\sum_{\zeta\in E_j} q(\eta,\zeta) - \bar{q}_{ij}\right)(\bar{g}_j - \bar{g}_i) = 0, \quad \eta\in E_i,\ 0\leqslant i\leqslant m, \quad \text{(B.19)}$$

then $\lambda_1 = \bar{\lambda}_1$.

Proof. For the given $g(\eta)$, let $\bar{g}_i = r_i$, which is the common value of $g(\eta)$ for $\eta\in E_i$. Starting from the eigenequation (B.16), we have for each $\eta\in E_i$,

$$-\lambda_1 g_i = -\lambda_1 g(\eta) = \Omega g(\eta) = \sum_\zeta q(\eta,\zeta)[g(\zeta) - g(\eta)]$$

$$= \sum_{j=0}^m \sum_{\zeta\in E_j} q(\eta,\zeta)[g(\zeta) - g(\eta)]$$

$$= \sum_{j=0}^m \min_{\eta\in E_i} \sum_{\zeta\in E_j} q(\eta,\zeta)[g(\zeta) - g(\eta)] \quad \text{(by (B.18) and (B.17))}$$

$$= \sum_{j\neq i} \bar{q}_{ij}(\bar{g}_j - g_i) \quad \text{(by (B.17))}, \qquad 0\leqslant i\leqslant m.$$

Since $(\bar{g}_i)$ is not a constant and $(\bar{q}_{ij})$ is reversible, this implies that $\lambda_1 \geqslant \bar{\lambda}_1$.

Conversely, let $(\bar{g}_i)$ be a strictly increasing eigenfunction of $\bar{\lambda}_1$ and define $\tilde{g}(\eta) = \bar{g}_i$ for all $\eta\in E_i$. Then the above proof, replacing g with $\tilde{g}$, shows that $\bar{\lambda}_1 \geqslant \lambda_1$. Therefore, we must have $\lambda_1 = \bar{\lambda}_1$. □

Applying the result to the Ising model on two sites, we get

$$\bar{q}_{01} = \bar{q}_{21} = 2e^{-\beta}, \qquad \bar{q}_{12} = \bar{q}_{21} = e^\beta.$$

As in Section B.2, the distance on $\{0,1,2\}$ is chosen to be

$$\rho(0,1) = \rho(1,2) = 1 \quad \text{and} \quad \rho(0,2) = 2.$$

Note that birth–death processes are always reversible and the eigenfunction of $\bar{\lambda}_1$ is strictly increasing (cf. Proposition 3.4 given in Section 3.7). Since the basic coupling $\tilde{\Omega}_b$ of $(\bar{q}_{ij})$ is order-preserving, the analogue of (B.12) for $(\bar{q}_{ij})$ with $\varphi(i,j) = \rho(i,j) = |\bar{g}_j - \bar{g}_i|$,

$$\tilde{\Omega}\rho(i,j) = -\bar{\lambda}_1\rho(i,j), \qquad i < j, \quad \text{(B.20)}$$

holds for this coupling. Now, because the distance is explicit, there is only one unknown variable $\bar{\lambda}_1$ in (B.20). We need only consider the coupling starting from $(0,1)$. Using the basic coupling, we get

$$2e^{-\beta}(0 - 2) = -2\bar{\lambda}_1.$$

That is, $\lambda_1 = \bar{\lambda}_1 = 2e^{-\beta}$ as expected. We point out that in the computation of (B.20), one may adopt different couplings (the classical coupling for instance), only the order preserving property is essential.

We are now ready to study the model with three sites. Label the state space as follows:

$$
\begin{aligned}
&\text{①} = (-1, -1, -1), &&\text{⑤} = (-1, 1, 1),\\
&\text{②} = (-1, -1, 1), &&\text{⑥} = (1, -1, 1),\\
&\text{③} = (-1, 1, -1), &&\text{⑦} = (1, 1, -1),\\
&\text{④} = (1, -1, -1), &&\text{⑧} = (1, 1, 1).
\end{aligned}
$$

Half of the transition rates are as shown in Figure B.4.

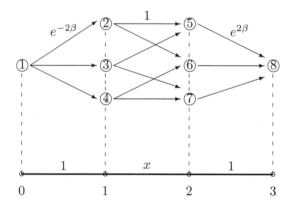

Figure B.4   Ising model on three sites

The eigenfunction g has four levels. The transition rates from level i to $i+1$ (or from i to $i-1$) are $e^{-2\beta}$, 1, and $e^{2\beta}$. Thus, we have a Q-matrix $(\bar{q}_{ij})$ on $\{0, 1, 2, 3\}$ as follows:

$$
\bar{q}_{01} = 3e^{-2\beta}, \quad \bar{q}_{12} = 2, \quad \bar{q}_{23} = e^{2\beta}.
$$

Symmetrically,

$$
\bar{q}_{32} = 3e^{-2\beta}, \quad \bar{q}_{21} = 2, \quad \bar{q}_{10} = e^{2\beta}.
$$

Otherwise, $\bar{q}_{ij} = 0$ for all $i \neq j$. By symmetry, we can define a distance on $\{0, 1, 2, 3\}$ as follows:

$$
\begin{aligned}
&\rho(0, 1) = \rho(2, 3) = 1, &&\rho(1, 2) = x > 0,\\
&\rho(i, j) = \rho(i, i+1) + \cdots + \rho(j-1, j), &&i < j.
\end{aligned}
$$

Conditions (B.18) and (B.19) are trivial for this example. Because there are two variables λ_1 and x only in (B.20), we need to compute two starting points of an order-preserving coupling.

When the starting point is $(0, 2)$, we adopt the coupling by (inner) reflection,

$$3e^{-2\beta}(x - 2 - x) = -\bar{\lambda}_1(2 + x).$$

That is,

$$6e^{-2\beta} = \bar{\lambda}_1(2 + x). \tag{B.21}$$

When the starting point is $(0, 1)$, we adopt the classical coupling,

$$\left(3e^{-2\beta} + e^{2\beta}\right)(0 - 1) + 2(x + 1 - 1) = -\bar{\lambda}_1 \times 1.$$

That is,

$$3e^{-2\beta} + e^{2\beta} - 2x = \bar{\lambda}_1. \tag{B.22}$$

Combining (B.21) with (B.22), we obtain the solution

$$\lambda_1 = \bar{\lambda}_1 = \frac{1}{2}\left\{3e^{-2\beta} + 4 + e^{2\beta} - \sqrt{\left(3e^{-2\beta} + 4 + e^{2\beta}\right)^2 - 48e^{-2\beta}}\right\}.$$

The eigenvalues of this example are 0, λ_1, $3e^{-2\beta} + e^{2\beta}$, $1 + e^{2\beta}$ (multiplicity 2), $3 + e^{2\beta}$ (multiplicity 2), and

$$\frac{1}{2}\left\{3e^{-2\beta} + 4 + e^{2\beta} + \sqrt{\left(3e^{-2\beta} + 4 + e^{2\beta}\right)^2 - 48e^{-2\beta}}\right\}.$$

B.4   Modification: four sites

Consider the Ising model on four sites. Label the state space as follows:

$$\begin{array}{ll}
①= (-1, -1, -1, -1), & ⑨= (-1, 1, 1, -1), \\
②= (1, -1, -1, -1), & ⑩= (-1, 1, -1, 1), \\
③= (-1, 1, -1, -1), & ⑪= (-1, -1, 1, 1), \\
④= (-1, -1, 1, -1), & ⑫= (1, 1, 1, -1), \\
⑤= (-1, -1, -1, 1), & ⑬= (1, 1, -1, 1), \\
⑥= (1, 1, -1, -1), & ⑭= (1, -1, 1, 1), \\
⑦= (1, -1, 1, -1), & ⑮= (-1, 1, 1, 1), \\
⑧= (1, -1, -1, 1), & ⑯= (1, 1, 1, 1).
\end{array}$$

The structure of the transition rates is marked in Figure B.5. The transition rates from level 0 to level 1 are all equal to $e^{-2\beta}$; the ones from level 3 to level 4 are all equal to $e^{2\beta}$. The rates from level 1 to level 2 are all equal to 1, except the dot directions, which have rates $e^{-2\beta}$. Similarly, the rates from level 2 to level 3 are all equal to 1, except the dot directions, which have rates $e^{2\beta}$. Once again, by exchanging the rates $e^{\pm2\beta}$ with $e^{\mp2\beta}$, we get the rates for the opposite transitions from level 4 to level 0.

To analyze this model, first we claim that for the eigenfunction g with lowest level 0 at site ①, there are five levels only. It is clear that the states

②, ③, ④, and ⑤ should be located at the same level, and similarly for the states ⑫, ⑬, ⑭, and ⑮. It is also clear that states ⑥, ⑧, ⑨, and ⑪ should be located at the same level, but it is not obvious why the states ⑦ and ⑩ should be. Suppose that $g(⑥) < g(⑦)$. That is, the potential at ⑦ is higher than that at ⑥. Thus, if we start from ⓪, then the level of ⑦ should be located on the right of the level of ⑥. However, by symmetry, if we start from ⑯, then the level of ⑦ should be located on the left of the level of ⑥. This contradiction shows that ⑥ and ⑦ should be located at the same level. Similarly, the same conclusion holds for state ⑩.

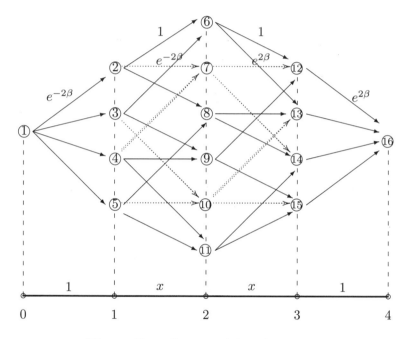

Figure B.5 Ising model on four sites

Note that the partial ordering used here is different from the ordinary one, for which the states ⑥ and ⑮ are not comparable.

Next, we discuss condition (B.18) given in Proposition B.3. Note that the sums of transitions from ⑥ and ⑦ are different. For this, we modify all of the transition rates from ⑦ and ⑩ to be 1, the same as the transition rates from the other states at level 2. This is based on the following observation. Since $g(⑫) - g(⑥) = g(⑥) - g(②)$, we have

$$\sum_{\zeta \in E_1 \cup E_3} \big(q(⑦, \zeta) - 1\big)\big(g(\zeta) - g(⑦)\big) = 0.$$

Therefore, the modification does not make any change in the solution to the eigenequation. This leads to condition (B.18). Actually, in the present situation, $g(\widehat{7}) = 0$ by symmetry, so we can use any symmetric rates we want for the transitions from level 2 to level 1 and level 3. Thus, we obtain a Q-matrix on $\{0, 1, \ldots, 4\}$ as follows:

$$\bar{q}_{01} = 4e^{-2\beta}, \quad \bar{q}_{12} = 2 + e^{-2\beta}, \quad \bar{q}_{23} = 2, \quad \bar{q}_{34} = e^{2\beta},$$

and by symmetry,

$$\bar{q}_{43} = 4e^{-2\beta}, \quad \bar{q}_{32} = 2 + e^{-2\beta}, \quad \bar{q}_{21} = 2, \quad \bar{q}_{10} = e^{2\beta}.$$

A distance on $\{0, 1, \ldots, 4\}$ is defined in Figure B.5, where only one variable, x, is used. For the same reason explained above, (B.19) holds.

We now compute $\bar{\lambda}_1$. When the starting point is $(0, 4)$, we adopt the coupling by reflection

$$4e^{-2\beta}(2x - 2x - 2) = -\bar{\lambda}_1(2 + 2x).$$

That is,

$$4e^{-2\beta} = \bar{\lambda}_1(1 + x). \tag{B.23}$$

When the starting point is $(0, 1)$, we adopt the classical coupling

$$\left(4e^{-2\beta} + e^{2\beta}\right)(0 - 1) + \left(2 + e^{-2\beta}\right)(x + 1 - 1) = -\bar{\lambda}_1.$$

That is,

$$4e^{-2\beta} + e^{2\beta} - \left(2 + e^{-2\beta}\right)x = \bar{\lambda}_1. \tag{B.24}$$

Solving the equations (B.23) and (B.24), we finally obtain

$$\lambda_1 = \bar{\lambda}_1 = \frac{1}{2}\left\{5e^{-2\beta} + 2 + e^{2\beta} - \sqrt{\left(5e^{-2\beta} + 2 + e^{2\beta}\right)^2 - 16\left(e^{-4\beta} + 2\right)}\right\}.$$

Remark B.4. Note that the three examples discussed so far are all of birth–death type. In general, if the reduced $(\bar{q}_{ij})$ is a birth–death Q-matrix, then we can apply Theorem 5.7 to estimate λ_1 for the original process.

Finally, we mention that some different approaches to estimating the convergence rate are presented in P. Diaconis and D.W. Stroock (1991), J.A. Fill (1991), F.R.K. Chung, A. Grigorýan and S.T. Yau (1997), R. Bhattacharya and E.C. Waymire (2001), N. Madras and D. Randall (2002), D.B. Wilson (2004). For infinite-dimensional path spaces, see S.Z. Fang (1994) and F.Z. Gong and L.M. Wu (2000). For higher eigenvalues, see M.S. Ashbaugh (1999) and M. Levitin and L. Parnovski (2002), and references within.

References

D. R. Adams and L. I. Hedberg. *Function Spaces and Potential Theory*. Springer, 1996.

S. Aida. Uniform positivity improving property, Sobolev inequality and spectral gaps. *J. Funct. Anal.*, 158, no.1:152–185, 1998.

S. Aida and I. Shigekawa. Logarithmic Sobolev inequalities and spectral gaps: Perturbation theory. *J. Funct. Anal.*, 126, no.2:448–475, 1994.

D. G. Aldous and J. A. Fill. *Reversible Markov Chains and Random Walks on Graphs*. In Preparation, www.stat.Berkeley.edu/users/aldous/book.html, 2004.

L. Ambrosio et al. *Optimal Transportation and Applications*. LNM 1813, L.A. Caffarelli S. Salsa (eds.), Springer, 2003.

W. J. Anderson. *Continuous-Time Markov Chains*. Springer Series in Statistics, 1991.

E. D. Andjel. Invariant measures for zero range process. *Ann. Prob.*, 10(3):527–547, 1982.

M. S. Ashbaugh. Isoperimetric and universal inequalities for eigenvalues. In *Theory and Geometry, London Math. Soc. Lecture Note Ser.*, eds: E. B. Davies and Yu. Safarov, Cambridge Univ. Press, Cambridge, 273:95–139, 1999.

D. Bakry. L'hypercontractivité et son utilisation en théorie des semigroupes. *Lectures Notes in Mathematics* **1581**, D. Bakry, R. D. Gill and S. A. Molchanov (eds.), *"Lectures on Probability Theorey"*, Springer-Verlag, 1992.

D. Bakry and M. Emery. Diffusions hypercontractives. *LNM*, 1123:177–206, 1985.

D. Bakry, T. Coulhon, M. Ledoux, and L. Saloff-Coste. Sobolev inequalities in disguise. *Indiana Univ. Math. J.*, 44(4):1033–1074, 1995.

D. Bakry and M. Ledoux. Lévy-Gromov's isoperimetric inequality for an infinite dimensional diffusion generator. *Invent. math.*, 123:259–281, 1996.

D. Bakry and Z. M. Qian. Some new results on eigenvectors via dimension, diameter and Ricci curvature. *Adv. Math.*, 155(1):98–153, 2000.

J. Barta. Sur la vibration fondamentale d'une membrane. *C. R. Acad. Sci. Paris*, 204:472–473, 1937.

F. Barthe and C. Roberto. Sobolev inequalities for probability measures on the real line. *Studia Math.*, 159(3):481–497, 2003.

M. Bebbington, P. Pollett, and X. G. Zheng. Dual constructions for pure-jump Markov processes. *Markov Processes and Related Filelds*, 1:513–558, 1995.

P. H. Bérard. *Spectral Geometry: Direct and Inverse Problem*. LNM. vol. 1207, Springer-Verlag, 1986.

P. H. Bérard, G. Besson, and S. Gallot. Sur une inégalité isopérimétrique qui généralise celle de Paul Lévy-Gromov. *Invent. Math.*, 80:295–308, 1985.

H. Berestycki, L. Nirenberg, and S. R. S. Varadhan. The principal eigenvalue and maximum principle for second-order elliptic operators in general domains. *Comm. Pure and Appl.*, XLVII:47–92, 1994.

R. Bhattacharya and E. C. Waymire. Iterated random maps and some classes of Markov processes. In *Handbook of Statistics* eds: D. N. Shanbhag and C. R. Rao, Elsevier Sci. B. V., 19:145–168, 2001.

I. Bihari. A generalization of a lemma of Bellman and its application to uniqueness problem of differential equations. *Acta Math. Hung.*, 7:81–94, 1956.

S. G. Bobkov. A functional form of the isoperimetric inequality for the Gaussian measure. *J. Funct. Anal.*, 135(1):39–49, 1996.

S. G. Bobkov. An isoperimetric inequality on the discrete cube, and an elementary proof of the isoperimetric inequality in Gauss space. *Ann. Probab.*, 25(1):206–214, 1997.

S. G. Bobkov and F. Götze. Discrete isoperimetric and Poincaré inequalities. *Prob. Th. Rel. Fields*, 114:245–277, 1999a.

S. G. Bobkov and F. Götze. Exponential integrability and transportation cost related to logarithmic Sobolev inequalities. *J. Funct. Anal.*, 163:1–28, 1999b.

C. Boldrighini, A. DeMasi, A. Pellegrinotti, and E. Presutti. Collective phenomena in interacting particle systems. *Stoch. Proc. Appl.*, 25:137–152, 1987.

P. Bougerol and J. Lacroix. *Products of Random Matrices with Applications to Schrödinger Operators.* Birkhäuser Boston, Inc., 1985.

K. Burdzy and W. S. Kendall. Efficient Markovian couplings: examples and counterexamples. *Ann. Appl. Probab.*, 10(2):362–409, 2000.

K. R. Cai. Estimate on lower bound of the first eigenvalue of a compact Riemannian manifold. *Chin. Ann. of Math.*, 12(B)(3):267–271, 1991.

E. A. Carlen, S. Kusuoka, and D. W. Stroock. Upper bounds for symmetric Markov transition functions. *Ann. Inst. Henri Poincaré*, 2:245–287, 1987.

I. Chavel. *Eigenvalues in Riemannian Geometry.* Academic Press, 1984.

I. Chavel. *Isoperimetric Inequalities: Differential Geometric and Analytic Perspectives.* Cambridge University Press, 2001.

J. Cheeger. A lower bound for the smallest eigenvalue of the Laplacian. *Problems in Analysis, a Symposium in Honor of S. Bochner,* Princeton Univ. Press, Princeton, pages 195–199, 1970.

J. W. Chen. The positive recurrence of a finite-dimensional Brusselator model. *Acta Math. Sci.*, 15(2):121–125, 1995.

L. H. Y. Chen. Poincaré-type inequalities via stochastic integrals. *Z. Wahrsch. Verw. Gebiete.*, 69:251–277, 1985a.

L. H. Y. Chen and J. H. Lou. Characterization of probability distributions by Poincaré-type inequalities. *Ann. Inst. H. Poincaré Sect. B (N.S.)*, 23:91–110, 1987.

M. F. Chen. Couplings of Markov chains (in Chinese). *J. Beijing Normal Univ.*, 4: 3–10, 1984.

M. F. Chen. Infinite-dimensional reaction–diffusion processes. *Acta Math. Sin. New Ser.*, 1(3):261–273, 1985b.

M. F. Chen. Couplings of jump processes. *Acta Math. Sinica, New Series*, 2(2): 123–136, 1986a.

M. F. Chen. *Jump Processes and Interacting Particle Systems* (in Chinese). Beijing Normal Univ. Press, 1986b.

M. F. Chen. Existence theorems for interacting particle systems with noncompact state spaces. *Sci. Sin.*, 30(2):148–156, 1987.

M. F. Chen. Probability metrics and coupling methods. *Pitman Research Notes in Math.*, 200:55–72, 1989a.

M. F. Chen. A survey on random fields (in Chinese). *Advances in Math.*, 18(3): 294–322, 1989b.

M. F. Chen. Stationary distributions of infinite particle systems with noncompact state spaces. *Acta Math. Sci.*, 9(1):7–19, 1989c.

M. F. Chen. Ergodic theorems for reaction–diffusion processes. *J. Statis. Phys.*, 58(5/6):939–966, 1990.

M. F. Chen. Exponential L^2-convergence and L^2-spectral gap for Markov processes. *Acta Math. Sin. New Ser.*, 7(1):19–37, 1991b.

M. F. Chen. On three classical problem for Markov chains with continuous time parameters. *J. Appl. Prob.*, 28:305–320, 1991c.

M. F. Chen. Uniqueness of reaction–diffusion processes. *Chin. Sci. Bulleten*, 36(12): 969–973, 1991d.

M. F. Chen. *From Markov Chains to Non-Equilibrium Particle Systems*. World Scientific, Singapore, Second Edition: 2004, 1992a.

M. F. Chen. Stochastic model of economic optimization (in Chinese). *Chin. J. Appl. Probab. and Statis.*, (I): 8(3), 289–294; (II): 8(4), 374–377, 1992b.

M. F. Chen. Optimal Markovian couplings and applications. *Technical Report*, No.215, 1993, Carleton Univ. and No.147, 1993, C. V. Volterra, Univ. of Roma II. *Acta Math. Sin. New Ser.* 10(3), 1994, pages 260–275, 1994a.

M. F. Chen. Optimal Markovian couplings and application to Riemannian geometry. in *Prob. Theory and Math. Stat.*, eds. Grigelionis B et al, VPS/TEV, pages 121–142, 1994b.

M. F. Chen. On ergodic region of Schlögl's model, *Carr Reports in Math. and Phys.* No.13, 1993. In M. Röckner Z. M. Ma and J. A. Yan, editors, *Proc. Intern. Conf. on Dirichlet Forms and Stoch. Proc.*, pages 87–102. Walter de Gruyter, 1994c.

M. F. Chen. Estimation of spectral gap for Markov chains. *Acta Math. Sin. New Ser.*, 12(4):337–360, 1996.

M. F. Chen. Coupling, spectral gap and related topics. *Chin. Sci. Bulletin*, (I): 42(16), 1321–1327; (II): 42(17), 1409–1416; (III): 42(18), 1497–1505, 1997a.

M. F. Chen. Trilogy of couplings and general formulas for lower bound of spectral gap. in *"Probability Towards 2000,"* edited by L. Accardi and C. Heyde, *Lecture Notes in Statistics*, Vol. **128**, 123–136, Springer-Verlag, 1998a.

M. F. Chen. Estimate of exponential convergence rate in total variation by spectral gap. *Acta Math. Sin. Ser.* (A), **41**(1) (Chinese ed.), 1–6; *Acta Math. Sin. New Ser.* **14**(1):9–16, 1998b.

M. F. Chen. Analytic proof of dual variational formula for the first eigenvalue in dimension one. *Sci. Sin.* (A), 42(8):805–815, 1999a.

M. F. Chen. Nash inequalities for general symmetric forms. *Acta Math. Sin. Eng. Ser.*, 15(3):353–370, 1999b.

M. F. Chen. Eigenvalues, inequalities and ergodic theory (II). *Advances in Math.*, 28(6):481–505, 1999c.

M. F. Chen. Single birth processes. *Chin. Ann. of Math.*, 20B(1):77–82, 1999d.

M. F. Chen. Equivalence of exponential ergodicity and L^2-exponential convergence for Markov chains. *Stoch. Proc. Appl.*, 87:281–297, 2000a.

M. F. Chen. Logarithmic Sobolev inequality for symmetric forms. *Sci. Sin.* (A), 43(6):601–608, 2000b.

M. F. Chen. Explicit bounds of the first eigenvalue. *Sci. China* (A), 43(10):1051–1059, 2000c.

M. F. Chen. The principal eigenvalue for jump processes. *Acta Math. Sin. Eng. Ser.*, 16(3):361–368, 2000d.

M. F. Chen. Explicit criteria for several types of ergodicity. *Chin. J. Appl. Prob. Stat.*, 17(2):1–8, 2001a.

M. F. Chen. Variational formulas and approximation theorems for the first eigenvalue. *Sci. China* (A), 44(4):409–418, 2001b.

M. F. Chen. *Ergodic Convergence Rates of Markov Processes — Eigenvalues, Inequalities and Ergodic Theory* [Collection of papers, 1993—2001]. http://math.bnu.edu.cn/~chenmf/main_eng.htm, 2001c.

M. F. Chen. Variational formulas of Poincaré-type inequalities in Banach spaces of functions on the line. *Acta Math. Sin. Eng. Ser.*, 18(3):417–436, 2002a.

M. F. Chen. A new story of ergodic theory. in "Applied Probability," 25–34, eds. R. Chan et al., *AMS/IP Studies in Advanced Mathematics*, 26, 2002b.

M. F. Chen. Ergodic convergence rates of Markov processes — eigenvalues, inequalities and ergodic theory. in *Proceedings of "ICM 2002,"* Higher Education Press, Beijing, III:41–52, 2002c.

M. F. Chen. Capacitary criteria for Poincaré-type inequalities. *preprint*, 2002d.

M. F. Chen. Variational formulas of Poincaré-type inequalities for birth–death processes. *Acta Math. Sin. Eng. Ser.*, 19(4):625–644, 2003a.

M. F. Chen. Variational formulas of Poincaré-type inequalities for one-dimensional processes. *IMS Lecture Notes – Monograph Series, Volume 41, Probability, Statistics and Their Applications:* Papers in Honor of Rabi Bhattacharya. K. Athreya, M. Majumdar, M. Puri, and E. Waymire (eds), pages 81–96, 2003b.

M. F. Chen. Ten explicit criteria of one-dimensional processes. *Proceedings of the Conference on Stochastic Analysis on Large Scale Interacting Systems,* Advanced Studies in Pure Mathematics 39, Mathematical Society of Japan, pages 89–114, 2004a.

M. F. Chen. Stochastic model of economic optimization. *Proceedings of the First Sino-German Conference on Stochastic Analysis*—A Satellite Conference of ICM 2002, in press, 2004b.

M. F. Chen, W. D. Ding, and D. J. Zhu. Ergodicity of reversible reaction–diffusion processes with general reaction rates. *Acta Math. Sin. New Ser.*, 10(1):99–112, 1994.

M. F. Chen, L. P. Huang, and X. J. Xu. Hydrodynamic limit for reaction–diffusion processes with several species. in *"Probability and Statistics, Nankai Series of Pure and Applied Mathematics,"* edited by S. S. Chern and C. N. Yang, World Scientific, 1991.

M. F. Chen and S. F. Li. Coupling methods for multi-dimensional diffusion processes. *Ann. Prob.*, 17(1):151–177, 1989.

M. F. Chen and Y. Li. Stochastic model of economic optimization. *J. Beijing Normal Univ.*, 30(2):185–194, 1994.

M. F. Chen and Y. G. Lu. Large deviations for Markov chains. *Acta Sci. Sin.*, 10(2):217–222, 1990.

M. F. Chen, E. Scacciatelli, and L. Yao. Linear approximation of the first eigenvalue on compact manifolds. *Sci. Sin.* (A), 45(4), 2002.

M. F. Chen and F. Y. Wang. On order preservation and positive correlations for multidimensional diffusion processes. *Prob. Th. Rel. Fields*, 95:421–428, 1993a.

M. F. Chen and F. Y. Wang. Application of coupling method to the first eigenvalue on manifold. *Sci. Sin.* (A), 23(11), 1993 (Chinese Edition), 37(1), 1994 (English Edition), 37(1), 1993b.

M. F. Chen and F. Y. Wang. Estimation of the first eigenvalue of second-order elliptic operators. *J. Funct. Anal.*, 131(2):345–363, 1995.

M. F. Chen and F. Y. Wang. General formula for lower bound of the first eigenvalue on Riemannian manifolds. *Sci. Sin.*, 40(4):384–394, 1997a.

M. F. Chen and F. Y. Wang. Estimation of spectral gap for elliptic operators. *Trans. Amer. Math. Soc.*, 349:1239–1267, 1997b.

M. F. Chen and F. Y. Wang. Cheeger's inequalities for general symmetric forms and existence criteria for spectral gap. Abstract: *Chin. Sci. Bulletin*, **43(18)**, 1516–1519. *Ann. Prob.* 2000, **28(1)**, 235–257, 1998.

M. F. Chen and Y. Z. Wang. Algebraic convergence of Markov chains. *Ann. Appl. Prob.*, 13(2):604–627, 2003.

M. F. Chen, Y. H. Zhang, and X. L. Zhao. Dual variational formulas for the first Dirichlet eigenvalue on half-line. *Sci. China* (A) 33(4) (Chinese ed.), 371–383; 46(6) (English ed.), 847–861, 2003.

R. R. Chen. An extended class of time-continuous branching processes. *J. Appl. Prob.*, 34(1):14–23, 1997b.

F. R. K. Chung. *Spectral Graph Theory*. CBMS, **92**, AMS, Providence, Rhode Island, 1997.

F. R. K. Chung, A. Grigoryán, and S. T. Yau. Eigenvalues and diameters for manifolds and graphs. In *Tsing Hua Lectures on Geometry and Analysis*, Inter. Press, Cambridge, Mass., 8(5):79–106, 1997.

K. L. Chung. *A New Introduction to Stochastic Processes* (in Chinese). World Scientific, Singapore, 1995.

Y. Colin de Verdière. *Spectres de Graphes*. Publ. Soc. Math. France, 1998.

J. B. Conrey. The Riemann hypothesis. *Notices of AMS*, pages 341–353, 2003.

M. Cranston. Gradient estimates on manifolds using coupling. *J. Funct. Anal.*, 99: 110–124, 1991.

M. Cranston. A probabilistic approach to gradient estimates. *Canad. Math. Bull.*, 35:46–55, 1992.

E. B. Davies. A review of Hardy inequality. *Operator Theory: Adv. and Appl.*, 110: 55–67, 1999.

D. A. Dawson and X. G. Zheng. Law of large numbers and central limit theorem for unbounded jump mean field models. *Adv. Appl. Math.*, 12:293–326, 1991.

A. DeMasi and E. Presutti. *Lectures on the Collective Behavior of Particle Systems*. LNM 1502, Springer, 1992.

J. D. Deuschel. Algebraic L^2 decay of attractive critical processes on the lattice. *Ann. Prob.*, 22:1:264–283, 1994.

J. D. Deuschel and D. W. Stroock. *Large Deviations*. Academic Press, New York, 1989.

P. Diaconis and L. Saloff-Coste. Logarithmic Sobolev inequalities for finite Markov chains. *Ann. Appl. Prob.*, 6(3):695–750, 1996.

P. Diaconis and D. W. Stroock. Geometric bounds for eigenvalues of Markov chains. *Ann. Appl. Prob.*, 1(1):36–61, 1991.

W. D. Ding, R. Durrett, and T. M. Liggett. Ergodicity of reversible theorems reaction–diffusion processes. *Prob. Th. Rel. Fields*, 85(1):13–26, 1990.

W. D. Ding and X. G. Zheng. Ergodic theorems for linear growth processes with diffusion. *Chin. Ann. Math.*, 10(B)(3):386–402, 1989.

B. Djehiche and I. Kaj. The rate function for some measure-valued jump processes. *Ann. Prob.*, 23(3):1414–1438, 1995.

R. L. Dobrushin. Prescribing a system of random variables by conditional distributions. *Theory Prob. Appl.*, 15:458–486, 1970.

W. Doeblin. Exposé de la théorie des chaines simples constantes de Markov à un nombre dini d'etats. *Rev. Math. Union Interbalkanique*, 2:77–105, 1938.

D. Down, S. P. Meyn, and R. L. Tweedie. Exponential and uniform ergodicity of Markov processes. *Ann. Prob.*, 23:1671–1691, 1995.

D. C. Dowson and B. V. Landau. The Fréchet distance between multivariate normal distributions. *J. Multivariate Anal.*, 12:450–455, 1982.

R. Durrett. Ten lectures on particle systems, St. Flour Lecture Notes. *LNM*, 1608, 1995.

R. Durrett and S. Levin. The importance of being discrete (and spatial). *Theoret. Pop. Biol.*, 46:363–394, 1994.

R. Durrett and C. Neuhauser. Particle systems and reaction–diffusion equations. *Ann. Prob.*, 22(1):289–333, 1994.

E. M. Dynkin. Commutative harmonic analysis I. General survey. Classical aspects. *Encyclopaedia Math. Sci.*, Springer-Verlag, 15:167–259, 1991.

Y. Egorov and V. Kondratiev. *On Spectral Theory of Elliptic Operators.* Birkhäuser, Berlin, 1996.

J. F. Escobar. Uniqueness theorems on conformal deformation of metrics, Sobolev inequalities, and an eigenvalue estimate. *Comm. Pure and Appl. Math.*, XLIII: 857–883, 1990.

S. Z. Fang. Inégalité du type de Poincaré sur l'espace des chemins Riemanniens. *C. R. Acad. Sci. Paris Série* I, 318:257–260, 1994.

J. F. Feng. The hydrodynamic limit for the reaction diffusion equation—an approach in terms of the GRP method. *J. Theor. Prob.*, 9(2):285–299, 1996.

S. Feng. Large deviations for empirical process of interacting particle system with unbounded jumps. *Ann. Prob.*, 22(4):2122–2151, 1994a.

S. Feng. Large deviations for Markov processes with interaction and unbounded jumps. *Prob. Th. Rel. Fields*, 100:227–252, 1994b.

S. Feng. Nonlinear master equation of multitype particle systems. *Stoch. Proc. Appl.*, 57:247–271, 1995.

S. Feng and X. G. Zheng. Solutions of a class of nonlinear master equations. *Stoch. Proc. Appl.*, 43:65–84, 1992.

J. A. Fill. Eigenvalue bounds on convergence to stationarity for nonreversible Markov chains, with an application to the exclusion process. *Ann. Appl. Prob.*, 1:62–87, 1991.

E. Fischer. Über quadratische Formen mit reellen Koeffizienten. *Monatsh. Math. Phys.*, 16:234–249, 1905.

M. Fukushima, Y. Oshima, and M. Takeda. *Dirichlet Forms and Symmetric Markov Processes.* Walter de Gruyter, 1994.

M. Fukushima and T. Uemura. Capacitary bounds of measures and ultracontractivity of time changed processes. *J. Math. Pure et Appliquées*, 82(5):553–572, 2003.

T. Funaki. Singular limit for reaction–diffusion equation with self-similar Gaussian noise. In S. Kusuoka D. Elworthy and I. Shigekawa, editors, *Proceedings of Taniguchi symposium, New Trends in Stochastic Analysis*, pages 132–152. World Sci., 1997.

T. Funaki. Singular limit for stochastic reaction–diffusion equation and generation of random interfaces. *Acta Math. Sin. Eng. Ser.*, 15, 1999.

F. R. Gantmacher. *The Theory of Matrices*, Volume: Two. Chelsea Publ. Com., 1989.

V. L. Girko. *Theory of Random Determinants*. Kluwer Acad. Publ., 1990.

C. R. Givens and R. M. Shortt. A class of Wasserstein metrics for probability distributions. *Michigan Math. J.*, 31:231–240, 1984.

F. Z. Gong and F. Y. Wang. Functional inequalities for uniformly integrable semigroups and application to essential spectrums. *Forum Math.*, 14:293–313, 2002.

F. Z. Gong and L. M. Wu. Spectral gap of positive operators and applications. *C. R. Acad. Sci. Paris Série* I, 331:293–313, 2000.

B. L. Granovsky and A. I. Zeifman. The decay function of nonhomogeneous birth–death processes, with application to mean-field models. *Stoch. Process. Appl.*, 72: 105–120, 1997.

D. Griffeath. Coupling methods for Markov processes. in *"Studies in Probability and Ergodic Theory, Adv. Math., Supplementary Studies,"* Academic Press 2, pages 1–43, 1978.

A. Grigor'yan. Isoperimetric inequalities and capacities on Riemannian manifolds. *Operator Theory: Advances and Applications*, 109:139–153, 1999.

M. Gromov. Paul Lévy's isoperimetric inequality. *preprint I.H.E.S.* [see Appendix C in Gromov (1999)], 1980.

M. Gromov. *Metric Structures for Riemannian and Non-Riemannian Spaces*. Progress in Mathematics. 152. Birkhäuser, Boston·Basel·Berlin, 1999.

L. Gross. Logarithmic Sobolev inequalities. *Amer. J. Math.*, 97:1061–1083, 1976.

L. Gross. Logarithmic Sobolev inequalities and contractivity properties of semigroups. *Lecture Notes in Mathematics* **1563**, E. Fabes et al. (eds.), *"Dirichlet Forms,"* Springer-Verlag, 1993.

A. Guionnet and B. Zegarlinski. Lectures on logarithmic Sobolev inequalities. In M. Ledoux J. Azéma, M. Émery and M. Yor, editors, *LNM* 1801, pages 1–134. Springer-Verlag, 2003.

E. Haagerup. Random matrices, free probability and the invariant subspace problem relative to a von Neumann algebra. in *Proceedings of "ICM 2002,"* Higher Education Press, Beijing, I:273–290, 2002.

K. Hamza and F. C. Klebaner. Conditions for integrability of Markov chains. *J. Appl. Prob.*, 32:541–547, 1995.

D. Han. Existence of solution to the martingale problem for multispecies infinite dimensional reaction–diffusion particle systems. *Chin. J. Appl. Prob. Statis.*, 6(3): 265–278, 1990.

D. Han. Ergodicity for one-dimensional Brussel's model (in Chinese). *J. Xingjiang Univ.*, 8(3):37–38, 1991.

D. Han. Uniqueness of solution to the martingale problem for infinite-dimensional reaction–diffusion particle systems with multispecies (in Chinese). *Chin. Ann. Math.*, 13A(2):271–277, 1992.

D. Han. Uniqueness for reaction–diffusion particle systems with multispecies (in Chinese). *Chin. Ann. Math.*, 16A(5):572–578, 1995.

D. Han and X. J. Hu. *Mathematical Models of Economics and Finance — Theory and Practice* (in Chinese). Shanghai Jiaotong Univ. Press, Shanghai, 2003.

G. H. Hardy. Note on a theorem of Hilbert. *Math. Zeitschr.*, 6:314–317, 1920.

H. Hennion. Limit theorems for products of positive random matrices. *Ann. Probab.*, 25(4):1545–1587, 1997.

R. Holley. A class of interaction in an infinite particle systems. *Adv. Math.*, 5: 291–309, 1970.

R. Holley. Recent results on the stochastic Ising model. *Rocky Mountain J. Math.*, 4(3):479–496, 1974.

Z. T. Hou and Q. F. Guo. *Time-Homogeneous Markov Processes with Countable State Space* (in Chinese). Beijing Sci. Press (1978). English translation (1988): Beijing Sci. Press and Springer-Verlag, 1978.

Z. T. Hou, Z. M. Liu, H. J. Zhang, J. P. Li, J. Z. Zhou, and C. G. Yuan. *birth–death Processes* (in Chinese). Hunan Sci. Press, Hunan, 2000.

E. P. Hsu. *Stochastic Analysis on Manifolds*. Amer. Math. Soc., Providence, R. I., 2002.

L. K. Hua. *Introduction to Advanced Mathematics,* Volume: The Remainder (in Chinese). Sci. Press, Beijing, 1984a.

L. K. Hua. The mathematical theory of global optimization on planned economy (in Chinese). *Kexue Tongbao,* (I): 1984, No.12, 705–709. (II) and (III): 1984, No.13, 769–772. (IV), (V), and (VI): 1984, No.16, 961–965. (VII): 1984, No.18, 1089–1092. (VIII): 1984, No.21, 1281–1282. (IX): 1985, No.1, 1–2. (X): 1985, No.9, 641–645. (XI): 1985, No.24, 1841–1844, 1984b.

L. K. Hua and S. Hua. The study of the real square matrix with both left positive eigenvector and right positive eigenvector (in Chinese). *Shuxue Tongbao*, 8:30–32, 1985.

L. P. Huang. The existence theorem of the stationary distributions of infinite particle systems (in Chinese). *Chin. J. Appl. Prob. Statis.*, 3(2):152–158, 1987.

C. R. Hwang, S. Y. Hwang-Ma, and S. J. Sheu. Accelerating diffusions. preprint, 2002.

N. Ikeda and S. Watanabe. *Stochastic Differential Equations and Diffusion Processes*. 2nd, ed. North-Holland, addr Kodansha, Tokyo, 1988.

S. R. Jarner and G. O. Roberts. Polynomial convergence rates of Markov chains. *Ann. Appl. Prob.*, 12:224–247, 2002.

M. R. Jerrum and A. J. Sinclair. Approximating the permanent. *SIAM J. Comput.*, 18:1149–1178, 1989.

F. Jia. Estimate of the first eigenvalue of a compact Riemannian manifold with Ricci curvature bounded below by a negative constant (in Chinese). *Chin. Ann. Math.*, 12A(4):496–502, 1991.

I. S. Kac and M. G. Krein. Criteria for discreteness of the spectrum of a singular string (in Russian). *Izv. Vyss. Učebn. Zaved. Mat.*, 2:136–153, 1958.

V. A. Kaimanovich. Dirichlet norms, capacities and generalized isoperimetric inequalities for Markov operators. *Potential Analysis*, 1:61–82, 1992.

L. S. Kang et al. *Nonnumerically Parallel Algorithms* (in Chinese), volume 1. Press of Sciences, Beijing, 1994.

W. S. Kendall. Nonnegative Ricci curvature and the Brownian coupling property. *Stochastics*, 19:111–129, 1986.

G. Kersting and F. C. Klebaner. Sharp conditions for nonexplosions and explosions in Markov jump processes. *Ann. Prob.*, 23(1):268–272, 1995.

H. Kesten and F. Spitzer. Convergence in distribution of products of random matrices. *Z. Wahrs.*, 67:363–386, 1984.

C. Kipnis and C. Landim. *Scaling Limits of Interacting Particle Systems*. Springer-Verlag, Berlin, 1999.

A. A. Konstantinov, V. P. Maslov, and A. M. Chebotarev. Probability representations of solutions of the Cauchy problem for quantum mechanical solutions. *Russian Math. Surveys*, 45(6):1–26, 1990.

I. Kontoyiannis and S. P. Meyn. Spectral theory and limit theory for geometrically ergodic Markov processes. *Ann. Appl. Prob.*, 13:304–362, 2003.

S. Kotani and S. Watanabe. Krein's spectral theory of strings and generalized diffusion processes. *Lecture Notes in Math.*, 923:235–259, 1982.

P. Kröger. On the spectral gap for compact manifolds. *J. Diff. Geom.*, 36:315–330, 1992.

P. Kröger. On explicit bounds for the spectral gap on noncompact manifolds. *Soochow J. Math.*, 23:339–344, 1997.

A. Kufner and L. E. Persson. *Weighted Inequalities of Hardy Type*. World Scientific, Singapore, 2003.

C. Landim, S. Sethuraman, and S. R. S. Varadhan. Spectral gap for zero range dynamics. *Ann. Prob.*, 24:1871–1902, 1996.

G. F. Lawler and A. D. Sokal. Bounds on the L^2 spectrum for Markov chain and Markov processes: a generalization of Cheeger's inequality. *Trans. Amer. Math. Soc.*, 309:557–580, 1988.

M. Ledoux. Concentration of measure and logarithmic Sobolev inequalities. In *Séminaire de Probabilités* 33. LNM 1709, pages 120–216. Springer-Verlag, 1999.

M. Ledoux. The geometry of Markov diffusion generators. *Annales de la Faculté des Sciences de Toulouse*, IX:305–366, 2000.

M. Ledoux. Logarithmic Sobolev inequalities for unbounded spin systems revised. In *Séminaire de Probabilités* 35. LNM 1755, pages 167–194. Springer-Verlag, 2001.

W. Leontief. Quantitive input–output relation in the economic system of the United States. *Review of Economics and Statistics*, XVIII(3):105–125, 1936.

W. Leontief. *The structure of the American Economy* 1919–1939. Oxford Univ. Press, New York, 1951.

W. Leontief. *Input–Output Economics*. 2nd ed., Oxford Univ. Press, 1986.

M. Levitin and L. Parnovski. Commutators, spectra trace identities, and universal estimates for eigenvalues. *J. Funct. Anal.*, 192:425–445, 2002.

P. Lévy. *Problèmes Concrets d'Analyse Fonctionnelle*. Gauthier–Villars, Paris, 1951.

P. Li. *Lecture Notes on Geometric Analysis*. Seoul National Univ., Korea, 1993.

P. Li and S. T. Yau. Estimates of eigenvalue of a compact Riemannian manifold. *Ann. Math. Soc. Proc. Symp. Pure Math.*, 36:205–240, 1980.

Y. Li. Uniqueness for infinite-dimensional reaction–diffusion processes (in Chinese). *Chin. Sci. Bulleten*, 22:1681–1684, 1991.

Y. Li. Ergodicity of a class of reaction–diffusion processes with translation invariant coefficients (in Chinese). *Chin. Ann. Math.*, 16A(2):223–229, 1995.

A. Lichnerowicz. *Géométrie des Groupes des Transformations*. Dunod, Paris, 1958.

T. M. Liggett. An infinite particle system with zero range interactions. *Ann. Prob.*, 1:240–253, 1973.

T. M. Liggett. *Interacting Particle Systems*. Springer-Verlag, 1985.

T. M. Liggett. Exponential L_2 convergence of attractive reversible nearest particle systems. *Ann. Prob.*, 17:403–432, 1989.

T. M. Liggett. L_2 rates of convergence for attractive reversible nearest particle systems: the critical case. *Ann. Prob.*, 19(3):935–959, 1991.

T. M. Liggett and F. Spitzer. Ergodic theorems for coupled random walks and other systems with locally interacting components. *Z. Wahrs.*, 56:443–468, 1981.

T. Lindvall. *Lectures on the Coupling Method.* Wiley, New York, 1992.

T. Lindvall. On Strassen's theorem, on stochastic domination. *Electr. Comm. Probab.*, 4:51–59, 1999.

T. Lindvall and L. C. G. Rogers. Coupling of multidimensional diffusion processes. *Ann. Prob.*, 14(3):860–872, 1986.

J. S. Lü. The optimal coupling for single-birth reaction–diffusion processes and its applications (in Chinese). *J. Beijing Normal Univ.*, 33(1):10–17, 1997.

S. L. Lu and H. T. Yau. Spectral gap and logarithmic Sobolev inequality for Kawasaki dynamics and Glauber dynamics. *Commun. Math. Phys.*, 156:399–433, 1993.

J. H. Luo. On discrete analog of Poincaré-type inequalities and density representation. Unpublished, 1992.

C. Y. Ma. *The Spectrum of Riemannian Manifolds* (in Chinese). Press of Nanjing Univ., Nanjing, 1993.

Z. M. Ma and M. Röckner. *Introduction to the Theory of (Non-symmetric) Dirichlet Forms.* Springer-Verlag, 1992.

N. Madras and D. Randall. Markov chain decomposition for convergence rate analysis. *Ann. Appl. Prob.*, 12(2):581–606, 2002.

Y. H. Mao. L^p-Poincaré inequality for general symmetric forms. *preprint*, 2001a.

Y. H. Mao. General Sobolev type inequalities for symmetric forms. *preprint*, 2001b.

Y. H. Mao. The logarithmic Sobolev inequalities for birth–death process and diffusion process on the line. *Chin. J. Appl. Prob. Statis.*, 18(1):94–100, 2002a.

Y. H. Mao. Nash inequalities for Markov processes in dimension one. *Acta. Math. Sin. Eng. Ser.*, 18(1):147–156, 2002b.

Y. H. Mao. Oral communication. 2002c.

Y. H. Mao. Strong ergodicity for Markov processes by coupling methods. *J. Appl. Prob.*, 39:839–852, 2002d.

Y. H. Mao. Convergence rates in strong ergodicity for Markov processes. *preprint*, 2002e.

Y. H. Mao. Ergodic degree for continuous-time Markov chains. *Sci. China* (A) 33(5) (Chinese ed.), 409–420, 2003.

Y. H. Mao. On empty essential spectrum for Markov processes in dimension one. *Acta Math. Sin. Eng. Ser., in press*, 2004.

Y. H. Mao and S. Y. Zhang. Comparison of some convergence rates for Markov process (in Chinese). *Acta. Math. Sin.*, 43(6):1019–1026, 2000.

Y. H. Mao and Y. H. Zhang. Exponential ergodicity for single birth processes. *J. Appl. Prob.*, 41(4), in press, 2004.

F. Martinelli. Lectures on Glauber dynamics for discrete spin models. In *LNM* 1717, pages 93–191. Springer-Verlag, 1999.

V. G. Maz'ya. On certain integral inequalities for functions of many variables. *J. Soviet Math.*, pages 205–237, 1973.

V. G. Maz'ya. *Sobolev Spaces.* Springer-Verlag, 1985.

M. L. Mehta. *Random Matrices.* 2nd ed., Academic Press, New York, 1991.

S. P. Meyn and R. L. Tweedie. Stability of Markovian processes (III):Foster-Lyapunov criteria for continuous-time processes. *Adv. Appl. Prob.*, 25:518–548, 1993a.

S. P. Meyn and R. L. Tweedie. *Markov Chains and Stochastic Stability*. Springer-Verlag, London, 1993b.

L. Miclo. Relations entre isopérimétrie et trou spectral pour les chaînes de Markov finies. *Prob. Th. Rel. Fields*, 114:431–485, 1999a.

L. Miclo. An example of application of discrete Hardy's inequalities. *Markov Processes Relat. Fields*, 5:319–330, 1999b.

R. A. Minlos. Invariant subspaces of the stochastic Ising high temperature dynamics. *Markov Processes Relat. Fields*, 2:263–284, 1996.

R. A. Minlos and A. G. Trishch. Complete spectral decomposition of the generator for one-dimensional Glauber dynamics (in Russian). *Uspekhi Matem. Nauk*, 49: 209–211, 1994.

T. S. Mountford. The ergodicity of a class of reversible reaction–diffusion processes. *Prob. Th. Rel. Fields*, 92(2):259–274, 1992.

B. Muckenhoupt. Hardy's inequality with weights. *Studia Math.*, XLIV:31–38, 1972.

A. Mukherjea. Tightness of products of i.i.d. random matrices. *Prob. Th. Rel. Fields*, 87:389–401, 1991.

J. Nash. Continuity of solutions of parabolic and elliptic equations. *Amer. J. Math.*, 80:931–954, 1958.

R. B. Nelssen. *An Introduction to Copulas, Lecture Notes in Stat.*, volume 139. Springer-Verlag, 1999.

C. Neuhauser. An ergodic theorem for Schlögl models with small migration. *Prob. Th. Rel. Fields*, 85(1):27–32, 1990.

E. Nummelin. *General Irreducible Markov Chains and Non-Negative Operators*. Cambridge Univ. Press, 1984.

E. Nummelin and P. Tuominen. Geometric ergodicity of Harris recurrent chains with applications to renewal theory. *Stoch. Proc. Appl.*, 12:187–202, 1982.

I. Olkin and R. Pukelsheim. The distance between two random vectors with given dispersion matrices. *Linear Algebra Appl.*, 48:257–263, 1982.

B. Opic and A. Kufner. *Hardy-type Inequalities*. Longman, New York, 1990.

V. I. Oseledec. A multiplicative ergodic theorem. Lyapunov characteristic number for dynamical systems. *Trans. Moscow Math. Soc.*, 19:197–231, 1968.

A Perrut. Hydrodynamic limits for a two-species reaction–diffusion process. *Ann. Appl. Prob.*, 10, no. 1:163–191, 2000.

J. G. Propp and D. B. Wilson. Exact sampling with coupled Markov chains and applications to statistical mechanics. *Random Structures Algorithms*, 9:223–252, 1996.

S. T. Rachev and L. Ruschendorf. *Mass Transportation Problems*. Vol. 1: *Theory*, Vol. 2: *Applications*. Springer Series in Statistics. Probability and Its Applications, 1998.

M. M. Rao and D. Ren, Z. *Theory of Orlicz Spaces*. Marcel Dekker, Inc. New York, 1991.

G. O. Roberts and J. S. Rosenthal. Geometric ergodicity and hybrid Markov chains. *Electron. Comm. Probab.*, 2:13–25, 1997.

G. O. Roberts and R. L. Tweedie. Geometric L^2 and L^1 convergence are equivalent for reversible Markov chains. *J. Appl. Probab.*, 38(A):37–41, 2001.

M. Röckner and F. Y. Wang. Weak Poincaré inequalities and L^2-convergence rates of Markov semigroups. *J. Funct. Anal.*, 185(2):564–603, 2001.

J. S. Rosenthal. Quantitative convergence rates of Markov chains: A simple account. *Elec. Comm. Prob.*, 7(13):123–128, 2002.

O. S. Rothaus. Diffusion on compact Riemannian manifolds and logarithmic Sobolev inequalities. *J. Funct. Anal.*, 42:102–109, 1981.

O. S. Rothaus. Analytic inequalities, isoperimetric inequalities and logarithmic Sobolev inequalities. *J. Funct. Anal.*, 64:296–313, 1985.

L. Saloff-Coste. Lectures on finite Markov chains. *Lectures on probability theory and statistics* (Saint-Flour, 1996), LNM, **1665**. Springer-Verlag:301–413, 1997.

R. Schoen and S. T. Yau. *Differential Geometry*. Science Press, Beijing, China, (in Chinese), English Translation: *Lectures on Differential Geometry*, International Press (1994), 1988.

R. H. Schonmann. Slow drop-driven relaxation of stochastic Ising models in the vicinity of the phase coexistence region. *Commun. Math. Phys.*, 161:1–49, 1994.

T. Shiga. Stepping stone models in population, genetics and population dynamics. in *"Stochastic Processes in Physics and Engineering,"* edited by S. Albeverio et al., pages 345–355, 1988.

A. J. Sinclair. *Algorithms for Random Generation and Counting: A Markov Chain Approach*. Birkhäuser, 1993.

A. D. Sokal and L. E. Thomas. Absence of mass gap for a class of stochastic contour models. *J. Statis. Phys.*, 51(5/6):907–947, 1988.

J. S. Song. Continuous time Markov decision processes with non-uniformly bounded transition rates. *Sci. Sin.* Ser. A, 12(11):1281–1291, 1988.

V. Strassen. The existence of probability measures with given marginals. *Ann. Math. Statist.*, 36:423–439, 1965.

D. W. Stroock. *An Introduction to the Theory of Large Deviations*. Springer-Verlag, 1984.

W. G. Sullivan. The L^2 spectral gap of certain positive recurrent Markov chains and jump processes. *Z. Wahrs.*, 67:387–398, 1984.

S. Z. Tang. The existence and uniqueness of reaction–diffusion processes with multi-species and bounded diffusion rates (in Chinese). *Chin. J. Appl. Prob. Stat.*, 1: 11–22, 1985.

L. E. Thomas. Bounds on the mass gap for finite volume stochastic Ising models at low temperature. *Comm. Math. Phys.*, 126:1–11, 1989.

H. Thorrison. *Coupling, Stationarity, and Regeneration*. Springer-Verlag, 2000.

P. Tuominen and R. L. Tweedie. Subgeometric rates of convergence of f-ergodic Markov chains. *Adv. Appl. Prob.*, 26:775–798, 1994.

R. L. Tweedie. Criteria for ergodicity, exponential ergodicity and strong ergodicity of Markov processes. *J. Appl. Prob.*, 18:122–130, 1981.

S. S. Vallender. Calculation of the Wasserstein distance between probability distributions on line. *Theory Prob. Appl.*, 18:784–786, 1973.

E. van Doorn. *Stochastic Monotonicity and Queuing Applications of Birth–Death Processes*, volume 4. Lecture Notes in Statistics, Springer-Verlag, 1981.

E. van Doorn. Conditions for exponential ergodicity and bounds for the decay parameter of a birth–death process. *Adv. Appl. Prob.*, 17:514–530, 1985.

E. van Doorn. Representations and bounds for zeros of orthogonal polynomials and eigenvalues of sign-symmetric tri-diagonal matrices. *J. Approx. Th.*, 51:254–266, 1987.

E. van Doorn. Quasi-stationary distributions and convergence to quasi-stationarity of birth–death processes. *Adv. Appl. Prob.*, 23:683–700, 1991.

E. van Doorn. Representations for the rate of convergence of birth–death processes. *Theory Probab. Math. Statist.*, 65:36–42, 2002.

N. Varopoulos. Hardy–Littlewood theory for semigroups. *J. Funct. Anal.*, 63:240–260, 1985.

C. Villani. *Topics in Mass Transportation, Graduate Studies in Math.*, volume 57. AMS, Providence, RI, 2003.

D. Voiculescu, K. Dykema, and A. Nica. *Free Random Variables*. CRM Monograph Series, vol. 1, AMS, Providence, R. I., 1992.

Z. Vondraçek. An estimate for the L^2-norm of a quasi continuous function with respect to a smooth measure. *Arch. Math.*, 67:408–414, 1996.

F. Wang. Latała–Oleszkiewicz inequalities. *Ph.D. thesis,* Beijing Normal Univ., 2003a.

F. Wang and Y. H. Zhang. *F*-Sobolev inequality for general symmetric forms. *Northeastern J. Math.*, 19(2):133–138, 2003.

F. Y. Wang. Gradient estimates for generalized harmonic functions on manifolds. *Chinese Sci. Bull.*, 39(22):1849–1852, 1994a.

F. Y. Wang. Gradient estimates in $\mathbb{R}^d$. *Canad. Math. Bull.*, 37(4):560–570, 1994b.

F. Y. Wang. Ergodicity for infinite-dimensional diffusion processes on manifolds. *Sci. Sin. Ser* (A), 37(2):137–146, 1994c.

F. Y. Wang. Uniqueness of Gibbs states and the L^2-convergence of infinite-dimensional reflecting diffusion processes. *Sci. Sin. Ser* (A), 32(8):908–917, 1995.

F. Y. Wang. Estimation of the first eigenvalue and the lattice Yang–Mills fields. *Chin. J. Math.* 1996, 17A(2) (in Chinese), 147–154; *Chin. J. Contem. Math.*, 1996, 17(2) (in English), 119–126, 1996.

F. Y. Wang. Logarithmic Sobolev inequalities for noncompact Riemannian manifolds. *Probab. Th. Relat. Fields*, 109:417–424, 1997.

F. Y. Wang. Functional inequalities for empty essential spectrum. *J. Funct. Anal.*, 170:219–245, 2000a.

F. Y. Wang. Functional inequalities, symmegroup properties and spectrum estimates. *Infinite Dim. Anal., Quantum Probab. and related Topics*, 3(2):263–295, 2000b.

F. Y. Wang. Sobolev type inequalities for general symmetric forms. *Proc. Amer. Math. Soc.*, 128(12):3675–3682, 2001.

F. Y. Wang. Coupling, convergence rate of Markov processes and weak Poincaré inequalities. *Sci. in China* (A), 45(8):975–983, 2002.

F. Y. Wang. A generalization of Poincaré and log-Sobolev inequalities. *Potential Analysis, in press*, 2004a.

F. Y. Wang. *Functional Inequalities, Markov Processes, and Spectral Theory*. Science Press, Beijing, 2004b.

F. Y. Wang and M. P. Xu. On order preservation for couplings of multidimensional diffusion processes. *Chinese J. Appl. Probab. Stat.*, 13(2):142–148, 1997.

Y. Z. Wang. Convergence rate in total variation for diffusion processes. *Chin. Ann. Math.* Chinese ed. 20(2), 261–266; English ed., 20(2):323–329, 1999.

Y. Z. Wang. Algebraic convergence of diffusion processes on the real line (in Chinese). *J. Beijing Normal Univ.*, 39(4):448–456, 2003b.

Y. Z. Wang. Algebraic convergence of diffusion processes in $\mathbb{R}^n$ (in Chinese). *Acta Math. Sin.* (A), in press, 2004.

Z. K. Wang. *birth–death Processes and Markov Chains* (in Chinese). Science Press, Beijing, 1980.

Z. K. Wang and X. Q. Yang. *Birth and Death Processes and Markov Chains*. Springer-Verlag, Berlin, 1992.

L. N. Wasserstein. Markov processes on a countable product space, describing large systems of automata (in Russian). *Problem Peredachi Informastii*, 5:64–73, 1969.

A. Wedestig. Some new Hardy type inequalitities and their limiting inequalities. *J. Ineq. Pure and Appl. Math.*, 4(3):Art. 61, 2003.

E. Wigner. Characterictic vector of boardered matrices with infinite dimensions. *Ann. Math.*, 62:548–564, 1955.

D. B. Wilson. Mixing times of lozenge tiling and card shuffling Markov chains. *Ann. Appl. Probab.*, 14(1), 2004.

L. M. Wu. Essential spectral radius for Markov semigroups (I): discrete time case. *Prob. Th. Rel. Fields*, 128:255–321, 2004.

X. J. Xu. Hydrodynamic limits (in Chinese). *Ph.D. thesis,* Beijing Normal Univ., 1991.

S. J. Yan and M. F. Chen. Multidimensional q-processes. *Chin. Ann. Math.*, 7(B):1: 90–110, 1986.

S. J. Yan and Z. B. Li. Probability models of non-equilibrium systems and the master equations (in Chinese). *Acta Phys. Sin.*, 29:139–152, 1980.

D. G. Yang. Lower bound estimates of the first eigenvalue for compact manifolds with positive Ricci curvature. *Pacific. J. Math.*, 190(2):383–398, 1999.

H. C. Yang. Estimate of the first eigenvalue for a compact Riemannjian manifold. *Sci. Sin.* (A), 33(1):39–51, 1990.

X. Q. Yang. *Constructions of Time-Homogeneous Markov Processes with Denumerable States* (in Chinese). Hunan Sci. Press, Hunan, 1986.

E. Zeidler. *Nonlinear Functional Analysis and Its Applications*, volume I. Springer-Verlag, 1986.

A. I. Zeifman. Upper and lower bounds on the rate of convergence for nonhomogeneous birth and death processes. *Stoch. Process. Appl.*, 59:157–173, 1995.

H. J. Zhang, X. Lin, and Z. T. Hou. Uniformly polynomial convergence for standard transition functions (in Chinese). In *"Birth–Death Processes"* by Hou, Z. T. et al. (2000), Hunan Sci. Press, Hunan, 2000.

S. Y. Zhang. Existence and application of optimal Markovian coupling with respect to nonnegative lower semi-continuous functions. *Acta Math. Sin. Eng. Ser.*, 16(2): 261–270, 2000a.

S. Y. Zhang and Y. H. Mao. Exponential convergence rate in Boltzman–Shannon entropy. *Sci. Sin.* (A), 44(3):280–285, 2000.

Y. H. Zhang. Conservativity of couplings for jump processes (in Chinese). *J. Beijing Normal Univ.*, 30(3):305–307, 1994.

Y. H. Zhang. Sufficient and necessary conditions for stochastic comparability of jump processes. *Acta Math. Sin. Eng. Ser.*, 16(1):99–102, 2000b.

Y. H. Zhang. Strong ergodicity for continuous-time Markov chains. *J. Appl. Prob.*, 38:270–277, 2001.

D. Zhao. Eigenvalue estimate on a compact Riemannian manifold. *Science in China,* Ser. A, 42(9):897–904, 1999.

J. L. Zheng. Phase transitions of Ising model on lattice fractals, martingale approach for q-processes (in Chinese). *Ph.D. thesis,* Beijing Normal Univ., 1993.

X. G. Zheng and W. D. Ding. Existence theorems for linear growth processes with diffusion. *Acta Math. Sin. New Ser.*, 7(1):25–42, 1987.

J. Q. Zhong and H. C. Yang. Estimates of the first eigenvalue of a compact Riemannian manifold. *Sci. Sin.* Ser. A, 27(12):1265–1273, 1984.

Author Index

Subject Index

Special Symbols

$\hat{\alpha}$, 95
η, 150
φ-optimal coupling, 31
λ_0, 62, 63, 71, 73, 77, 93–96, 141
$\lambda_0(A)$, 74, 75, 79
λ_1, 2, 3, 11, 43, 50, 58, 67, 69–71, 74, 75, 83, 93–96, 105, 121, 150, 197
$\lambda_1(B)$, 75
ρ-optimal coupling, 18, 28
ρ-optimal coupling operator, 29, 30, 51
σ, 150

$C_0(E)$, 74, 131
$\mathrm{Cap}\,(K)$, 74, 132
$\mathscr{D}_w(L)$, 34, 110, 196
$\mathrm{Ent}(f)$, 12, 69, 82, 114, 125, 150
F-Sobolev inequality, 140, 141, 144
L^1-exponential convergence, 13,15, 100, 155, 156
L^2-algebraic convergence, 13, 14, 153, 157
L^2-exponential convergence, 12,13, 15, 100, 157, 158, 160
L^p-exponential convergence, 14
(MO), 22
(MP), 19
q-pair, 19
q-process, 19
Q-matrix, 19, 89, 164, 202
Q-process, 20, 90
$\mathrm{Var}(f)$, 10, 12, 149
W_p-distance, 28

A

algebraic convergence, 145
algebraic form, 195, 196
analytic method, 44, 106
approximation procedure, 10, 95, 96, 116, 117, 120
asymptotically stable, 178

B

basic coupling, 18, 23
birth–death process, 9, 15, 29, 30, 39, 62, 87, 88, 91, 92, 95, 100–102, 106, 111, 113, 122, 124, 126, 134, 145, 170, 175, 176, 184, 203
Brusselator model, 165, 174

C

Cheeger's constant, 10, 12, 68
classical coupling, 22, 25
classical variational formula, 4, 9, 43
closed (lower semicontinuous) function, 31
collapse time, 183
conservative, 19
consumption matrix, 188
coupling, 7, 17, 24, 45, 47, 171, 197
coupling by inner reflection, 23
coupling by reflection, 25, 29, 47
coupling method, 7, 45, 109
coupling of marching soldiers, 23, 25, 47, 173